Once a River

ONCE A RIVER

Bird Life and Habitat Changes on the Middle Gila

AMADEO M. REA

Bird Sketches by Takashi Ijichi

University of Arizona Press
TUCSON, ARIZONA

About the Author . . .

AMADEO M. REA's field of expertise is the taxonomy and distribution of birds of western North America. His interest in natural history began during a childhood spent on a ranch in El dorado County, California. Five years of teaching on the Gila River Reservation at St. John's Indian School, Komatke, gave him the opportunity not only to pursues ornithological studies but also to tap the wealth of biological data in the oral history of the Pima Indians. He earned his Ph.D. at the University of Arizona, Tucson, and in 1977 became curator of birds and mammals at San Diego's Natural History Museum.

The University of Arizona Press
www.uapress.arizona.edu

Century Collection edition 2016

Printed in the United States of America
21 20 19 18 17 16 7 6 5 4 3 2

ISBN-13: 978-0-8165-0799-3 (cloth)
ISBN-13: 978-0-8165-3481-4 (Century Collection paper)

This book was set in 11/12 ITC Garamond Light on a V-I-P.

Library of Congress Cataloging-in-Publication Data
Rea, Amadeo M.
Once a River.

Bibliography: p.
Includes index.
1. Birds—Gila River Valley (N.M. and Ariz.)—
Ecology. 2. Riparian ecology—Gila River Valley
(N.M. and Ariz.) 3. Desertification—Gila River Valley
(N.M. and Ariz.) 4. Pima Indians—Ethnozoology.
5. Indians of North America—Gila River Valley (N.M. and
Ariz.) I. Title.
OL683.G5R4 1983 598.252'6323'097917 82-23815

♾ This paper meets the requirements of ANSI/NISO Z39.48-1992
(Permanence of Paper).

to
Allan
who raised questions
and
Sally
who sometimes wondered why

Contents

Foreword *xi*
Acknowledgments *xiii*
Introduction *1*

I. Changes on the Middle Gila

Chapter 1. The Pima People, Past and Present *7*
Chapter 2. Changes in Bird Habitats: Historic Accounts *16*
Chapter 3. Major Modern Habitats *41*
Chapter 4. Riparian Analogs *61*
Chapter 5. Changes in Bird Life *76*
Chapter 6. Interrelationship of Birds and Reservation Habitats *98*

II. Species Accounts

Contents *113*
Plan of the Species Accounts *115*
The Species *125*

Reference Material

Appendix A: Piman Orthography *255*
Appendix B: Riverine Pima Ethnotaxonomy of Birds *257*
Bibliography *259*
General Index *275*
Index of Piman Words and Expressions *285*

Illustrations

Figures

1.1 Map of the Gila River Indian Reservation 8
1.2 a. Map of the northwest section of the reservation 12
b. Aerial photograph of the same area 13
1.3 Climatological data, 1963–1972 14
2.1 a. A view of Pima and Maricopa villages and fields sketched by John R. Bartlett in the summer of 1852 26
b. Same view, 1982 27
2.2 a. The Gila River channel at Sacaton in 1905 28
b. Modern condition of the Gila channel 29
2.3 a. Komatke around 1914 32
b. Komatke in 1982 32
2.4 A dead mesquite bosque: New York Thicket, June 1978 35
2.5 Modern Maricopa Wells, near Pima Butte 39
3.1 The Sierra Estrella 43
3.2 Bajada plant community 44
3.3 An arroyo dissecting the bajada above Santa Cruz Village 45
3.4 The saltbush-bursage community 47
3.5 Traditional Pima rancheria-type settlement 48
3.6 Pima field at Komatke 48
3.7 Fence row on traditional Pima ranch 49
3.8 Ditch banks on Pima ranch 49
3.9 Tamarisk grove about former Pima house, Komatke 51
3.10 Second-growth mesquite bosque 51
3.11 Modern mechanized farm near Sacate 52
3.12 Sewage pond west of Sacaton 52
3.13 Barehand Lane pond in April 1981 54
3.14 a. Barehand Lane marsh (December 1974) fifteen months after burning 55
b. Barehand Lane marsh in April 1981 55
3.15 Salt-Gila confluence grove 58
3.16 Channel above the Salt-Gila confluence with cattails and willows 58
3.17 Rapid regeneration of willows: Salt River channel at 91st Avenue 59
3.18 Chandler boundary cottonwoods 59
4.1 Blue Point Cottonwoods 64
4.2 The Gila River at The Buttes–Cochran area 65
4.3 Map of The Buttes–Cochran area 69
6.1 Habitat changes on the Gila 109

Distribution Maps

Turkey Vultures (summering) 128
Turkey Vultures (wintering and northern migrants) 129
American Kestrels (breeding) 138
Snowy Egrets (breeding and nonbreeding) 144
Great Horned Owls 167
Poorwills (breeding) 172
Say's Phoebes (breeding adults and juvenals) 185
Say's Phoebes (wintering) 186
Horned Larks (to end of 1940s) 192
Horned Larks (1970s) 193
Common Ravens 200
Verdins (fall specimens) 203
Cactus Wrens (fall specimens) 206
Northern Orioles (breeding) 243
Great-tailed Grackles (1960s and 1970s) 249

Tables

1. Comparison of breeding birds of Blue Point Cottonwoods and the Gila River Indian Reservation 62
2. Comparison of wintering birds of Blue Point Cottonwoods and the Gila River Indian Reservation 67
3. Comparison of breeding birds from two sections of the Gila River: The Buttes–Cochran area and the Gila River Reservation 70
4. Species extirpated from Gila River Reservation 80
5. Population declines associated with habitat deterioration 82
6. Species colonizing Arizona during the twentieth century 88
7. Species with range extensions within the Southwest during historic times 88
8. Species formerly wintering later on the Gila River 93
9. Years of appearance of some erratic winter visitants 95
10. Taxa involved in the 1972–1973 flight year 96
11. Casuals on lower Salt, winter 1977 97
12. Field ornithologists working on or near the reservation 116
13. Abbreviations for collections 119
14. Nonbreeding southwestern *Ardea thula* specimens 145

Foreword

Since time immemorial, canaries have been used in coal mines to detect the presence of poisonous gases. A dead canary means danger.

Amadeo Rea, ornithologist, writes about "extirpated" birds in chronicling the death of a river. The Great Blue Heron, Whistling Swan, Snow Goose, Elf Owl, Vermilion Flycatcher, and Summer Tanager are among the more than two dozen species of birds that no longer winter or breed within the present boundaries of southern Arizona's Gila River Indian Reservation. They are nature's canaries. Their absence says "danger."

The human onslaught against rivers in the American Southwest is no longer a secret. Philip Fradkin has called the mightiest of them all, the Colorado, *A River No More* (New York: Alfred A. Knopf, 1981). The Santa Cruz, marking the eastern fringe of the Sonoran Desert and upon which Tucson depends for its very sustenance, not only sports a dry surface throughout most of its length but is polluted in its underground meanderings by industrial contaminants. Such is grist for the mill of newspaper feature writers.

Amadeo Rea's study, however, is possibly the first to offer an in-depth look at the results of human activities on a narrowly defined stretch of a southwestern stream, the Middle Gila, where it formerly coursed from east to west down the lengthwise center of the Gila River Indian Reservation. While his immediate concern is with birds ("avifauna" in the jargon of science), the story that unfolds is one of the destruction of a habitat leading to a place that was "once a river." Cutting, chopping, damming, gouging, bulldozing, leveling, pumping, ripping, poisoning, and grazing are only a few of the forces employed to fuel insatiable human appetites, all at work on the Gila. All in the name of "bigger is better"; all in the name of Progress.

To make his point, Rea takes us to the data from prehistoric archaeology, where bones from great water birds bespeak a once salubrious clime. And with him we see the Middle Gila through the eyes of Spanish missionaries and explorers, as well as through the books and journals of mid-nineteenth century Anglo Americans who found themselves tracking the edges of the flowing stream. Too, he relies heavily on the careful observations of earlier twentieth century naturalists, both ornithologists and botanists, who also were concerned with this piece of riverine habitat.

Most important, however, are Rea's own studies which began in 1963 and which have continued with varying degrees of intensity to 1982. These include not only qualitative studies of the Gila River Indian Reservation's bird life and bird habitat, but of nearby control areas as well. All this is supplemented, moreover, with data from Pima Indian consultants, people who have been the region's most sustained bird watchers and the most pained witnesses to the demise of their river. These data are not casual. They are a systematic offering of the Pima knowledge of the avian world and how that world has traditionally been viewed by them. The inclusion of such information in a study of this kind is a first for the Southwest.

Once a River should become a kind of model handbook for ecologists interested in desert environments. It will become the checklist for bird watchers along the Gila River. It will provide anthropologists and others involved in studies of cultural and linguistic domains with insights into Pima ethno-ornithology and suggest leads for similar research among additional cultural groups. It will forever be the baseline report for future studies of Middle Gila ecology.

Most significantly, Amadeo Rea's little book should help make all of us ask hard questions of ourselves about the land in which we live and to which we must look for support for future generations. He directs his immediate challenge to the governing body of the Gila River Indian Reservation. But all of us *Homo sapiens,* whether Indian or non-Indian, share high stakes in the eventual outcome.

BERNARD L. FONTANA

Acknowledgments

I wish to acknowledge with thanks all those who have helped in any way to bring this work to completion. Its preparation has been so long and diverse that some of the many people who assisted in so many different ways may be omitted from formal acknowledgments. Such oversights do not indicate any lack of gratitude.

Two people have assisted me most importantly throughout the course of this work. Allan R. Phillips has provided general guidance since its inception. Sally Giff Pablo has provided facilities and countless other courtesies during my studies, making fieldwork possible.

R. Roy Johnson and James M. Simpson loaned freely of their data from current studies at Blue Point Cottonwoods and from their several decades of study of the Salt River Valley.

Collecting on the reservation was assisted for several years each by Joseph V. Kanseah, Francisco Valenzuela, Albert Pablo, and Marvin Pablo; for shorter periods, assistance was supplied by a number of other Indian students. Many students at St. John's Indian School helped with preparations. In particular, Irma Barehand and Mary Wallen volunteered evenings during 1968–1969 to prepare specimens.

The bird sketches, so kindly supplied by Takashi Ijichi were inspired by Hohokam pottery designs.

Wade C. Sherbrooke and Gary P. Nabhan took some of the habitat photographs, and Daniel Gottlieb worked diligently to produce satisfactory prints from my own negatives.

Most of the plant identifications are by Gary P. Nabhan. Richard S. Felger and Reid Moran edited the botanical taxonomy and nomenclature.

Special thanks are due to the Pima consultants who put up with my years of checking and rechecking ethnotaxonomies and historical data: Ruth Giff and the late Joseph Giff, Sylvester Matthias, and the late David and Rosita Brown.

Fathers Celestine Chinn, O.F.M., and the late Bonaventure Oblasser, O.F.M., shared recollections of their early days on the reservation when the Gila still ran.

To help solve taxonomic problems of birds resident on or migratory through the reservation, it was necessary to collect fresh specimen material (particularly of species with postmortem color changes) from many parts of the West. The following people assisted me in the field: in the Salt River Valley of Arizona, J. M. Simpson and R. R. Johnson; on the Fort Apache Reservation, Arizona, Timothy Nozie, Ralph Thomas, Benedict Lupe, Arnold Lupe, and Nelson Lupe, Sr.; at Zuni, New Mexico, Keith Peywa, Albert Peywa, Jr., and Octavius Seotewa; at Acoma Pueblo, New Mexico, Gary Valley, Wilfred Antonio, and Michael Antonio; at Mescalero, New Mexico, Joseph V. Kanseah, Charles Belin, and Avery Belin; in California, William F. Bowman and Aldena J. Stevens; in Nevada, George T. Austin and Alex W. Stevens; in the Pacific Northwest, Jon Jantzen, Anthony de la Torre, Ruth Horn, Allan R. Phillips, Robert M. Chandler, Takashi Ijichi, Donald Payne, and the late Donald F. Martin.

The following provided food, lodging, and facilities during my field work: Sally Pablo, Komatke, Arizona; Mary V. Riley, Fort Apache, Arizona; Mr. and Mrs. Nelson Lupe, Sr., White Mountain, Arizona; Robert and Charmion McKusick, Globe, Arizona; Rose Peywa, Zuni, New Mexico; Mary Valley, Acomita, New Mexico; Amador and Arlene Martinez, Mescalero, New Mexico; Benjamin and Josephine Romero, Barona Indian Reservation, California; Roger F. Pearson, Los Angeles, California; Aldena Stevens, Placerville, California; and the de la Torre family in Washington.

Salome (Bixby) R. Demaree deserves special thanks for supplying over the years critical specimens and other data from the lower Salt.

Asylum during the first draft writing was provided by Richard R. Purcell and the Sacred Heart Community at Covered Wells, Papago Indian Reservation, Arizona. Dr. and Mrs. Allan R. Phillips kindly offered hospitality in Mexico City during my month's study of Gilman's manuscript and the Phillips Collection.

The following critically read the manuscript and provided many helpful comments: R. Roy Johnson, Gary P. Nabhan, Henry F. Dobyns, Allan R. Phillips, James M. Simpson, William F. Bowman, Wade C. Sherbrooke, and Reid Moran. Opinions on particular problems of avian taxonomy and nomenclature were provided by Kenneth C. Parks and Allan R. Phillips. All final decisions, however, remain matters of my own judgment.

Deanne Demeré typed the manuscript. Marie Hoff Cox drafted most of the maps. Sarah Feuerstein volunteered many weeks of her time preparing the index. Mary Trotter volunteered typing the index.

For various collecting permits and for their continued cooperation I thank the Arizona Game and Fish Department and the United States Fish and Wildlife Service. Special acknowledgment goes to Alexander Lewis, Sr., former Governor of the Gila River Tribal Communities, who early in this study authorized me to collect vertebrate specimens on the reservation. My 1977–1979 field work among the Pima Bajo and other Piman speakers was supported by National Science Foundation grant NSF-BNS-77-08582.

I am grateful to the University of Arizona Press, and particularly Marie Webner, for effecting publication of this work.

A. M. R.

Introduction

> What can we gain by sailing to the moon if we are not able to cross the abyss that separates us from ourselves? This is the most important of all voyages of discovery, and without it all the rest are not only useless but disastrous. Proof: the great travellers and colonizers of the Renaissance were, for the most part, men who perhaps were capable of things they did precisely because they were alienated from themselves. In subjugating primitive worlds they only imposed on them, with the force of cannons, their own confusion and their own alienation.
>
> Thomas Merton
> *The Wisdom of the Desert*

European man and his introduced Eurasian animals have made profound changes in the American biota in post-Conquest times (A. Crosby, 1972). Many changes, often dramatic and irreversible, occurred so soon after contact that the aboriginal or precontact conditions of the New World ecosystems can now only be surmised (Sauer, 1966). In places, tropical forests became savannas, thorn scrub invaded grasslands, watersheds were overgrazed, permanent watercourses dried, and entire species of plants and animals were extirpated because of altered habitats or excessive predation by man and introduced predators or diseases (Bennett, 1968; Budowski, 1956; Cook, 1949; Gordon, 1957; Humphrey, 1958; Johannessen, 1963; Warner, 1968). Certain ecosystems are especially susceptible to upset by human-induced influences. The arid regions of the world, such as the Sonoran Desert, have proven to be delicately balanced systems (Hastings and Turner, 1965; Dobyns, 1981).

When the first European explorers entered the deserts of the American Southwest they found them crossed with rivers wherever they went: the Fuerte, Mayo, Yaqui, Sonora, San Miguel, Sonoita, San Pedro, Santa Cruz, Gila, Verde, Colorado. These desert streams and rivers had a number of things in common. They were perennial though water levels fluctuated, rising highest from the snowmelts in late spring and summer, spreading nutrient-rich silts over the floodplains, rejuvenating the soil for the summer growth of annuals. Numerous backwaters and constantly changing side channels of the larger rivers were filled with emergent vegetation: cattails, rushes, seepwillow. Most impressively, each river was a green avenue of bottom-land timber—cottonwood, willow, sycamore, and ash—running for hundreds of miles through an otherwise dry desert.

These narrow ribbons of forest along a river are called riparian woodlands, from Latin *riparius,* 'relating to a riverbank.' Riparian biotic communities are ecotones—meeting places or interfaces between two major vegetation types—in this case xeric desert and mesic broad leafed woodlands (even though here the gallery forest is linear). Both humans and native animals found the desert riparian woodlands areas of the highest food resources: annual seeds, perennial fruits, insects and aquatic invertebrates, fish of various sizes. And of course, *water,* the commodity most limited in the desert, was here available in abundance for both plants and animals. A look at the distribution of native human populations at the time of European contact shows concentrations in the floodplains along these riparian woodlands. The highest known breeding bird densities in the United States have been recorded in the desert riparian habitat (Carothers et al., 1974). By 1968 only 279,600 acres (113,150 ha) of riparian woodland were left in Arizona, 100,700 acres (40,750 ha) along the Gila (Babcock, 1968), and of this only 6,000–8,000 acres (2,000–3,000 ha) of cottonwood remained (Barger and Ffolliott, 1971).

Water tables once were high in the lower terraces of these desert waterways. Dense thickets of gray arrowweed and "cane" or commonreed grew close to the river. Where spring floodwaters receded from banks or where lagoons dried, stands of annuals such as wild sunflowers grew quickly in the rich silt. Great areas of the floodplains were covered by extensive bosques or woodlands of mesquite, the trees reaching enormous sizes. The summer crops of sweet mesquite pods provided ample food for man and beast.

Today this has almost all changed in the American Southwest. The larger rivers of northwestern Mexico (the Fuerte, Mayo, Yaqui, and Sonora) have been tamed. Dams now regulate their flow, and the streams no longer dump their enriching silt annually on the floodplains. Some no longer even reach the sea in surface flow. The once-huge Colorado, draining half the southern Rocky Mountains, is no longer muddy red, as its name implies, and no longer huge. A series of dams, beginning with Hoover Dam built in 1935, "controls" the river. Its waters are drawn off to supply metropolitan areas such as Los Angeles. By the time the remaining water has been recycled, it reaches the Colorado River delta so saline as to kill most plants. The once-productive delta bottom lands of the native Yuman peoples are now barren, gaunt communities of exotic saltcedar, eschewed by most native animals. The Verde River has fared a bit better. True to its name, it is still a green ribbon for some stretches of its length down to its confluence with the Salt. The north-flowing San Pedro, a tributary of the Gila, has been drastically changed. Its beaver dams, ciénagas, and fine growths of grasses are gone (Miller, 1961) and the channel severely damaged, but surface flow persists (1978), with a riparian community of mesquite, saltcedar, and cottonwoods (Reichhardt et al., 1979).

But other desert waterways have not survived even this well. The lower Salt, Agua Fria and Hassayampa, and at least the lower 250 miles (400 km) of the Gila and the entire Santa Cruz are *dead.* There is no surface flow except during floods immediately following violent storms, when torrents of water rush down the denuded gulleys and are gone in a few days. The recharge of groundwater is negligible.

On nearly all these desert streams the delicate fabric of the ecological community has been destroyed or warped beyond recognition and restoration. Emergent vegetation has all but disappeared. The riparian forest is largely gone or replaced by feral saltcedar. Other weedy species proliferate. The water table of the floodplains, formerly but a few feet below the surface, now averages hundreds of feet underground. In places the water table is so low that even the deep-rooted mesquite cannot reach moisture—witness the skeletons of the dead bosques at San Xavier del Bac, Casa Grande Ruins National Monument, and New York Thicket (Fig. 2.4).

This book examines the microcosm of the Gila River Indian Reservation and focuses on minute details of specific avian habitats and of ecosystem deterioration. The story it tells, however, is really the history of all of southern Arizona and may, some day, be the story of all the arid West. This abuse of arid lands produces a disease called *desertification.*

According to Sheridan (1981:4), "approximately 10% of the U.S. land mass is in a state of severe or very severe desertification. The actual acres *threatened* by severe desertification, however, are almost twice that amount."

Most earmarks of desertification annotated by Sheridan are found to an extreme degree on the middle Gila:

Reduction of surface streamflow
Declining groundwater tables
Salinization of topsoil and water
Desolation of native vegetation
Unnaturally high soil erosion

The causes for the watershed deterioration of the Gila and its tributaries and the desertification of the floodplains are multiple:

Overgrazing of arid and adjacent semiarid uplands
Excessive woodcutting in watersheds and mesquite bosques
Overtrapping of beaver and loss of beaver dams on watersheds and lowlands
Gullying streambanks and hillsides by the trampling of cattle and other Old World herbivores

Cutting unprotected wagon roads and stripping these bare by horses, cattle, and wheels

Cultivating fragile desert and upland soils with methods suited to wet woodland agriculture

Developing water-control technologies unsuited to alluvial soils and short, intense precipitation

Pumping "fossil" groundwater far in excess of natural recharge

Some of these causes are discussed in Chapter 2. These and others have been treated in greater detail by Henry F. Dobyns (1978, 1981) and by Charles Bowden (1977).

While the causality for deterioration and desertification is complex and the relative contributions of the various factors, all human induced, will long be debated, what is more important, I think, are the underlying ethics and cultural axioms that produced these causes. *These continue to be operative.*

How can biologists, ecologists, and taxonomists at this late stage of the destructive process hope to comprehend the desert riparian of the Southwest as an ecosystem? R. Roy Johnson (1977), student of riparian communities, capsulizes the problems and the program:

> The dearth of relative and detailed historic data on riparian habitats is lamentable. We begin investigations on a complex and closely interwoven ecosystem without a reliable baseline. Much of our knowledge to date focuses on the aftermath of decades of abuse. But armed with this hindsight, we are beginning to collect, organize, and apply hard data to difficult problems. What percentage of breeding birds are dependent on riparian habitat? How much grazing is too much? How many species of plants occur only in riverine ecosystems? We have begun the long and expensive task of quantifying our generalities about the riparian habitat, so that we can offer valid alternatives to managers in their attempts to preserve the scant remnants of what was once a vast network of thriving and varied habitats.

In a minutely detailed manner this book attempts to present what is known of the interactions on the middle Gila of human cultures, water regimes, plant communities, and birds in the recent historic past, and especially in the 1960s and 1970s.

PART I

Changes on the Middle Gila

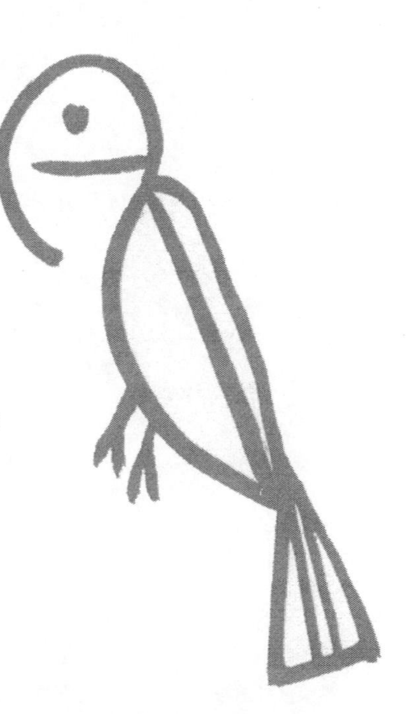

CHAPTER 1

The Pima People, Past and Present

The Gila River was once a well-defined stream meandering across a Lower Sonoran Desert floodplain with here and there marshes, lagoons, and oxbows. Its gallery forest of native cottonwoods and willows formed a green ribbon that travelers could trace for hundreds of miles through the desert. Other living streams—the San Pedro, Santa Cruz, Salt, Agua Fria—added their own waters to the middle Gila. Villages of agricultural Indians, early historic as well as prehistoric, dotted the fertile floodplains. These streams with their woods, lagoons, and grasslands, all abounding in birds and other forms of wildlife, are a thing of the past. The rivers are dead. Their biotic communities are gone. Their fragile watersheds have collapsed during decades of abuse. By the time the middle Gila had its first resident ornithologist (1907), the water regime had been altered, the channel was broad and unstable, the marshes and most of the timber had been scoured away by years of floods, and the grasslands had entirely disappeared. The native villages Father Kino found in the 1690s all along the Gila from the mouth of the San Pedro River to Yuma had been reduced to about half a dozen settlements.

This study of bird life and habitat change takes a close look at what was once the most mesic part of the Gila—its middle in south central Arizona, or what is now the Gila River Indian Reservation (Fig. 1.1). For several reasons, the reservation is ideal for such a study. When the first Spanish explorers pushed their way northward into the area they called Pimería Alta, they found Piman Indians dwelling along 150 miles (250 km) of the prime Gila River bottoms. These indigenous people have preserved an oral history extending back at least a century and a half and a lexicon of avian ethnotaxonomy reaching well over three centuries into the past. Later, when Europeans began traveling overland to California, the Pima Villages were a major outpost until

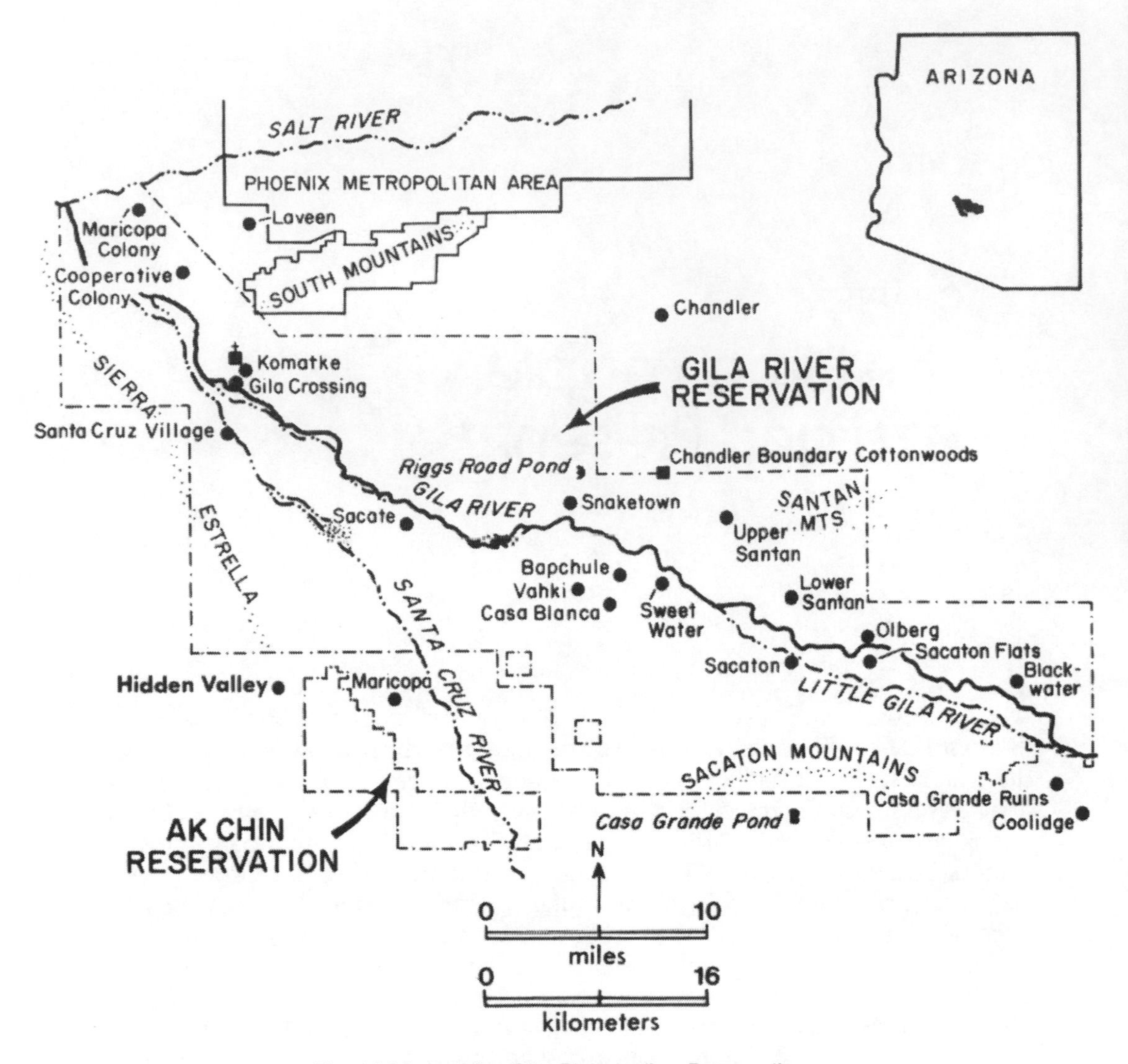

Fig. 1.1. Map of the Gila River Indian Reservation.

after the American establishment of Phoenix (1869) and the Southern Pacific Railroad (1878). Subsequently, various Spanish and American itinerants left accounts of the local habitats and sometimes of the fauna. Still later, M. French Gilman, a government teacher and biologist, worked on the reservation during a critical period (1907–1915) of riparian habitat destruction. These various historical sources provide a baseline to compare with my own investigations of habitats and bird life from 1963 to 1981.

The Riverine Pimans (often called "Pimo" by nineteenth-century Anglo travelers) are the northern part of a large subbranch of the Uto-Aztecan language family. This subbranch, Tepiman, extends from northern Jalisco,

Mexico, to the Gila River (Kroeber, 1934). Included are the Tepecano, the Northern and Southern Tepehuán, the various Pima Bajo groups, the extinct Sobaipuri of the San Pedro and Santa Cruz rivers, various dialectical groups comprising the Papago of southern Arizona, and the Gileños or Gila River Pima. All the lowland Pimans call themselves *áw' awtam* (or some variant such as *ó'otam* or ó'od'ham).

Among themselves these northern Pimans distinguish two essentially ecological groups: The *Táwhano Áw'awtam* (Desert People), known to Americans as the Papago, and the *Ákimul Áw'awtam* (River People), usually called today simply the Pima. The Pima/Papago tribal distinction is of administrative convenience to Americans. A third group, the *Kohadt* or *Kwahatk,* living downstream (north) from Picacho where the Santa Cruz River went underground, are culturally and geographically intermediate between the *Tawhano* and *Ákimul Áw'awtam* but dialectically are nearest Pima. They live today in villages on the northern Papago Reservation or are intermarried with Gila Pima. This book deals with the ecology of the heartland of the River People.

In the other directions the Gila Pima were bounded by non-Uto-Aztecan peoples. To the east were the nomadic Apache, dwelling for the most part in the Gila watershed. Living partly by raiding, they were a rather constant threat to agriculturists of the fertile middle Gila valley. To the north were the Hualapai, the Yavapai, and the various Puebloan groups. To the west were Yuman speakers. Five Yuman groups from the Colorado and lower Gila rivers amalgamated through time, eventually becoming known as the Maricopas (Ezell, 1963). They sought refuge with the Pima at least as long ago as the early half of the 1800s (Russell, 1908:93, 196; Spier, 1933:1, 18, 26, 40). Since that time the Maricopa settlements have been immediately to the west of the Pima villages, where they have preserved their ethnic identity to this day. "Their settlements [originally] were on both sides of the river from Sacate and Pima Butte to Gila Crossing as the western limit. On mesquite gathering and fishing expeditions they were accustomed to camp along the slough (Santa Cruz River) at the northeastern foot of the Sierra Estrella, in the Gila-Salt confluence and on the Salt as far upstream as Phoenix, but they had no settlements there [then]" (Spier, 1933:18).

Both the Pima and the Maricopa traditionally are irrigation agriculturalists, farming the Gila floodplain with hot-season crops such as cotton, melons, beans, and corn, and winter crops such as wheat and alfalfa. That the Pima were *ditch* irrigationists prehistorically has been contested (Winter, 1973). If they were not, at least they enthusiastically adopted ditch irrigation, together with wheat, after initial Hispanic contact near the close of the seventeenth century. The Jesuit Kino, the first European to explore the northern part of Pimería Alta, found Pimans not only on the San Pedro (the Sobaipuri) but also spread down the Gila to what is today Gila Bend.

Still earlier the Hohokam, one of the great cultures of the Southwest, flourished in this area, subsisting by means of agriculture, gathering, and limited hunting (Haury, 1976). Beginning perhaps as early as 300 B.C., but certainly by A.D. 300, these Mesoamerican people plied the lower Salt and middle Gila valleys with an extensive system of huge canals, further helping to make this area an enormous desert oasis. The Hohokam apparently disappeared around A.D. 1450, ungratefully leaving little solid evidence on which archaeologists could build theories of the culture's demise.

The one thing the Hohokam did leave in abundance, however, was painted ceramics and potsherds. These have designs in a dark orange-red or even reddish purple pigment painted on a slip that is usually warm buff or tan. Both positive and negative representations appear. Zoomorphic figures are common and reveal a great deal about Hohokam inspirations and surroundings. One of the longest inhabited and subsequently most thoroughly excavated sites is the Hohokam metropolis of Snaketown, near modern Bapchule (Fig. 1.1). Over a third of the Snaketown life-form sherds depict birds and of these about 40 percent are unmistakably water birds with long bills, necks and legs. Often these are shown eating reptiles or fish. Aquatic birds were clearly a conspicuous element in the Hohokam environment. The Japanese *sumi-e* artist Takashi Ijichi analyzed the technique and style of Hohokam pottery painting, then produced the brush paintings that appear throughout this book.

Some two and a half centuries later when the Spaniards reached Pimería Alta, they found only Pima rancheria settlements along the Santa Cruz, San Pedro, and middle Gila rivers. Some scholars see a connection between the Hohokam and the modern Pima, but there is scant support for such an idea. Riverine Pima ties—linguistically, esthetically, mythologically, ethnohistorically—are intimately connected to the other Piman groups to the south, of which they are just the northernmost extension. The Piman response to both Hispanic and Anglo explorers was the same when asked about the great ruins found in the Salt-Gila Valley: they disassociated themselves completely from them.

The Gila River Indian Reservation, established in 1859, consists today of 371,933 acres (150,521 ha) lying in the Sonoran Desert of the lower Sonoran life-zone. The reservation encloses a section of the middle Gila River for approximately 65 miles (104 km), running from Coolidge and Florence northwest to the Phoenix metropolitan area. The reservation averages 35 miles (56 km) wide. The floodplain elevation ranges from 1,455 feet (444 m) at the Blackwater area to 941 feet (287 m) at the Salt-Gila confluence.

The northwestern section of the reservation (Fig. 1.2) is bounded by the Salt River, historically a stream of greater volume than the Gila (Font 1931:47, 54; see also Pfefferkorn, 1949:28). The third river on the reservation is the Santa Cruz. When it was still a living river, the Santa Cruz flowed north from Sonora for about a hundred miles (160 km) and then sank into the sand north of Tucson, to emerge again on the reservation, where it flowed into the Gila.

The human population of the reservation consists of approximately 9,000 Pima Indians and about 300 Maricopa Indians, the latter living primarily in a single colony at the northwest corner of the reservation. The Pima traditionally have been farmers, cultivating the Gila floodplain by canal irrigation at least since the time of first Hispanic contact in the late seventeenth century.

Approximately half of the reservation is mountain and bajada country and the rest disturbed floodplains. In 1976 the tribe was leasing some 20,500 acres (8,340 ha) of agricultural land to non-Indians. The Gila River Tribal Farm cultivated 16,000 acres (6,500 ha). An additional 9,300 acres (3,785 ha) were farmed by individual Indians. Thus, a total of 45,800 acres (18,625 ha) of the reservation floodplains, about one-quarter of the lowlands, were under cultivation by the mid 1970s.

I have done no fieldwork on the Ak Chin Reservation (see Fig. 1.1). Ecologically, it is all disturbed farmland, almost entirely mechanized. Politically, it is under the jurisdiction of the Gila River Tribal Communities with headquarters at Sacaton. Likewise, I have done no work on the Salt River Indian Reservation. This is a politically independent community of Riverine Pima and Maricopa people.

The average annual temperature for the Gila River Reservation is 69°F (20.6°C). From 1963 to 1972 (Fig. 1.3) the average annual highest temperature was 115°F (46.1°C) and the average annual lowest temperature was 18.9°F (−7.28°C). The average annual precipitation for this period was 8.56 inches (217 mm) with extremes from 5.96 inches (145 mm) in 1973 to 12.44 inches (316 mm) in 1965.

The precipitation pattern (Fig. 1.3) is biseasonal. Summer rains begin about mid-summer and storms are generally short but intense. These come from the south or are convective in origin. The greater part of the annual precipitation falls at this time. The winter period is relatively dry, with storms originating from the northwest. The precipitation at this time is usually gentle.

Weather data are from stations at Sacaton (elevation 1285 feet; 392 m), Casa Grande Ruins National Monument (1,419 feet; 433 m), and Laveen (1,115 feet; 340 m) (U.S. Dept. Commerce, Weather Bureau, 1964–1973).

I began my studies of the bird life and habitats in August 1963 and was in residence on the reservation until the fall of 1969. From 1970 through 1975, I made several visits a month to the study area. From 1976 to 1982 I visited at critical seasons. Altogether I spent approximately 1,000 days in the field, collecting over 1,500 avian specimens on or near the reservation.

This work has several functions. Firstly, it provides a regional account of the bird life of the Gila River Indian Reservation and adjacent areas. As such it describes the distribution, status, and movements of some 244 full species that occur on the reservation or have occurred there within the historic past. Secondly, it focuses on the migration of birds through this one section of central Arizona. The subspecies taxonomy is one of the primary tools in the study of bird movements. Thirdly, this account is a study of avifaunal changes within the past century.

I consider distribution, taxonomy, and change to be tightly interrelated facets of an avifaunal study. Changes within historic time are usually overlooked because of a paucity or complete lack of historical data. The Gila River Reservation proves to be an excellent area to study historic changes in the Southwest. With European settlement of the Southwest, a chain of events (beginning probably with overtrapping beaver and overgrazing the Gila watersheds) led to the reduction of surface and subsurface water in the Gila, alteration of the vegetation, and change in the fauna. The result eventually was a complete loss of aquatic and riparian communities from most of the river.

Fortunately, the death of the riparian community has not been uniform. A few reasonably healthy reaches still persist in several locales in the low, hot desert not far from the reservation. One of these is Blue Point

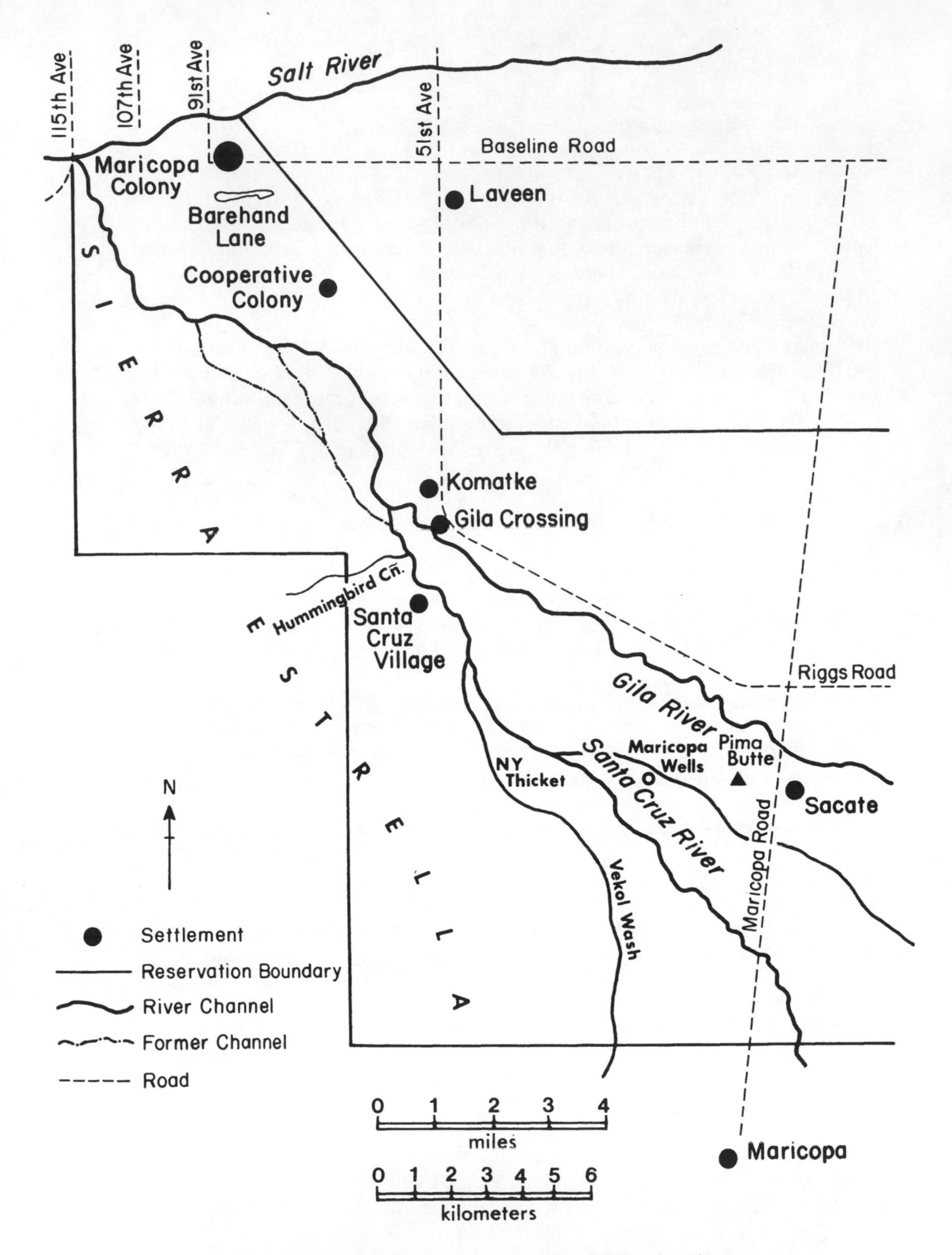

Fig. 1.2a. Map of the northwest section of the reservation (showing the confluence of the Salt, Gila, and Santa Cruz rivers).

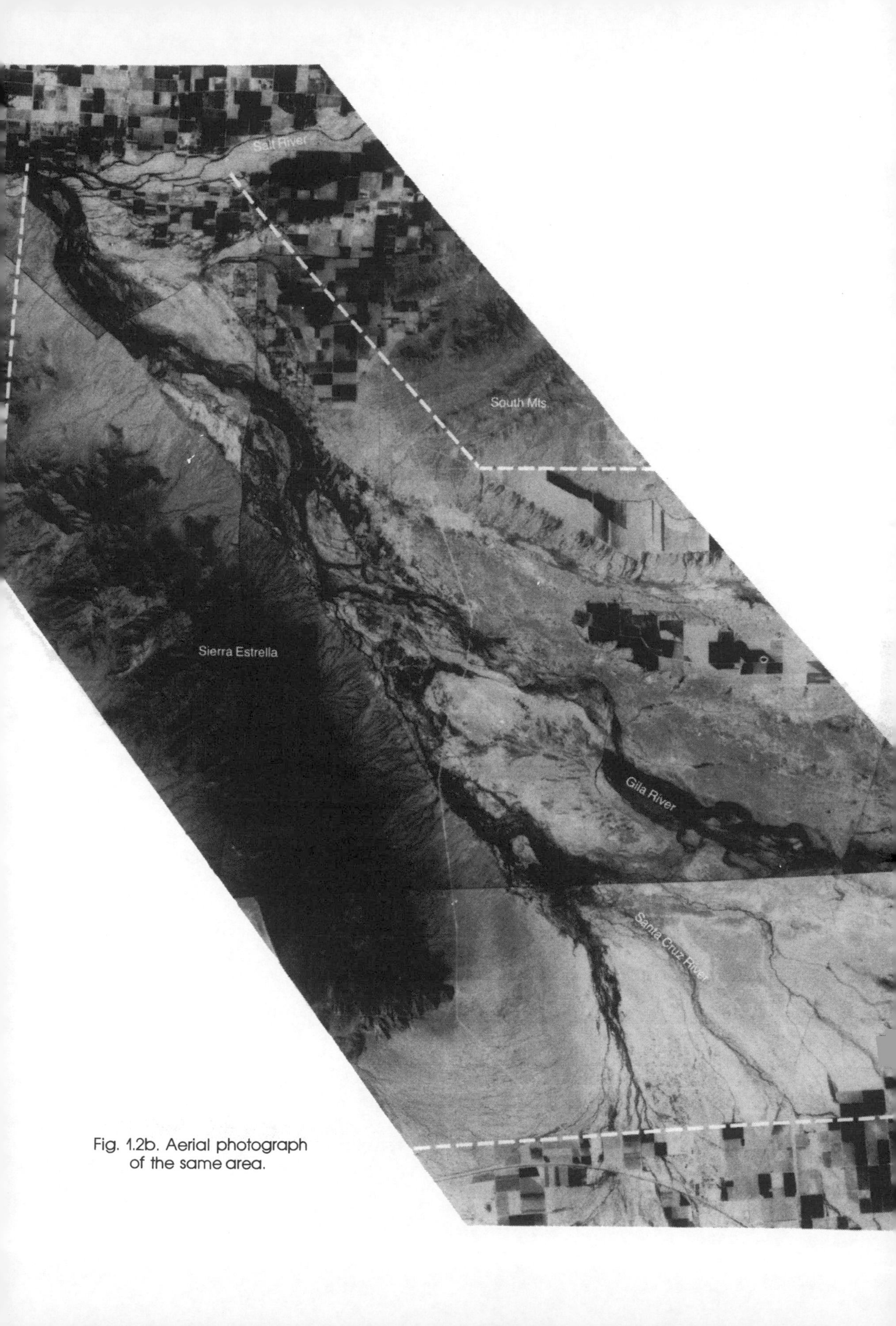

Fig. 1.2b. Aerial photograph of the same area.

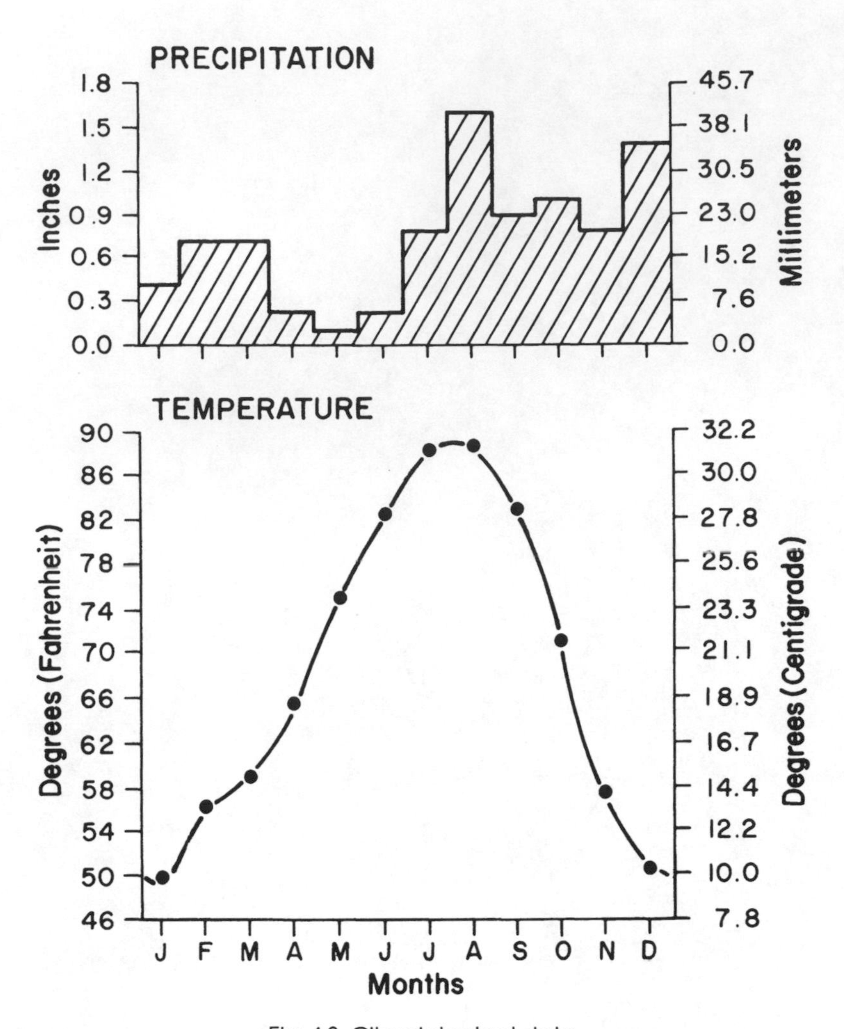

Fig. 1.3. Climatological data.

The precipitation data are the mean monthly total precipitation for the 10-year period, 1963–1972. The temperature data are the means of the monthly averages of daily minimum and maximum temperatures for the same 10-year period. Data are averaged from three weather stations on or near the reservation (see text). (Data from U.S. Department of Commerce, 1964–1973.)

Cottonwoods at the Salt-Verde River confluence, north of the reservation. The other is "The Buttes," a section of the Gila east of the reservation and above the Ashurst-Hayden Diversion Dam, between North and South Butte and the ghost town of Cochran. I have used these as modern analogs approximating aboriginal riparian conditions. (Of course, *all* desert communities in existence today are artifacts of decades of alteration—which makes the ecological studies that ignore this fact at best amusing.)

Change is an ongoing process. Study of change must not be limited to comparing the present with the past. If bird life is to be studied in the future, precise statements regarding present status (distribution, taxonomy, migration) are essential. This study should provide the indispensable baseline data for subsequent comparisons.

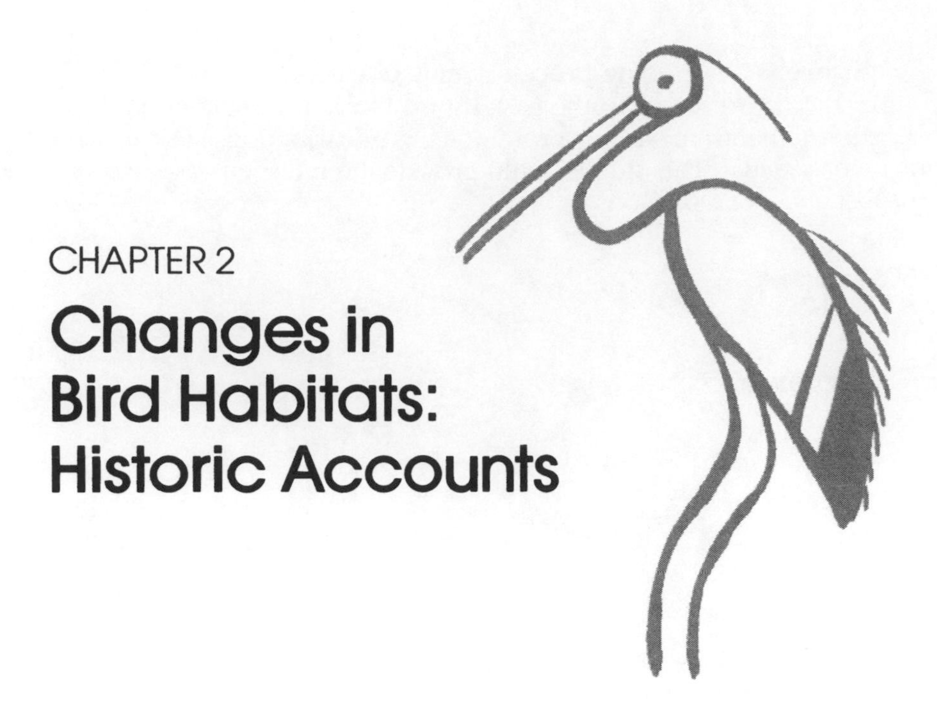

CHAPTER 2

Changes in Bird Habitats: Historic Accounts

Significant environmental changes have occurred on the middle and lower Gila River within historic times. The perennial flow of surface water was drastically reduced by the end of the nineteenth century. Only in exceptional places, such as about the junction of the Santa Cruz and Gila rivers, did surface flow continue until the 1950s. Groundwater levels dropped. As a result the entire riparian community of willows (*Salix*) and cottonwoods (*Populus*) was eliminated, being replaced along the river channel by exotic saltcedars (*Tamarix ramosissima*). Lowe (1964:30) summarized:

> Along the formerly great Gila River (the now dry bed of which stretches across the Sonoran Desert of western Arizona) there were extensive marshes, swamps, and floodplains with cattail (*Typha domingensis*), bulrush (*Scirpus olneyi*), giant reed (*Arundo donax*), commonreed (*Phragmites communis*), arrowweed (*Pluchea sericea*), and many trees. The dense vegetation of these well-developed riparian communities often stood 10–15 feet [3–4.5 m] high and supported a tremendous quantity and variety of wildlife.

This chapter documents the original habitat of the reservation portion of the river from selected historical Spanish and American accounts. (Literature citations of historical documents here give the original author with date of publication in English, rather than the date of the original manuscript.) For one who knows only the disturbed habitat as it is found in the twentieth century, it is difficult even to imagine that these earlier chroniclers are talking about the same desert stream we now know in central Arizona.

Hispanic Accounts

The first European to visit what today is designated the Gila River Indian Reservation was the Jesuit missionary Eusebio Kino. This was in November 1694. On his second *entrada,* November 1697, Kino (1919:171–172) recorded:

> Travelling always by the valleys of the Río de Quiburi [San Pedro River], we arrived at the large river, or Río de Hila. And following its bank and its very large cottonwood groves, after travelling three days' journey to the west, we arrived at the Casa Grande, and its neighboring rancherías. As we journeyed we always had on the right hand and in sight but on the other bank of the river, the very extensive Apachería.

Kino's last *entrada* was in March 1699. Speaking of the lower Gila River he said (Kino, 1919:195):

> All its inhabitants are fishermen, and have many nets and other tackle with which they fish all the year, sustaining themselves with the abundant fish and with their maize, beans, and calabashes.

At the Pima villages he noted (Kino, 1919:197):

> In some places they gave us so much and so very good fish that we gave it as a ration to the men, just as beef is given where it is plentiful.

It is an ironic footnote of history that the cattle industry Kino introduced into the Southwest indirectly led to the destruction of the Gila riparian community, which Kino was the first European to describe.

Captain Juan Manje, who accompanied Kino on his last visit to Pimería Alta, left his own account (1954:124; Hayden, 1965:8) of conditions:

> On the fourth [of March 1699] we continued toward the east and passing through the settlement of Encarnación, nine leagues [36 km] after this [area of Maricopa Wells], we slept in a pasture and broad fields, abundant in grass, where our horses ate well. We were informed that the other side of the river is much better for a cattle ranch and for horses; but I pass judgement on what I saw.

On this last *entrada,* Kino and Manje had with them eighty horses.

In 1744 another Jesuit, Father Jacobo Sedelmayr (1955:23), explored the Pima country for potential mission sites:

> Today there dwell in the basin of the Gila, not very distant from the Casa Grande, a branch of the Pima tribe divided into three rancherías. That farthest east is called Fuguissan [Tuquisan]; four leagues down stream in [is] Tussonimo. Still farther on the river runs entirely underground in hot weather, and where it emerges there is situated the great rancherías called Sudacsson [Sudacson]. All of these rancherías on either bank of the river and on the islands enjoy broad acres for the cultivation of crops. These Indians raise corn, beans, squash, and cotton which they use for clothing. Those of Sudacson raise wheat by irrigation.

Sedelmayr (1955:24) further described the area which is now the northwest part of the reservation:

> Leaving behind these Pima settlements and trekking down stream we came upon broad savannas of reed grass and clumps of willow and a beautiful spring with good land for pasture. We named the place Santa Teresa [Maricopa Wells]. Passing on down river another five or six leagues and keeping it always in view with its willows and cottonwoods, we came to its confluence with the Río de la Asunción [Salt], which in its turn is formed by the Salado and the Verde. A very pleasant country surrounds this fork of the rivers. Here the eye is regaled with creeks, marshes, fields of reed grass and an abundant growth of alders [*sic,* willows] and cottonwood.

The luxuriant habitat at the confluence described by Sedelmayr was completely eradicated by the 1950s. but during the 1960s and 1970s it was slowly redeveloping (on a very small scale) because of effluent released into the Salt River channel from the Phoenix sewage treatment plants (see Chapter 3).

The "Rudo Ensayo" originally written 1761–1762 by the Jesuit Juan Nentvig (1894:128–129) described the Pima country:

> Between these Casas Grandes the Pimas, called Gileños, inhabit both banks of the River Gila, occupying ranches on beautiful bottom land for ten leagues further [*sic*] down, which, as well as some islands, are fruitful and suitable for wheat, Indian corn, etc. So much cotton is raised and so wanting in covetousness is the husbandman that after the crop is gathered in more remains in the fields than is to be had for a harvest here in Sonora—this upon the authority of a missionary father who saw it with his own eyes in the year 1757.
>
> The most important of these ranches are, on this side, Tusonimo, and on the other Sudacason or the Incarnatión [*sic*], where the principal of their chiefs, called Tavanimo, lived, and further down, Santa Theresa [*sic*], where there is a very copious spring. Having passed out from among these ranches, the Gila, at a distance of ten or twelve leagues, receives the waters of the Assumption [Salt] River.

The last historic Jesuit to describe the Pima country was Ignaz Pfefferkorn (1949:28–29), who was stationed at the mission of Guevavi (below Tucson on the Santa Cruz River) in 1763:

> Following down the Gila beyond this, spread along both sides of the river, are the still unconverted Pimas. This tribe is separated into three populous communities, of which the largest inhabits a pleasant abundantly tree-covered country fourteen miles [22 km] long and irrigated by aqueducts, which are built from the river to the surrounding country with little difficulty because the land is so level.
>
> From the habitation of the Pimas it is approximately twelve miles [19 km] to the above-mentioned Río de la Asunción. The region where this river flows into the Gila is very beautiful, is entirely level, and is exceptionally good for all kinds of grain and plants. Both sides of these two rivers are inhabited by the Cocomaricopas.

In 1767 came the decree of Carlos III of Spain suppressing the Jesuits and expelling them from the New World. Shortly thereafter Franciscans took over the Piman missions in southern Arizona. (Franciscan missions were not established among the Riverine Pima until early in the twentieth century.) Within a year of the Jesuit expulsion Franciscan Father Francisco Garcés was stationed at Mission San Xavier del Bac. Garcés was an inveterate traveler. Alone or accompanied by whatever guides he could coerce from friendly tribes, he penetrated as far as the Hopi Mesas in Arizona and the Tulare country of southern California. He made numerous trips between 1768 and 1775 to visit the Gila Pimas. He described the vegetation of the Gila (Garcés, 1968:23): "En la ribera del Gila se hallan álamos, sauces, y mesquites. Este río por lo común es escaso de pastos, pero en las rancherías de San Andrés, y despobladas y en las inmediaciones del Sutaquison hay algunos zacates y en todo el hay abundancia de chamizo y carrizo."

"On the Gila River there are cottonwoods, willows, and mesquites. This river for the most part is scarce in pasturage, but in the rancherías de San Andrés, and the uninhabited areas and in the immediate vicinity of Sutaquison there is some grass and everywhere there is an abundance of chamise (saltbush) and carrizo (cane)" [my translation].

Garcés had a penchant not only for traveling but also for writing rather detailed accounts of the country and its people. In 1775 he accompanied the Anza Expedition to Los Angeles. Garcés' diary account (1900:107–109) of the Pima villages, translated by the ornithologist Elliott Coues, reads:

> In all these pueblos they raise large crops of wheat, some of corn, cotton, calabashes, etc., to which end they have constructed good irrigating canals, surrounding the fields in one circuit common (to all) and divided (are) those of different owners by particular circuits. Go dressed do these Indians in blankets of cotton which they fabricate and others of wool, either of their own sheep or obtained from Moqui. Not is this portion of the river abounding in pasturage but in the last pueblo called Sutaquison there is abundance, even to maintain a presidio, as has reported Señor Capitan Don Bernardo de Vrrea, having passed personally to inspect the situation most fit for founding missions.

The expedition consisted of 240 people, 695 horses and mules, and 355 head of cattle (Font, 1930:23). The abundant pasturage Anza encountered here was enough for 1,050 head of livestock.

The other Franciscan accompanying the Anza Expedition to California was Pedro Font. On 30 October 1775 the group, arriving from the south, halted at a lagoon before reaching the Gila River. Font (1930:32) surmised that the lagoon "appears to be formed by the water which runs into it from the plains during the rainy season, or from the Gila River itself when it overflows and wanders from its channel." This may well be what was later known as the "Little Gila."

Font complained of everything encountered in the villages. In his eyes the Pima were black and ugly, especially the women, their smell apparently aggravated by coarse food; the soil arose in clouds as the troops marched through, the water was salty, the pasturage scant, the fields poor except along the river banks, and the timber was not very large. Font (1930:33) concluded his entry for the day: "In short, in all this land of Papaguería [*sic*] which we passed through I did not see a single thing worthy of praise."

Early in November Captain Anza halted the march at some lagoons or pools south of the Gila River for several days' recuperation. Anza (1930:22) records, "It is unfortunate that these lakes have such bad water for they have an abundance of good pasturage." Some saddle animals became sick, but Anza reported that the people took the precaution of bringing domestic water from the Gila three leagues (about 7.5 miles or 12 km) away. But Font drank the water and suffered with the animals.

Font (1930:43–44) was not impressed with the Gila River. "This stream is large only in the season of floods, and now it carries so little water that when an Indian waded in and crossed it the water only reached half way up his legs. This is the reason why they had not yet planted, according to what they have told me, for the river was so low the water could not enter the ditches." Anza (1930:19) was told that the water would rise by the middle of the month.

The earlier Spanish itinerants make no mention of fences. But the introduction of domestic herbivores and the intensification of agriculture (Doelle, MS) apparently changed things. In 1775, Font (1930:43) wrote: "In the afternoon I went with Father Garcés, accompanied by the Pápago [*sic*] governor of Cojat [Kohadk], to visit the pueblo [of Uturituc or Juturitucam, between Sacaton and Sweetwater, *fide* Bolton] and see the fields. The latter are fenced in with poles and laid off in divisions, with very good irrigating ditches, and are very clean. They are close to the pueblo and on the banks of the river." In 1795, Fr. Diego Bringas (1977:89–90) reported of the Gileños:

> Their pueblos are solidly built, and in some of them I have seen houses with brick walls. Their fields are solidly and cleverly fenced and diligently worked. They know how to make use of the advantages offered them by the Gila River near whose banks they live. It covers the fields at flood season. This is one of the factors which promises a successful foundation in a terrain which with the assistance of improved methods will yield an abundant harvest. This land is good for all kinds of seeds and plantings.

Bringas, Apostolic Visitor, hoped to establish Franciscan missions among the Gila Pima. In 1796–1797, he wrote to his potential sponsor, the King of Spain, detailed accounts of this northern section of Pimería Alta (Bringas, 1977:90–91):

> The natural products of the country as far as to the banks of the Gila are kinds of bushes: chamizo (fourwing saltbush), ocotillo, hediondilla (creosote [bush]), and others which are unknown. Pasturage is scarce except for some points right beside the river, and in a *ciénega* which is to the west of the pueblos. As for trees, the banks of the river are covered with cottonwoods and willows which are the only timber for construction, although at a distance of 25 leagues [about 100 km] almost directly north there is an abundance of pine. Mesquites, creosote bushes, and saguaros are found in the open country, as well as quail, rabbits, hares and deer. The river abounds in fish of various species. As for harmful

animals, none are seen but the coyote. The fruits which are actually grown and cultivated by the inhabitants are various: beans, maize, wheat, watermelons, melons, squash, and cotton. If you wish to considèr others which can be grown there, every species of grain, tree and legume would do well because of the mild climate and even temperature. The river can fertilize these beautiful tracts of land with its waters. These can easily be conducted anywhere for farming. Even the gentiles steal a portion of its water by means of a poorly built dam which feeds a main ditch along their fields and distributes the water to small fields cultivated by each family.

But as editors Matson and Fontana point out (*in* Bringas, 1977:2), Spain and England became involved in war in 1790 and the Bringas report was destined never to be delivered to the king. Mexican revolutionaries and Spanish royalists began dividing up sides, even among the clergy, and war broke out in 1810, leaving little political or economic interest in Pimería Alta.

Early Anglo Accounts

The United States acquired from France, by the 1803 Louisiana Purchase, an immense parcel of land west of the Mississippi, ending somewhat vaguely at the Rocky Mountains. Western boundaries were technicalities for statesmen to worry about. Following the Mexican War of Independence in 1821, the already weak Hispanic influence began to wane in what was to become southern Arizona. Floods of Anglo-Americans began to traverse the country, often passing through the Pima Villages. The earliest of these were trappers (who seldom left reliable written accounts) and United States military personnel on reconnaissance into Mexican territory. The military personnel often kept journals of their Arizona travels. Beaver attracted the trappers. They occurred in exploitable numbers from the headwaters of the Gila to its mouth at the Colorado River (G. Davis, 1982; Dobyns, 1981).

At this time there were two major transcontinental routes to California—few attempting a more northern continental crossing through the Rocky Mountains and the Sierra Nevada. The more difficult of the two was to follow the Gila directly west from its headwaters in western New Mexico. A longer but somewhat easier route was to come west at a lower latitude to the Mexican village of Tucson, then head northwest to the Gila River, a course followed almost exactly by Interstate Highway 10 today. Both routes passed through the Pima Villages.

Few itinerants followed the Gila downstream from the Pima Villages to its confluence with the Salt. Most left the western (Maricopa) villages about the area of Sacate and Maricopa Wells (earlier called Santa Teresa) and traveled a *jornada* of 45 miles (72 km) south of the Sierra Estrella to Gila Bend, where they again struck the river. The *jornada* was a day's travel without available food or water for pack animals; understandably, some travelers preferred to make the trip at night. In spite of the hardships, the *jornada* shortened the road to California by about 90 miles (145 km). As a result few Europeans ever reached the great expanse of marshes at the confluence. The exceptions who left written accounts (quoted earlier) were Kino, Sedelmayr, and Pfefferkorn. Just 3 miles (4.8 km) below this confluence, the now dry Agua Fria River (together with the New River) enters the Gila from the north. This river drains the east side of the Bradshaw Mountains about Prescott and the west side of the Black Hills about Jerome. Still farther to the west the Gila received water from the Hassayampa River draining the southwest side of the Bradshaws. The confluences of all these desert streams (Santa Cruz, Salt, Agua Fria, Hassayampa) contributed to a vast marshland the likes of which we can scarcely imagine now. These rivers are all now dry and there is no record of their biotic communities, except those tantalizing hints mentioned in passing by the early Jesuits.

Lieutenant Colonel William H. Emory, while on military duty with the Kearny Expedition reconnoitering the Southwest, usually detailed at each night's camp the geological, faunal, and floral features encountered during the day's march. However, when the expedition, heading for action in California, passed through the Pima Villages in November 1846, Emory (1848:81) was more concerned with documenting the archaeology and ethnography. The troops camped somewhat up river from the Villages on 10 November:

> Near our encampment, a corresponding range draws in from the southeast, giving the river a bend to the north. At the base of this chain is a long meadow, reaching for many miles south, in which the Pimos [*sic*] graze their cattle; and along the whole day's march were remains of zequias [*acequias*], pottery, and other evidences of a once densely populated country. About the time of the noon halt, a large pile, which seemed the work of human hands, was seen to the left. It was the remains of a three-story mud house [Casa Grande ruins].

The following day Emory (1848:85) recorded, "The bed of the Gila, opposite the [Maricopa] village, is said to be dry; the whole water being drawn off by the sequias of the Pimos for irrigation; but the ditches are larger than necessary for this purpose, and the water which is not used is returned to the bed of the river with little apparent diminution in its volume." His overall impression of the Pima (Emory, 1848:86) was, "This peaceful and industrious race are in possession of a beautiful and fertile basin. Living remote from the civilized world, they are seldom visited by whites, and then only by those in distress, to whom they generously furnish horses and food."

Captain Abraham R. Johnston (1848:559), a member of the same expedition, likewise kept a travel account:

> Leaving the Casa Grande, I turned towards the Pimos, and traveling at random over the plain, now covered with mesquite, the piles of earth and pottery showed for hours in every direction. I also found the remains of a sicia [*acequia*], which followed the range of houses for miles. It had been very large. When I got to camp I found them on good grass, and in communication with the Pimos, who came out with a frank welcome. Their answer to Kit Carson, when he went up and asked for provisions was, "Bread is to eat, not to sell; take what you want."

A third member of the Kearny Expedition to keep a detailed journal was Henry Smith Turner. At the camp downstream from Casa Grande ruins, about 8 miles (13 km) above the Pima Villages, Turner (1966:108) wrote: "We have traveled today through a very extensive valley, capable of being irrigated for cultivation—it has occurred to some of us it would be a suitable place for Mormons to establish themselves. We have had a delightful day—a little too warm for comfort. We find the grass quite green and good—infinitely better than we had any expectation of finding." Before making the *jornada* to Gila Bend the expedition halted at what was to become known as Maricopa Wells, southwest of the present Sacate. "We are encamped on a spot where the grass is excellent, wood sufficiently abundant, and water a short distance off. We have left the Gila several miles to the north of us, and will not see it again for 2 days. We are now encamped in a slough [Santa Cruz River] that empties into the Gila" (Turner, 1966:109–110).

Marshall's discovery of gold in the mill tailings at Coloma, El Dorado County, California, led to the 1849 Gold Rush. The Pima were on the major route to the Mother Lode. "It was reported that some sixty thousand Anglo travelers passed through their villages during the westward migration" (Spicer, 1962:147).

John Durivage (1937:217), passing through the villages in June 1849, recorded that "The Gila at this point is narrow, not more than 100 yards [28 m], and flows at the rate of six miles [10 km] an hour. The meadow in which we camped [west of Sacate] is three miles [4.8 km] from the Gila and has a number of springs in it. The water is nearly all saline, and one spring is strongly impregnated with sulphur."

George Evans with a contingent of forty-niners arrived at the Gila from Tucson, camping 22 August 1849 about eight miles (13 km) above the Pima Villages (Evans, 1945:152–153): "Here we found a little grass, and our poor animals are once more feeding with a good appetite... The Gila River opposite our present camp is a deep, narrow, and rapid stream of warm, muddy water, with the banks covered with a dense growth of wild willow and weeds, tall cottonwoods, and the low willow tree, known as the water willow [seepwillow, *Baccharis glutinosa*]. A dam has been constructed and by small canals the water is conveyed over the bottoms and thrown into the fields. Here they raise nearly every kind of grain that is usually found in a warm climate."

A month later John W. Audubon, the younger of the famous naturalist's two sons, passed through the Villages (1906:157): "The river bottom here forms a great flat, which was, I think, once irrigated; at all events, it is cut up by a great many lagoons, nearly all muddy, but the water is not so salt in those that do not run, as to be undrinkable; in some places the water is so impregnated that as the water evaporates, a cake of pure salt is deposited, and the Indians on being asked for it, brought us five or six pounds [2–3 kg] in a lump. It was pure white when broken, but on the surface a sediment covered it. The country is nearly flat, and on the light sandy soil there is found grass, in some places very sparse and thin, and in others pretty good."

Robert Eccleston (1950:207) approached the Pima Villages from the south on 18 November 1849."It was not long before the road came close to the long-looked-for Gila. I rode in to see it, as the cottonwood, willow, etc., obstruct the view, and found a swift stream about 40 feet [12 m] wide, not as clear as I expected to find it, but perhaps this may have been caused by the late rain."

Following the peace treaty of Guadalupe Hidalgo in 1848 between the United States and the Mexican Republic a survey was required to determine the new boundary from San Diego to El Paso. The survey began late in 1850, with the Mexican contribution being nominal at best. John R. Bartlett, the U.S. Commissioner, and his crew of surveyors, draftsmen, naturalists and artists, spent two weeks of early July 1852 at the villages of the Maricopas and Pimas, surveying and making illustrations (Fig. 2.1a; see also Hine, 1968). There were enough horses and mules on the survey to haul 160 tons of freight from Louisiana to California. A good deal of pasturage was needed on this desert trek.

Bartlett (1854) left the most elaborate nineteenth-century account of the Pima Villages. Coming up the Gila from Yuma, they traveled the *jornada* from Gila Bend to the villages at night, arriving 30 June.

> At daylight we passed the southern end of a range of mountains [Sierra Estrella] which extend to the Gila, terminating near the mouth of the Salinas [Salt] River; and at half past six we reached some waterholes [Maricopa Wells], about a mile from the first Coco-Maricopa Village, thus making the journey of forty-five miles [72 km] in thirteen hours.... It was indeed a pleasant sight to find ourselves once more surrounded by luxuriant grass. Although we had met with a little salt grass [*Distichlis*] in one or two places on the march, which no animal would eat if he could get anything else, we had not seen a patch of *good* grass since leaving our camp at San Isabel [*sic*], fifty-six miles [90 km] from San Diego [California].... As it would yet be several hours before we could look for the wagons and the remainder of the party, we turned the mules out to luxuriate on the rich pasture before them, and creeping under some mezquit [*sic*] bushes, soon fell asleep.... The water here is found in several holes from four to six feet [1.2 – 1.8 m] below the surface, which were dug by Colonel Cooke on his march to California. In some of these holes the water is brackish, in others very pure. The Gila passes about two miles [3.2 km] to the north; for one half of which distance the grass extends, the other half being loose sand. [Bartlett, 1854(2):210 – 211]

This area is upstream from the New York Thicket. Today it is devoid of grass, except sparse salt grass and ephemeral growth (Fig. 2.1b).

Bartlett's men were anxious to push on to a place with shade trees on the river where they could bathe. "After crossing a deep arroyo of sand [Vekol or Greene's Wash?], which is filled by the river at its floods, and pushing our way through a thick underbrush of willows, we at length reached the bank of the [Gila] river. There were many fine large cottonwood trees, beneath which we stopped, and which afforded us a good shade from the scorching rays of the sun, but there was not a blade of grass to be seen, and, what was worse, *the Gila was dry!* [Bartlett, 1854(2):214 – 215]." But Bartlett's men obtained water by digging two-foot [0.6-m] wells, and they found pasture some miles beyond the easternmost Pima village.

> The dryness of the river was produced by the water having been turned off by the Indians to irrigate their lands, for which the whole stream seemed barely sufficient. It is probable, however, that, with more economical management, it might be made to go much further. The valley or bottom-land occupied by the Pimos and Coco-Maricopas extends about fifteen miles [24 km] along the south side of the Gila, and is from two to four miles [3 to 6 km] in width, nearly the whole being occupied by their villages and cultivated fields. The Pimos occupy the eastern portion. There is no dividing line between them, nor anything to distinguish the villages of one from the other. The whole of this plain is intersected by irrigating canals from the Gila, by which they are enabled to control the waters, and raise the most luxuriant crops. At the western end of the valley is a rich tract of grass, where we had our encampment. This is a mile or more from the nearest village of the Coco-Maricopas. On the north side of the river there is less bottom land, and the irrigation is more difficult. There are a few cultivated spots here; but it is too much exposed to the attacks of their enemies for either tribe to reside upon it. [Bartlett, 1854(2): 215, 232–233]

As a result of the Gadsden Purchase of 1853, lands south of the Gila River (including the Villages) became territory of the United States Government. The first Pima reservation, consisting of 100 square miles (40.47 ha) of land along the Gila, was established in 1859. A number of events during the next fifty years not only profoundly affected the lives of the Pima and Maricopa Indians, but radically altered the entire biotic community. The upper Gila and its tributaries had luxuriant grasslands, as has been amply documented by Hayden (1965). Cattle, escaped from abandoned Mexican ranches in southeastern Arizona, increased to form enormous feral herds. In 1846, for instance, the Mormon Batallion was attacked by a herd of wild bulls southeast of Tucson. At the end of a battle lasting several hours, more than fifty bulls lay dead on the banks of the San Pedro. Three years later the Texas Argonauts encountered a wild herd here estimated at between five and fifteen thousand head (Harris, 1960). The Apache, long in control of the Gila headwaters, undoubtedly practiced some animal husbandry, but to what extent is unknown. Dobyns (1981:79–89) demonstrated on the basis of an 1830 report by the commander of the Tucson Mexican presidio that the Western Apache had "a significant scale of herding of large Old World domestic animals in the Gila-Salt River headwaters." He believed the numbers were sufficient to initiate erosion. However, when the dreaded Apache were finally contained in 1873, great areas of the upper and middle Gila became available to Anglo settlement. Starting approximately in the 1860s and continuing for the next half-century the Gila Basin and its headwaters sustained a period of drought lasting until a reversal of the trend in 1906 as shown in the tree-ring record (see Fritts, 1965).

At the turn of the century Arizona's resident ornithologist, Herbert Brown (1900), published an essay on how the two-pronged specter was affecting the bird life of the Territory:

> The general aridity of the country is such that vast tracts of land must, perforce, remain forever uninhabited. Cattle interests are, however, a dominant feature in the development of the country, and to these, more than all else combined, must

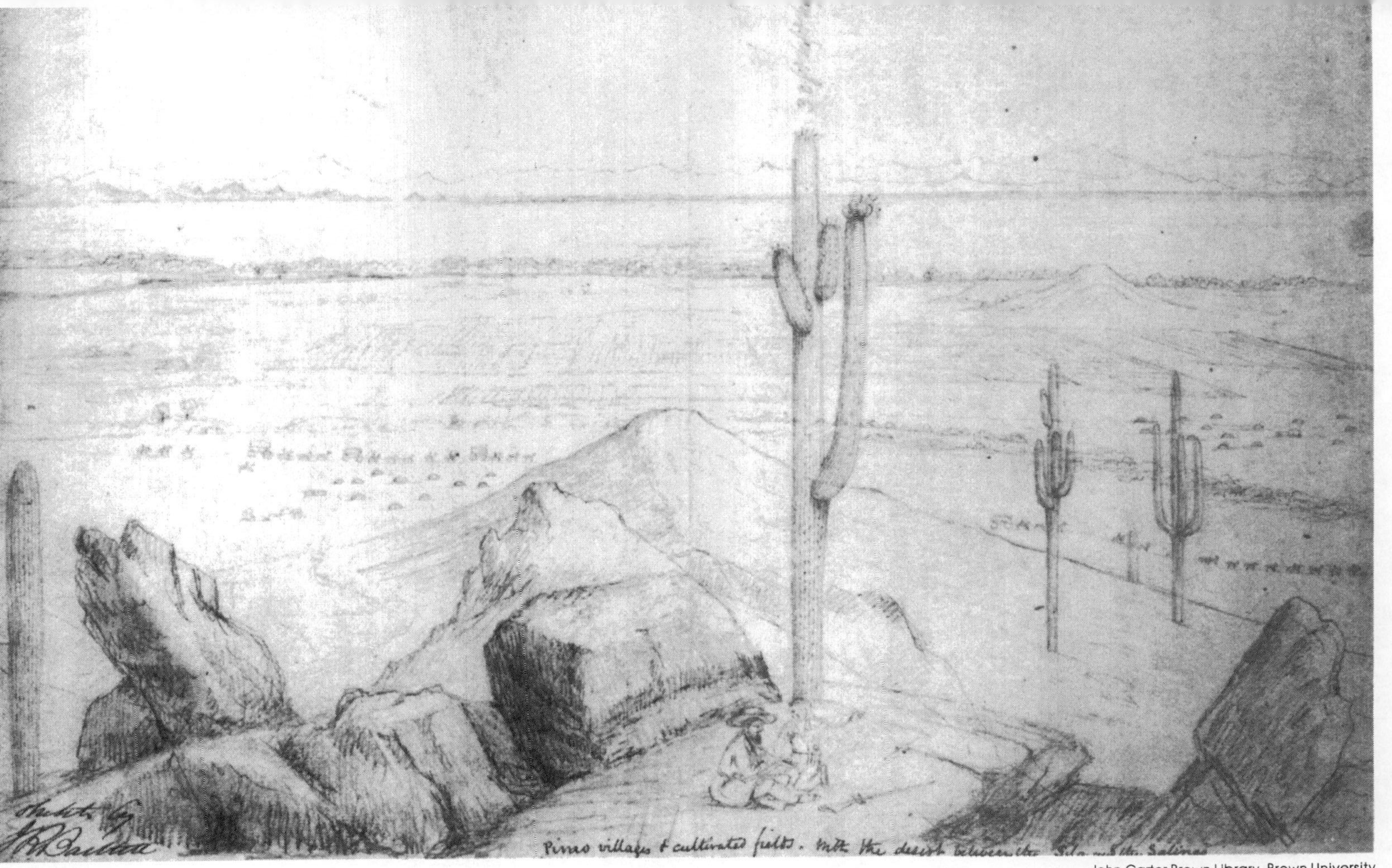

John Carter Brown Library, Brown University

Fig. 2.1a. A view of Pima and Maricopa villages and fields sketched by John R. Bartlett in the summer of 1852.

This sketch was made from the southeast end of the Sierra Estrella looking northeast across the lower Santa Cruz River (middle) and the Gila (running across the upper middle), with Pima Butte on right (above two distant saguaros). The lower parts of the South Mountains (horizon to left) appear dark as they do today. The Superstition Mountains would be on the horizon to the right. The fields and villages are south and west of the Gila.

At this time the only serious modification of the Gila River system had been excessive beaver trapping. The middle Gila and lower Santa Cruz are confined to narrow channels with well-defined banks unlike the broad sandy waste shown in the 1905 photograph at Sacaton (Fig. 2.2a). The gallery forest along the Gila is thick and mostly continuous, even though in places the Indians drew off all the surface flow for agriculture. To the left the Gila opens into a broad lagoonlike area. The water table in this area was particularly high. Serious disruption was to begin within the next 10–15 years.

Fig. 2.1b. Same view, 1982.

The same view 130 years after boundary survey commissioner John Bartlett sat on this point of the Sierra Estrella and sketched the villages and fields below him (now entirely gone). The several dark channels of the Santa Cruz River cross the picture in front of Pima Butte. These are mostly dead mesquites, fringed with a few living trees. Saltcedars engulf the channel to the left. The ruins of Maricopa Wells stagecoach station are mid-photograph between the Santa Cruz channels. The almost continuous cottonwood gallery forest along the Gila is gone. The famous grasslands of early travelers and the ciénaga of Santa Teresa have disappeared without a trace. (Photographed May 1982 by A. M. Rea.)

Fig. 2.2a. The Gila River channel at Sacaton in 1905.

A thin fringe of cottonwoods can be seen along the opposite bank of the river, but by this date years of drought and devastating floods had scoured away most of the riparian community. The formerly well-defined channel had become a bare sandy bed. M. French Gilman began studies here two years later. (Photo by H. L. Shantz, courtesy of University of Arizona Herbarium.)

be charged the obliteration of bird life in the so-called desert portions of the Territory. The stock business at one time promised enormous profits and because of this the country was literally grazed to death. During the years 1892 and 1893 Arizona suffered an almost continuous drouth, and cattle died by the tens of thousands. From 50 to 90 per cent of every herd lay dead on the ranges. The hot sun, dry winds and famishing brutes were fatal as fire to nearly all forms of vegetable life. Even the cactus, although girdled by its millions of spines, was broken down and eaten by cattle in their mad frenzy for food. This destruction of desert herbage drove out or killed off many forms of animal life hitherto common to the great plains and mesa lands of the Territory. Cattle climbed to the tops of the highest mountains and denuded them of every living thing within reach. The ranges were foolishly overstocked, and thus many owners of big herds were financially ruined by their covetousness, but under the most favorable circumstances it will be years before the life, once so common to the desert country, recovers from the shock.

But Herbert Brown was wrong—the country was never to recover. A point of no return had been reached in the arid ecosystem. Drought, coupled with overgrazing, reduced the surface flow of the Gila River (see Fig. 2.2). Writing in 1924, Hayden (1965:59) said:

Fig. 2.2b. Modern condition of the Gila channel.

This is the site, between Sacate and Pima Butte, of a former great grove of cottonwoods where Pimas from settlements both upriver and down would come to swim, fish, and picnic until the 1950s. On Bartlett's 1852 sketch the grove appears immediately to the right of Pima Butte. Non-native saltcedars are almost the exclusive tree cover. The understory is composed of disturbance plants: Russian thistle (tumbleweed), four-wing saltbush, cocklebur, and seepwillow. Water is ephemeral and there is no emergent vegetation in the channel. (Photo by A. M. Rea, May 1982.)

> The cause of this disaster is not difficult to find. A great change has taken place in the entire watershed of the Gila in eastern Arizona and New Mexico. A vast area which was once covered with a marvelous growth of grass has been damaged by livestock.

There were other factors contributing to the demise of the river system (Dobyns, 1978, 1981). Starting earlier in the century beaver, so important in flood control, were intensively and illegally exploited by American trappers in what was then Mexican territory. Since mines in the Gila headwaters required timbers as well as fuel for smelting, heavy demands were made on the forests. Pack animals in the dry, steep mining country set up erosional processes. Anglo farming methods, adapted to the temperate woodlands of northern Europe and northeastern North America, were ill suited to desert alluvial soils, and many of the fertile floodplains that had been farmed for centuries by natives suddenly became severely gullied (Duce, 1918; Bryan, 1925). In an already xeric ecosystem undergoing the stress of decades of drought, all these elements combined to produce an irrevocable disruption of the delicately balanced edaphic and biological network that formed the Gila watershed. The next hundred years is a story of decline.

Oral Historical Accounts

Shortage of water in the latter part of the nineteenth century led to economic and demographic instability of the Gileños. Many who could no longer earn a livelihood by farming turned to woodcutting as a source of income, and the older mesquite bosques along the river began to be felled to supply firewood to the new settlement of Phoenix.

In each village of the Pima and Maricopa, one man recorded the oral history of his settlement. Events were recalled by means of a calendar stick, a pole with notches or carved markings for each year which served the narrator as mnemonic devices. The calendar sticks of Gila Crossing and Salt River village for the year 1872–1873 (Russell, 1908:54) record the movement of people from the central part of the reservation, northwest of Casa Blanca, to the Salt River:

> For several years the Pimas had had little water to irrigate their fields and were beginning to suffer from actual want when the settlers on Salt river [*sic*] invited them to come to that valley. During this year a large party of Rso'tuk Pimas accepted the invitation and cleared fields along the river bottom south of their present location. Water was plentiful in the Salt and the first year's crop was the best they had ever known.

Twelve hundred Pima moved north (Spicer, 1962:148). Russell (1908:54) adds, "The motive of the Mormons on the Salt was not wholly disinterested, as they desired the Pimas to act as a buffer against the assaults of the Apaches, who were masters of the country to the north and east."

The following year the Gila Crossing calendar stick records that some Pima families searching for water went downstream, re-establishing the settlement at Komatke. Russell (1908:54) adds, "There is an unfailing supply of water at this place. The Gila, after flowing 75 miles [120 km] beneath the surface, rises to form a stream large enough to irrigate several hundred acres."

The predominantly Anglo settlement of Florence was established in 1868 just upstream from the reservation. According to Spicer (1962:149), "By 1887 the irrigation canal constructed to take water out of the Gila River for the White settlers utilized the whole flow. No water reached any of the Pima fields downstream." The only surface flow persisting on the reservation was from the confluence of the Santa Cruz to the confluence of the Salt.

Local damage was becoming obvious elsewhere in the Gila lowlands, as about Winkelman and its satellite communities at the mouth of the San Pedro, 45 miles (70 km) upstream (Granger, 1960):

> By 1890 the dangers of overgrazing by cattle on the hills which surrounded the valley became evident. With no grass roots to absorb and hold the rains which fell to the earth, floods washed down from the hills across the valley and into the San Pedro River, which broadened with every flood. Several times the store at Dudleyville was moved to prevent its being washed away by the river.

The Maricopa had lived above the Salt-Gila confluence at least since 1800 (Spier, 1933:18–20). Other peregrinating Yuman speakers (Halchidomas) began settling with them around 1833. The combined people lived in a series of villages, mostly along the Gila from Sacate and Pima Butte (sometimes called M Mountain today) to the Gila Crossing area. The names of some of their villages are stark reminders of former ecological conditions: Thick Cottonwood, Good Cottonwood, Cattails, and Reedy Place. The only Maricopa calendar stick recorded began in 1833, as do most of those of the Pima. It records that in 1869–1870 (the year, not incidentally, the Anglo settlement of Phoenix was established) the Maricopa abandoned their village on the Salt near Phoenix and settled above the Salt-Gila confluence, where the village is today. The remaining Maricopa from the traditional Sacate to Gila Crossing villages joined them about the 1880s, presumably also because of declining water (Spier, 1933:138–142).

The new horse, mule, and cattle industry, adopted by the Riverine Pima almost as enthusiastically as winter wheat agriculture, must have damaged the middle Gila pastures. The animals were herded by youngsters and driven into enclosures at night (ethnohistorical record; see also Dobyns 1981:80). Unlike free-ranging herds, close-herded animals caused intensive trampling. The annalist who kept the Blackwater village calendar stick records for the year 1869–1870 (Russell, 1908:53): "An unusually heavy rain occurred during the winter, which gullied the hills deeply." Overgrazing had brought about a turning point in the local erosional processes. This ecological change must have been significant to have been the event of the year recorded by the native historian.

After the floods of 1906 washed out their fields, the Komatke Pima at the base of the Estrellas relocated this village east of the Gila where the floodplain was broader (Fig. 2.3a). Native Pima consultant Sylvester Matthias (born 1911) described many small creeks and seepage water entering the Gila near Komatke during his boyhood. *Scirpus, Typha, Pluchea,* in addition to *Salix,* grew abundantly in these lagoonlike areas. Matthias tells of a great "lake" between Gila Butte and Gila Crossing on the Gila near where Lone Butte Housing is (incorrectly shown on Arizona Bureau of Mines/U.S.G.S. maps of 1923 and 1933 as being on the Indian Canal at Komatke): "It was circular, about one and a half or two miles long [2.5–3 km] and very clear, very clear. You could just see the fishes. Good divers could go in there. That's why we called it *má'as ákimur. Má'as* means clear. There was an ancient canal you could see, starting on the river near Pima Butte and running about thirty miles [48 km] across the desert to the ruins where the Maricopa Cemetery is [north edge Barehand Lane]." Bartlett's 1852 sketch of the westernmost villages (Fig. 2.1a) clearly shows this "lake" at the left. But by 1927 there began a marked decline in the Gila even in this part of the reservation. Also this year Matthias recalls that the saltcedar, *Tamarix ramosissima,* became established near Komatke. This Eurasian species disperses readily by seeds falling on exposed mud flats. Harris (1966:18) notes, "The natural flood regime of rivers in the Southwest, which is characterized by receding snowmelt flows in late spring and early summer, is therefore ideally suited to colonization." An explosive invasion must have followed the disintegration of the native riparian community. Today most of the Gila channel from the Colorado River to the Safford Valley is virtually pure saltcedar.

Fig. 2.3a. Komatke around 1914.

This photograph shows rather thick cottonwood rows lining the irrigation ditches in the village. The flats are well covered with saltbushes (*Atriplex*) and wolfberry (*Lycium*). The house walls and roof as well as the storage basket to the right are constructed of arrowweed (*Pluchea*). (Courtesy of Southwest Museum, Los Angeles, California. Photographer unknown.)

Fig. 2.3b. Komatke in 1982.

The 1914 house is still standing, though enlarged and reinforced. All the native cottonwoods are gone. The dark shade trees are exotic athel tamarisks (*Tamarix*). Scrubby mesquite growth has invaded the fallow fields behind the house. (Photo by A. M. Rea.)

Few Pima have committed their oral ethnohistory to writing. George Webb of Gila Crossing village is an admirable exception. Lack of water forced Webb to abandon local farming in 1938. His English prose (Webb, 1959) preserves the Pima narrative style, reminiscent of Isaiah writing in another desert:

> In the old days, on hot summer nights, a low mist would spread over the river and the sloughs. Then the sun would come up and the mist would disappear. The red-wing blackbirds would sing in the trees and fly down to look for bugs along the ditches. Their song always means that there is water close by as they will not sing if there is not water splashing somewhere.
>
> The green of those Pima fields spread along the river for many miles in the old days when there was plenty of water.
>
> Now the river is an empty bed full of sand.
>
> Now you can stand in that same place and see the wind tearing pieces of bark off the cottonwood trees along the dry ditches.
>
> The dead trees stand there like white bones. The red-wing blackbirds have gone somewhere else.

Webb astutely sums up the Pima plight in the early twentieth century: "Where everything used to be green, there were acres of desert, miles of dust, and the Pima Indians were suddenly desperately poor." Woodcutting to prevent starvation caused mature bosques to vanish. "Today there are no mesquite trees left on the Reservation that are not second growth. If you look at the base of any mesquite tree you will find a dry stump. That is where a much bigger tree once grew." But the construction of San Carlos (Coolidge) Dam in 1930 held promise of a return to the traditional Pima livelihood. "When the dam was completed there would be plenty of water. And there was. For about five years. Then the water began to run short again. After another five years it stopped altogether." Webb placed the blame on excessive groundwater pumping and on white competition for the precious surface flow.

On the lower Santa Cruz River, upstream from the Piman settlement of Santa Cruz, was an extensive mesquite bosque known as the New York Thicket. Local Pima Joseph Giff and Sylvester Matthias provided a description of the vegetation as it was in the earlier half of the twentieth century. The grass was high and the reeds (*Phragmites australis*) so thick one had to follow cow trails to get through. There were long sloughs standing all the time. The mesquites were very large and the Pima came here to cut house beams. Milkweed vine (*Sarcostemma heterophyllum*) and graythorn (*Ziziphus obtusifolia*) were abundant. There were many open meadows covered with salt grass (*Distichlis spicata*), dock (*Rumex* sp.), and so much yerba mansa (*Anemopsis californica*) that whole areas were completely white and could be smelled from a distance. There were few screwbean (*Prosopis pubescens*), only two or three cottonwoods in different places, but no willows or cattails. Really good water came from underground seepage. Many little streams supplied the Santa Cruz with most of its water. Giff adds, "You could hear the voice of the *aw' áwkukoi* [White-winged Doves] early in the morning

—all calling—like one voice. They nest in there, but go up in the *Hawshañ* [saguaros] at the foot of the Estrellas to eat."

Johnson A. Neff (1940a,b) of the U.S. Biological Survey, spent the summers of 1938 and 1939 in the New York Thicket studying the breeding of White-winged Doves. Neff (1940b) provided this account:

> Near the mouth of the Santa Cruz river southwest of Laveen is a great alluvial flat with intermittent heavy growths of mesquite that is called by the local Pima Indians the New York thicket. In some places it is aid to be fully six miles [10 km] in diameter. Nowhere is the growth continuous, however, mesquite thickets covering several acres being interspersed with openings of similar size. The thickets are connected by strands of mesquite growing along the banks of the numerous washes that meander in from the south. At the margins of the big thicket the growth of mesquite is young and short, but in older sections the mesquite and screwbean reach heights up to about 40 feet [12 m]. The entire thicket is on the Gila River Indian Reservation, a few miles east of the Sierra Estrella mountains, along the base of which is a narrow belt of giant cactus that furnishes food for the white-wings.

By the late 1970s, when I visited New York Thicket, about 90 percent of the mesquites were dead (Fig. 2.4). The groundwater table had fallen to about 100 feet (30 m), a depth not even the deep-rooted mesquites can readily reach. Between the dead trees were extensive areas of pale adobe soil devoid of vegetation. The living trees were mostly at the edge of the floodplain bordering the Sierra Estrella bajada. Evidently enough subsurface moisture drains from the bajada to keep some vegetation alive. Trees still living out in the bosque were found to be growing in the bottom of a Hohokam canal over 6.5 feet (2 m) wide. In spots seepweed (*Suaeda moquinii* [=*S. torreyana*]), provided fairly good coverage, together with tumbleweed (*Salsola iberica*) and an occasional *Atriplex* bush. The only grasses of importance were *Schismus barbatus* and occasional *Phalaris minor,* both introduced Mediterranean weeds. Even the salt grass has almost disappeared.

Early Twentieth-Century Accounts

The ornithological fieldwork of Breninger in 1901 and Gilman from 1907 to 1915 must be seen in the context of not only greatly reduced surface flow, but also lowering of the groundwater on the reservation with consequent gradual loss of riparian community.

Frank Russell, ethnographer of the River Pima, conducted his fieldwork in 1901–1902. Scattered throughout the ethnography, Russell (1908:69, 84–85, 116, 134, 147) made a number of passing references to riparian loss:

"The cottonwood occurs in a thin fringe, with here and there a grove along the Gila and Salt rivers."

"The once famous grassy plains that made the Pima villages a haven of rest for cavalry and wagon-train stock are now barren, and it is not until the mesquite leaves appear in April that the horses can browse upon food sufficiently nourishing to put them in good condition."

Fig. 2.4. A dead mesquite bosque: New York Thicket in June 1978.

Most of the mesquite trees in this once great lower Santa Cruz River bosque are now dead. The predominant plants remaining are winter ephemerals: Russian thistle or tumbleweed (*Salsola*), globe mallow (*Sphaeralcea*), tansey mustard (*Descurainia*), and *Schismus*, a Mediterranean grass. Some mesquites, seepweed (*Suaeda*), and graythorn bushes (*Zizyphus*) survive in favored spots. (Photographed by Wade C. Sherbrooke.)

"[Women] frequently smooth the hair with a brush which was formerly made of the roots of the 'Sacaton grass,' *Sporobolus wrightii,* but as this no longer grows along the river where the majority of the villages are situated, they now make use of maguey fibers. . . ."

"Reeds, *Phragmitis* [*sic*] *communis* [= *Phragmites australis*], were formerly common along the Gila, but continuous seasons of drought caused them to disappear."

"[Sleeping] mats were formerly made by the Pimas of the cane, *Phragmitis* [*sic*] *communis,* that grew in abundance along the Gila until the water supply became too scant for the maintenance of this plant."

The "Little Gila" mentioned by a number of early explorers from Kino onward, apparently was one of the last places for the water supply to give out on the eastern half of the reservation. Breninger's (1901:44) fieldwork centered here:

> A strip of land on both sides of the river is cultivated by the Indians, water for irrigation being drawn from the river and from a lake [*sic*]. The latter is of crater origin, and supplies an abundance of water. In consequence of this never-failing

> supply, a large area of ground covered with a dense willow growth is always flooded, and at the time of my observations teemed with bird-life. The crops along the ditch tapping the lake were luxuriant. The corn, the beans and the pumpkins sent forth such a pleasant freshness that it is not to be wondered at the tired, wing-sore birds after a night's flight, should be attracted by such a scene of peace and plenty. Away from irrigation was desert, dry and barren, supporting only such plant life as can withstand long droughts, and the heat of a long summer.

Breninger's supposition about the origin of the "lake," of course, was incorrect. This side channel may have been an ancient meander of the Gila or a Hohokam canal enlarged by floods (see Gilman's account, below).

Gilman's unpublished manuscript gives the first detailed floral and habitat description of the Gila River Reservation:

> The district covered [in ornithological explorations] was a narrow strip of territory on each side of the river for a distance of twenty miles [32 km]. My first base of operation was at Sacaton on the south side of the river, where was located the Agency and the Indian Training School. This point is about midway between the Indian village of Blackwater to the east and Casa Blanca to the west, each about ten miles [16 km] distant. The presence at Sacaton of cultivated fields and shade and orchard trees made it a favorable location for birds not often found in the open desert. Each of the Indian villages was also a center of such bird life for the same reasons. The last four years of my stay I spent at the Indian village of Santan on the north side of the river and directly opposite from Sacaton. On this side of the river the territory covered was from North Blackwater ten miles [16 km] eastward, to Skaw-Kwee or Snaketown, ten miles [16 km] to the west. On both sides of the river were located Indian villages a few miles apart, each with cultivated fields and irrigation ditches which had water in [them] whenever the Gila was not dry.
>
> On the south side of the river was a small sluggish stream containing water most of the year. It began in some swampy bottom land just east of Blackwater and emptied into the river a few miles west of Sacaton, running a distance of about fifteen miles [24 km]. It was commonly supposed the channel was an ancient irrigating canal which had been enlarged by floods of the Gila River. This stream was a favorite resort of ducks and geese in favored seasons and was frequented by waders and other water loving birds.
>
> The vegetation was abundant and varied according to location, the nature of the soil, its moisture and alkali contents, and the distance from the river, and also elevations and drainage. Nearest the river where occasional overflows were frequent were found growths of cottonwood and *Baccharis glutinosa.* Where it was moist without overflow grew willows and screwbean, *Prosopis odorata* [= *pubescens*]. Where moisture and alkali are abundant and combined, there are large patches of arrowweed, *Pluchea sericea,* and numerous bushes of *Allenrolfea occidentalis.* In dryer alkaline situations may be found *Sarcobatus vermiculatus* and some of the salt bushes, *Atriplex lentiformis, A. canescens, A. polycarpa,* or *A. linearis.* On good soil extending from the river back as far as moisture content of the soil will permit are found mesquite trees, *Prosopis velutina* and mixed with them when not too close together are the wild jujube, *Zizyphus lycioides.* Scattered among the mesquites and other growth are found

> several kinds of squawberries as they are locally termed: *Lycium fremontii, L. cooperi, L. torreyi, L. fremontii bigelovii, L. exsertum,* and a few others, the different varieties being found according to soil and location favorable to each. A few ash trees, *Fraxinus velutina,* were found scattered along irrigation ditches and the borders of fields next to fences were grown up in many cases with *Dondia torreyana* [= *Suaeda moquinii*] as well as some of the salt bushes.
>
> In the sand washes were found a few mesquites, palo verdes, *Parkinsonia torreyana* [= *Cercidium floridum*], ironwood, *Olneya tesota,* and catsclaw, *Acacia greggii.* As the hills were approached on either side there were found various species of cactus, as the giant cactus or saguaro, *Cereus giganteus; Echinocactus* [= *Ferocactus*] *lecontei; Echinocereus engelmannii;* the *Opuntias:* species *fulgida, bigelovii, versicolor, leptocaulis, arbuscula,* and some others. On the hills and at their rocky bases are found another palo verde, *Parkinsonia* [= *Cercidium*] *microphylla; Hyptis emoryi,* a sort of tree sage; ocotillo, *Fouquieria splendens,* and several small shrubs, as *Fagonia californica* [= *laevis*], *Encelia farinosa* and others. On suitable ground is found large areas of creosote bush *Larrea.* Scattered throughout this growth are specimens of crucifixion thorn, *Holocantha* [= *Castela*] *emoryi.*
>
> The abundance and variety of vegetation makes for a large bird population and the non-interference of the Pima Indians allows the birds to become quite tame [Gilman MS].

Gilman (1909a:49) detailed a bit more of the "Little Gila" vegetation:

> Half a mile [0.8 km] south of the Gila, and flowing parallel with it for about twenty miles [32 km] is a small stream called the Little River. Along its banks are a few cottonwoods, many willows and much water-mote (*Baccharis glutinosa*). Between the two streams, on the "Island," as it is called, are groves of cottonwoods, and a few Arizona ash trees (*Fraxinus velutina*). In places not cleared and cultivated by the Indians, is a dense growth of mesquite (*Prosopis velutina*), screwbean (*P. odorata*), and arrow-wood (*Pluchea sericea*), besides a number of scattered plants of squaw-berry *Lycium berlandieri*) and jujube (*Zizyphus lycioides*).

Conspicuously absent from Gilman's descriptions of vegetation is any mention of *Typha* or *Phragmites* (and he makes only one reference to a "tule marsh"), even from the species accounts of Red-wings, yellowthroats, and Song Sparrows. The nests of such marsh birds he records as placed in *Baccharis, Pluchea,* or *Salix.* The great flood of 1905 undoubtedly had removed the emergent vegetation and probably had scoured much of the drought-weakened riparian growth. Damaging floods continued on the Gila from 1905 to 1917 (Burkham, 1972). Figure 2.2a shows the Gila River near Sacaton during this period.

In 1917 Eckman, Baldwin, and Carpenter (1923) conducted a soil survey of the Middle Gila Valley, extending from 5 miles (8 km) upriver from Florence to the Maricopa-Pinal County Line (near the present village of Santa Cruz). Little remained of the former riverine habitats along this section of the river in 1917.

> The Gila River has a channel varying in width from less than one-fourth mile to a mile [0.4 to 1.6 km] or more. The banks of this stream are generally poorly defined and unstable. Shifting of the channels and cutting of the banks take place at each overflow, and the process is increasingly destructive as the stream bottoms are used more extensively for agriculture. The stream has an average gradient of about 10 feet per mile [1.9 m per km] through the area. During the rainy season great volumes of water fill the channel and overflow the banks, but during the dry season its sandy bed is bare. [Eckman et al., 1923:2088]

In his comprehensive survey of the erosion and sedimentation histories of various southern Arizona streams, Bryan (1922:72–73) summarized the Gila:

> Early reports indicate that up to about 1880 the Gila flowed in a relatively deep channel through its flood plain, overflowing it only in times of flood. There was also a considerable low-water flow. At present the channel is a sandy waste with many tortuous subsidiary channels, constantly shifting in position, and there is no low water flow except in favored places. There seems to be a greater proportionate load of sediment, which under present conditions is silting up the channel.

Eventually the decrease in surface and subsurface water in the Gila had an adverse effect on the remaining riparian community of the west end of the reservation, from the lower Santa Cruz to the Salt confluence, the last "favored place." Photographs of the Komatke area from 1923 to 1925 (Franciscan Archives, Mission San Xavier del Bac) show a thin but almost continuous line of cottonwoods on the north side of the river. The trees are about half a mile (0.8 km) above the river channel and are probably along irrigation canals. But streamside timber was already gone. Photographs taken there from 1939 to 1941 show a great reduction of the cottonwoods. There was only one when I arrived in 1963, but it has since disappeared (Fig. 2.3b).

Perspective

From these various sources a picture now emerges of the Gila at the time of contact. Aboriginally, the Gila at the Pima Villages was a stream of sufficient width to support tillable islands. Throughout most of the year the depth in most places was probably not usually greater than 3 feet (1 m) and the river was readily fordable on foot. At times the entire surface flow was diverted for irrigation or sank for part of its course into the sand. Such observations were probably coincident with seasonal periods of drought in the watersheds. The river gradient was shallow and the floodplain was so level that lagoons formed in places along the main channel. At least three such marshy areas existed along the Gila on what is now the reservation: near Sacaton, at the confluence with the Santa Cruz River, and at the mouth of the Salt River. The channel was well defined, and the banks were consolidated and well vegetated. Beaver abounded. Riparian timber, consisting mostly of cottonwood, willow, and ash,

Fig. 2.5. Modern Maricopa Wells, near Pima Butte.

The drainages of three major washes (Vekol, Greene's, and Santa Rosa) and the Santa Cruz River once formed a ciénaga where the water table was less than 3 feet (1m) below ground level. Early Hispanic and Anglo travelers found enough grass at this spot to pasture over a thousand head of livestock. Today, because of watershed damage and excessive groundwater pumping, the ciénaga and grasslands have totally disappeared and the mesquite bosques have died. Surviving vegetation included seepweed and wolfberry, and a non-native annual grass, *Schismus.* Along the base of Pima Butte a line of green mesquite is visible on the Santa Cruz channel. (Photo taken 26 April 1981 by A. M. Rea.)

was well developed. There was a dense understory of arrowweed, batamote, graythorn, and cane. The bottom lands were fertile, 6 to 8 miles (9.6 to 13 km) across and a strip several miles wide was cultivated by the Pima and Maricopa. Back from the fields and riparian growth were dense mesquite bosques.

Though spotty, grasslands and pasturage were well developed on at least two areas on or near the reservation. About 8 miles (13 km) above Sacaton and below Casa Grande Ruins was a grassland many miles long (Emory, Johnston, Turner, Evans). A second major grassland was west of Sacate and several miles south of the Gila, where a number of drainages (Vekol Wash, Greene's Wash, Santa Rosa Wash) running northwest converged with the Santa Cruz Wash, which once again became an above-surface stream approximately where it entered the reservation. This area, named Santa Teresa by Sedelmayr, later was known as Maricopa Wells. Itinerants found abundant and luxuriant pasture for as many as a thousand pack animals and cattle. The water table was only 4 to 6 feet (1.2 to 1.8 m) below the surface. Peculiar edaphic conditions impregnated some of the lagoons and wells with so many soluble chemicals that their brackish water affected animals and man (Garcés, Font, Anza; Reid, 1858). The wash today (Fig. 2.5) has neither water nor pasturage.

During the latter half of the nineteenth century the Gila River and its riparian communities were severely damaged by overgrazing upstream during successive years of drought, coupled with other factors that disrupted the watershed. By 1907 emergent vegetation was absent from the upper half of the reservation and the gallery forest persisted as but a mere fringe. Surface flow ceased and remnant emergent vegetation disappeared from the Santa Cruz – Gila confluence area in the 1950s.

CHAPTER 3

Major Modern Habitats

The reservation can be divided into three major physiographic sections: mountains, bajadas, and floodplains. Of these, the vegetation of the mountains and bajadas is relatively undisturbed, while the floodplains for the most part are greatly disturbed. The floodplains of three rivers occur on the reservation, and each has a different land use history. The lowlands are more diverse both from the viewpoint of overall differences in vegetational structure and in human modification, and hence receive relatively greater attention in this discussion.

The term habitat is used for broad divisions of the physiographic sections as proposed by Ricklefs (1973), Hanson and Churchill (1961), and others. In this work, habitat delimitation is based primarily on vegetative structure as utilized by birds (see Chapter 5) and is understandably somewhat arbitrary. Within each major habitat division are various avian microhabitats. Few if any bird species on the reservation have identical microhabitat requirements. Some habitats of the study area are unique in that only one exemplary site is represented.

Reservation habitats may be considered of two stability types. Those associated with surface water are "dynamic" in that they have been undergoing their major vegetational development within the past two decades or so (in the 1960s and 1970s). The future of these dynamic habitats is especially precarious because they are maintained by wastewater from human activities. In contrast, the "static" communities outside the floodplains are relatively stable, changing primarily in response to long-term climatological shifts.

Voucher specimens of most plant species mentioned in this chapter are deposited in the University of Arizona Herbarium and the San Diego Natural History Museum Herbarium. Specimens collected in the Sierra Estrella by Dr. E. Linwood Smith, marked with an asterisk, are in the Arizona State University Herbarium.

The Mountains

Several ranges of desert mountains occur along the Gila River, at least partially on the reservation. The Sierra Estrella, the largest and highest of the ranges, is the only one I have studied systematically. It runs from north-northwest to south-southeast for approximately 20 miles (32 km). The highest part of the range, Montezuma Peak, with an elevation of 4,510 feet (1,375 m), is more than 2,952 feet (900 m) above the surrounding floodplain. Much of the Sierra Estrella is precipitous and for this reason whole upper basins have been naturally protected from grazing by Eurasian herbivores. Small numbers of bighorn sheep (*Ovis canadensis*), mule deer (*Odocoileus hemionus*), and white-tailed deer (*Odocoileus virginianus*) persist in the Sierra Estrella. Native grass species surviving here include: tanglehead (*Heteropogon contortus*), sand dropseed (*Sporobolus cryptandrus*), big galleta (*Hilaria rigida*), Arizona cottontop (*Digitaria* [*Trichachne*] *californica*), three-awn (*Aristida glabrata*), slim tridens (*Tridens muticus*), *cane beardgrass (*Andropogon barbinodis*), *bush muhly (*Muhlenbergia porteri*), *sideoats grama (*Bouteloua curtipendula*), *woody melic (*Melica frutescens*), and *desert needlegrass (*Stipa speciosa*).

As with the bajadas, the mountains are floristically rich. Important woody species include: brittle bush (*Encelia farinosa*), wolfberry (*Lycium fremontii*), Arizona rosewood (*Vauquelinia californica*), ocotillo (*Fouquieria splendens*), prickly pear (*Opuntia violacea*), foothill paloverde (*Cercidium microphyllum*), desert agave (*Agave deserti*), Crossosoma (*Crossosoma bigelovii*), Bernardia (*Bernardia incana*), bush penstemon (*Keckiella [Penstemon] antirrhinoides microphylla*), desert hackberry (*Celtis pallida*), buckwheat brush (*Eriogonum fasciculatum*), *desert senna (*Cassia covesii*) and catclaw (*Acacia greggii*).

On the north-facing slopes of a few peaks in the Sierra Estrella are limited areas with Upper Sonoran life-zone vegetation. Distinctive vegetation, probably relictual, includes: one-seeded juniper (*Juniperus monosperma*), scrub live oak (*Quercus turbinella*), gooseberry (*Ribes quercetorum*), banana yucca (*Yucca baccata*), crucifixion thorn (*Canotia holacantha*), *desert oregano (*Aloysia wrightii*), wormwood (*Artemisia ludoviciana*), miner's lettuce (*Claytonia [Montia] perfoliata*), and virgin's bower (*Clematis* cf. *drummondii*). The junipers are mature trees, and no young trees have been found. The *Canotia*, while not as widespread in the sierra as *Vauquelinia*, forms locally tall and rather extensive stands. I have visited the area of oak, gooseberry, and juniper only twice (April 1967 and 5–6 May 1975), but visited the lower crests near Santa Cruz usually several times each year (see Fig. 3.1).

The vegetation on the more xeric, south-facing slopes of the mountains includes: white bursage (*Ambrosia dumosa*), brittle bush (*Encelia farinosa*), barrel cactus (*Ferocactus wislizenii*), foothill paloverde (*Cercidium microphyllum*), ratany (*Krameria parvifolia*), creosote bush (*Larrea tridentata*), teddy bear cholla (*Opuntia bigelovii*), fishhook cactus (*Mammillaria microcarpa*), and hedgehog cactus (*Echinocereus engelmannii*).

Fig. 3.1. The Sierra Estrella

The upper end of Hummingbird Canyon, above Santa Cruz Village, has been little disturbed floristically. The upper basin, above the cliff, has native grasses, agave, Arizona rosewood, and saguaros. Near a tank at the base of the cliff (middle of picture) grow bernardia, bush penstemon, desert hackberry, buckwheat brush, and bedstraw. In the foreground the predominant cover is tree sage, paloverdes, and brittle bush. Canyon Towhees, Rufous-crowned Sparrows, and Costa's Hummingbirds breed at these elevations. (Photo by Gary P. Nabhan, 25 November 1978.)

The Bajadas

In the Southwest, the sloping, well-drained alluvial piedmonts are called bajadas. These areas are extensive and well developed on the reservation, being found along the Sierra Estrella, South (Salt River) Mountains, Santan Mountains, and Sacaton Mountains. The bajadas are relatively diverse floristically (see Fig. 3.2). Important woody species include: foothill paloverde (*Cercidium microphyllum*), Mormon tea (*Ephedra fasciculata*), prickly pear (*Opuntia chlorotica*), buckhorn cholla (*O. acanthocarpa*), pencil cholla (*O. arbuscula*), saguaro (*Cereus giganteus*), ratany (*Krameria parvifolia*), creosote bush (*Larrea tridentata*), elephant tree (*Bursera microphylla*), triangleleaf bursage (*Ambrosia deltoidea*), white bursage (*A. dumosa*), and brittle bush (*Encelia farinosa*).

Courtesy of Environment Southwest

Fig. 3.2. Bajada community.

This plant community is floristically diverse and vertically stratified. An overstory is provided by saguaros, foothill paloverdes, and ironwood. The understory includes brittle bush, bursage, creosote bush, and buckhorn cholla. Though probably once rather heavily grazed, since the 1950s the reservation bajadas have received relatively little grazing pressure from non-native herbivores. (Former Komatke village site west of Gila, photographed by Wade C. Sherbrooke, June 1978.)

Fig. 3.3. An arroyo dissecting the bajada above Santa Cruz Village.

Dominant vegetation at this elevation includes beloperone, ocotillo, both types of paloverdes, and tree sage. This part of the arroyo is the main winter nesting area for Costa's Hummingbirds. (Photo by Gary P. Nabhan, 25 November 1978.)

Cutting through the bajadas, sometimes to a considerable depth, are arroyos or washes (Fig. 3.3) that are dry except following storms. Some of these arroyos, particularly those along the north slope of the Sierra Estrella, have the appearance of being large sections of dry rivers and attract certain migrant and breeding birds seldom found elsewhere on the reservation. Distinctive arroyo vegetation includes: desert ironwood (*Olneya tesota*), catclaw (*Acacia greggii*), blue paloverde (*Cercidium floridum*), chuparosa (*Justicia [Beloperone] californica*), canyon bursage (*Ambrosia ambrosioides*), desert lavender (*Hyptis emoryi*), burrobush (*Hymenoclea salsola pentalepis*), *yerba del venado (*Porophyllum gracile*), and bedstraw (*Galium stellatum*). Both mesquite (*Prosopis velutina*) and foothill paloverde (*Cercidium microphyllum*) occur commonly on arroyo banks.

The Floodplains

The floodplains formed by the three rivers occupy a major portion of the reservation. Virtually all the area is disturbed, having been cultivated to a greater or lesser extent since the beginning of the Hohokam period, perhaps as early as 300 B.C. The Pima and Maricopa continue to farm the floodplains. Today enormous areas are being converted to industrialized farming by non-Indians who lease reservation land. Eight major habitat divisions of the floodplains (described below) are important to bird life. As a contrast with traditional Pima farming methods, I have included in this analysis an additional man-modified habitat adjoining the reservation boundary.

The Gila and Santa Cruz River Channels

Most of the channel today is either barren or overgrown with pure stands of exotic winter-deciduous saltcedars (*Tamarix ramosissima*) (Fig. 2.2b). There are no exposed gravel beds remaining on either river. Along the lower 15 miles (24 km) of the Gila, from the Maricopa Highway near Sacate to the confluence of the Salt, are continuous saltcedar stands of varying widths. No other plants become established within this consociation. What few screwbeans (*Prosopis pubescens*) survive on the reservation are on the sandy lower terraces, just above the channel. Here also are found arrowweed (*Pluchea sericea*), burrobush (*Hymenoclea* spp.), and seepweed (*Suaeda moquinii*).

The Saltbush-Bursage Community

This distinctive habitat (Fig. 3.4) of ornithological significance is found on the upper terraces of the floodplain. Important plant species include: seepweed (*Suaeda moquinii*), four-wing saltbush (*Atriplex canescens*), quailbrush (*A. lentiformis*), desert saltbush (*A. polycarpa*), triangle-leaf bursage (*Ambrosia deltoidea*), pickleweed (*Allenrolfea occidentalis*), burrobush (*Hymenoclea salsola pentalepis*), wolfberries (*Lycium andersonii* and *L. fremontii*), and mesquite (*Prosopis velutina*).

A nearly continuous strip of this habitat extends along the north side of the Gila from Gila Crossing to Sweetwater and Santan, but elsewhere much of this important community has been destroyed for agricultural development. Some of these flats are so saline that the soil appears white.

Pima Rancherias

Within the floodplains the Pima rancherias or traditional farming communities (Fig. 3.5) form a distinctive habitat for birds (Rea, 1979). Historically, fences (first mentioned by Font in 1775) were constructed of double rows of poles thrust into the ground several meters apart, with the intervening space filled with mesquite and other spiny branches as the fields were cleared (Fig. 3.6). Some trees were allowed to remain in the fields (see Plate 11a in Russell, 1908). Barbed wire was introduced in 1900. Traditional Pima farms now often

Fig. 3.4. The Saltbush-Bursage community.

The predominant vegetation shown on this salt flat is desert saltbush (*Atriplex polycarpa*), quail-brush (*A. lentiformis*), triangle-leaf bursage (*Ambrosia deltoidea*), and several species of wolfberry (*Lycium* spp.). The three larger, still bare trees are mesquites, and the few dark trees in the distance are athel tamarisks planted around Pima houses. These flats are prime habitat for the Le Conte's Thrasher. The higher parts of the Sierra Estrella appear on the horizon. (Photographed by Wade C. Sherbrooke, March 1976.)

have fence rows (Fig. 3.7) grown up with mesquite (*Prosopis velutina*), graythorn, (*Condalia lycioides* [= *Zizyphus obtusifolia*]), wolfberry (*Lycium fremontii*), seepwillow (*Baccharis glutinosa*), arrowweed (*Pluchea sericea*), jimmy weed (*Haplopappus heterophyllus*), seepweed (*Suaeda moquinii*), and various suffrutescent *Atriplex* species. In only a few localities (particularly around Santan) do cottonwoods (*Populus fremontii*) and Mexican elderberry (*Sambucus mexicana*) remain as part of fence rows. Fields are usually small, consisting of one to several acres. Important crops include alfalfa (*Medicago sativa*), corn (*Zea mays*), Milo maize (*Sorghum bicolor*), and various other grains. Canal and ditch banks (Fig. 3.8) are often overgrown with Johnson grass (*Sorghum halapense*), seepweed (*Suaeda moquinii*), sunflower (*Helianthus annuus*), spiny aster (*Aster spinosus*), cocklebur (*Xanthium strumarium canadense*), tumbleweed (*Salsola iberica*), and *Amaranthus* sp. In the 1960s traditional Pima rancheria-type settlements were still found around most villages; but they are rapidly giving way to large-scale mechanized farming, where thousands of acres are cleared and leveled, and no trees or shrubs remain.

Fig. 3.5. Traditional Pima rancheria-type settlement.

Pima ranches typically consist of an adobe or sandwich house, small canals, fields, and fence rows. Around the house is a clump of planted athel tamarisk. The fence row is grown up with cottonwood, mesquite, saltcedar, and wild sunflowers. The fields beyond the house have reverted to a second-growth mesquite bosque. (Photographed March 1976 near Bapchule village by Wade C. Sherbrooke.)

Fig. 3.6. Pima field at Komatke.

The field is nearly surrounded by living mesquite fences (background), with a low brush fence of cut mesquite branches along one side (foreground). A large mesquite tree is left in the field. Crops in this field include Sudan grass, Pima pumpkin, and the short white Pima corn. These fields provide food and cover for a high diversity of breeding and wintering birds and abundant spring and fall migrants. (Photo by Gary P. Nabhan, fall 1976.)

Fig. 3.7. Fence row on traditional Pima ranch.

Mesquite, wolfberry, and graythorn have been permitted to grow up over most of the original barbed wire fence. In places thorny lower branches have been trimmed back and stacked to reinforce weaker parts of the fence. The field is in permanent pasture. At the time this photo was taken Bell's Vireos, Lucy's Warblers, Northern Orioles, Verdins, and Northern Cardinals were nesting in the fence rows. (Photographed in Komatke Village, May 1982, by A. M. Rea.)

Courtesy of Environment Southwest

Fig. 3.8. Ditch banks on Pima ranch.

The Pima Indians and before them the Hohokam plied the middle Gila floodplain with miles of canals and ditches which greatly increased the oasis effect of the river. Typical of rancherias, this unlined earthen ditch at Komatke supports spiny aster, cocklebur, quail-plant, and seepweed. A living fence of young mesquites follows the ditch. (Photographed by Wade C. Sherbrooke, June 1978.)

Formerly, the Pima planted the evergreen athel tamarisk (*Tamarix aphylla*), in rectangles about their homes (Fig. 3.9). This Eurasian species does not self-propagate here, unlike the congeneric saltcedars (Harris, 1966). These tamarisk clumps grow to considerable size, persisting long after the original adobe houses (*bit kik*) have disintegrated. Both the associated insects and the reduced temperatures within these tamarisk groves attract migrant birds, particularly those species arriving during the very hot months of July, August, and September.

Though grasslands and pastures are specifically mentioned by Emory (1848), Johnston (1848), Bartlett (1854), Garcés (1900), Kino (1919), Anza (1930), Sedelmayr (1955), Bringas (1977), and others, and the names of two modern villages (Sacaton and Sacate) are derived from Spanish words for grasses, there is nothing on the reservation today that could be called a grassland. The nearest ecological approximations are the alfalfa fields (Fig. 3.18), where certain sparrows winter and Horned Larks (*Eremophila alpestris*) and limited numbers of meadowlarks (*Sturnella neglecta*) may breed. Some very steep and inaccessible upper reaches of the Sierra Estrella still have native grasses intermixed with the forb-shrub cover.

The Mesquite Bosques

Modern Pima fields lying fallow quickly revert to even-aged bosques of mesquite (*Prosopis velutina*). All bosques I have found are second-growth (Fig. 3.10). Woodcutting, fires, and dropping of the water table have all contributed to the destruction of mature mesquite stands. Massive stumps marking former bosques can still be seen about Blackwater, Casa Grande Ruins, and on the terrace north of the Gila along the Maricopa Highway and in New York Thicket (Fig. 2.4; see also Fig. 2.5).

Various plants, most commonly four-wing saltbush (*Atriplex canescens*), *Haplopappus heterophyllus, Suaeda moquinii, Hoffmanseggia densiflora,* and graythorn (*Condalia lycioides* [= *Zizyphus obtusifolia*]) are sometimes found growing in second-growth bosques.

Mechanized Farms

Large-scale farms (Fig. 3.11) are being developed across the reservation, usually by off-reservation enterprises. The terrain is leveled and no natural vegetation of any sort persists. Fossil groundwater (a nonrenewable resource) is pumped from wells hundreds of feet deep into concrete-lined ditches. Monocultural crops, principally cotton, are raised. The Indians are actively encouraged and assisted by a federal governmental agency at Sacaton to adopt this style of agriculture.

Runoff Ponds

Water collecting temporarily in certain shallow ponds on the reservation is important in the ecology of migrating aquatic birds. These runoff ponds are

Fig. 3.9. Tamarisk grove about former Pima house, Komatke.

The field in the foreground is in permanent pasture. Other fields are second-growth mesquite. A number of vagrants were found in this grove including Williamson's Sapsucker, Eastern Phoebe, Willow Flycatcher, Brown Creeper, Varied Thrush, Philadelphia Vireo, and the eastern race of Summer Tanager. (Photographed by Wade C. Sherbrooke, June 1978.)

Fig. 3.10. Second-growth mesquite bosque.

A short trail has been cut in the foreground. Otherwise these bosques are virtually impenetrable to humans and livestock. (Photographed in Komatke village in 1976 by Gary P. Nabhan.)

Fig. 3.11. Modern mechanized farm near Sacate.

This area, west of Vahki, formerly had small Pima ranches and thorny fence rows where I collected for a number of years. About in the middle foreground was a mesquite thicket occupied by numbers of Bendire's Thrashers, Red-winged Blackbirds, Pyrrhuloxias, and Abert's Towhees. (Photographed by A. M. Rea, May 1982.)

Fig. 3.12. Sewage pond west of Sacaton.

Open water, emergent vegetation, and at least some shallow shoreline provide cover and food for ducks, coots, gallinules, and shorebirds. Migrants such as swallows that drink on the wing also use these ponds. Chain link fences, to keep people out of contaminated water, give resting birds protection from most terrestrial predators. (Photo by A. M. Rea, December 1981.)

usually located at the lower edge of fields, though some occur in the bed of the Gila. Since they are filled only during periods of peak irrigation, they seldom develop permanent vegetation. Some support narrow strips of saltcedars along their deeper edge. The shallow, muddy edges attract wading birds such as sandpipers, stilts, and avocets.

The largest such pond on the study area was the Riggs Road pond, a borrow pit remaining from the construction of Interstate Highway 10. This consisted of four fingers one-half mile (0.3 km) long, originally as much as 2 yards (2 m) deep. The pit was filled apparently by the early fall rains and runoff of 1972 and dried at the end of July 1973.

The pond I visited most frequently is along Interstate Highway 10 in the Casa Grande area, 16.7 miles (27 km) southeast of Bapchule. When the reservation boundary was marked late in 1975 the pond proved to be off the reservation. Nevertheless, data from this pond are included in this study. Most of this pond is less than a foot (one-third m) deep, and when filled to maximum it takes about ten minutes to walk the edges.

Government-constructed sewer effluent ponds are also important in attracting migrating and wintering ducks. These differ from irrigation runoff ponds in having steep banks and deeper, permanent water. One such pond west of the Sacaton Agency (Fig. 3.12) supports a growth of cattail (*Typha domingensis*) and tumbleweed (*Salsola iberica*).

Barehand Lane Marsh

This is one of the two most dynamic habitats on the reservation. It is dependent, as is the Salt, on wastewater. The marsh is between Co-op and Maricopa colonies. It appears clearly on the USGS Phoenix 15′ quadrangle (1952). When I discovered the swampy area in March 1966, it had probably had a number of years of aquatic development. A naturally low area in the floodplain, extending from east to west at about the 1000-foot contour, drains water from the irrigated fields east of 75th Avenue to about 79th Avenue. Drainage is toward the Gila River. The surrounding area not under cultivation is upper terrace covered with *Atriplex polycarpa,* a very low wolfberry (*Lycium californicum*), a taller *Lycium parishii,* Engelmann hedgehog (*Echinocereus engelmannii*), and an occasional mesquite tree.

At the east (upper) end tail water collected in a shallow, open lagoon, which originally had small clumps of cattails (*Typha domingensis*) (see Fig. 3.13). The drainage channel itself, over a mile (a kilometer and a half) long and averaging 250 feet (75 m) wide, must originally have supported mesquite trees with some elderberry (*Sambucus mexicana*), then later became choked with saltcedars. Some of these mesquite trees, eventually killed by the waterlogged soil, were of considerable size. There were no cottonwoods or willows in the entire marsh. In places most of the water was confined to a small stream running through the largely impenetrable thicket. Where the stream spread out and saltcedars had not gained a foothold, extensive clumps of cattail became established. Each year, from 1966 to 1974, the cattail stands approximately doubled in area and the saltcedars became more dense.

Fig. 3.13. Barehand Lane pond in April 1981.

This is one of two open ponds at the east end of the marsh. Cattails line some edges. Dominant plants surrounding the pond include saltcedar, Russian thistle, jimmy weed (*Haplopappus heterophyllus*), white horse-nettle (*Solanum elaeagnifolium*), and quail-brush (*Atriplex lentiformis*). (Photo by A. M. Rea.)

The extreme western end of Barehand Lane was a soggy meadow of perhaps several acres (a hectare) enclosed by mesquite trees. The common emergent vegetation here was: smartweed (*Polygonum* sp.), buttercup (*Ranunculus scleratus*), water speedwell (*Veronica anagallis-aquatica*), two species of dock (*Rumex violascens* and *R. crispus*), and cocklebur (*Xanthium strumarium canadense*). A few head of cattle were often run here, which perhaps prevented this large open area from being converted into a saltcedar and cattail thicket.

During the next few years there was always open water at the eastern lagoon except during brief periods when all the fields were fallow and not being irrigated. The soil surrounding the lagoon was highly saturated with salts that accumulated downfield from evaporating tail water. The north side of this open marsh usually supported rank growth of weedy vegetation such as sunflowers (*Helianthus annuus*), spiny aster (*Aster spinosus*), and cocklebur (*Xanthium strumarium canadense*). Sedges (*Carex* sp.) grew at the water's edge. A clump of bulrush (*Scirpus acutus*) was established by the summer of 1972. This doubled in area by the following summer, until the *Scirpus* and

Fig. 3.14a. Barehand Lane marsh fifteen months after burning.

The marsh was completely destroyed in December 1974. The water was confined to a small channel (mid-foreground) and a larger canal paralleling the marsh. In three months elderberries sent up suckers from roots and stumps, and the new bushes were blooming when this picture was taken. The recolonizing plants at this time are dock, sow thistle (*Sonchus* sp.), quail-plant, and saltcedars. (Photographed by Wade C. Sherbrooke, 27 March 1976.)

Fig. 3.14b. Barehand Lane marsh in April 1981.

This shows six years of plant regeneration at the same section of the marsh appearing in the previous picture. The ground is once again saturated, and there is a dense cover of dock (*Rumex* spp.), salt-marsh fleabane (*Pluchea oderata*), nightshade (*Solanum nodiflorum*), and rabbitfoot grass (*Polypogon monspeliensis*), all species tolerating high soil salinity. Taller vegetation consists of cattail, mesquite, saltcedar, and (immediately behind the camera) elderberry. The Sierra Estrella appears on the horizon. (Photographed by A. M. Rea.)

Typha coalesced to form a large contiguous stand of emergent vegetation. Yet there was always an area of open water, usually from knee to hip deep, even during the maximum growth of cattail and bulrush. This open water attracted swallows and nighthawks, which came to drink on the wing. Teal and other dabbling ducks used the open marsh.

Various attempts were made to dry the east marsh and to reclaim a small section of the field where the water collected, but usually the drainage ditch silted in again and the marsh re-formed before the aquatic vegetation suffered. For instance, from 28 January to 29 March 1970 the marsh was ditched and dry, but it reflooded by 5 April. It was ditched again the following fall but silted in within five weeks. In November 1973 the open marsh was in good condition, but by 26 December 1973 it had been ditched, and the emergent vegetation burned and plowed under. By 20 May 1974 both the east and west sides of 75th Avenue were deeply ditched and there was no longer standing water except at the two extreme ends of the marsh. The size of the various cattail clumps had roughly doubled during the past year, but with the heat of summer they began to yellow and die. Finally by December 1974 the irrigation effluent was turned into a side canal and all the brushy vegetation was burned (Fig. 3.14a). A few blackened saltcedar and mesquite snags sticking starkly into the air were all that remained to mark the former site of Barehand Marsh. At the far west end a few mesquite trees and saltcedars escaped burning. By June 1975 some elderberries had sent up suckers from the burned stumps and the bare soil was still moist. The largest of the former cattail stands I measured was 490 × 220 feet (148 × 67 m). The ground where the cattails had been was nearly white with shells of two species of snails and a freshwater clam (*Corbicula* sp.).

In 1976 the only plants making a successful comeback were elderberry, buttercup, and water speedwell. By 1979 much of the western marsh was colonized by nightshade (*Solanum nodiflorum*), salt-marsh fleabane (*Pluchea odorata*) and coarse grasses. During the growing seasons of 1980 and 1981 the ground along the channel continued to become more waterlogged, some cattail thickets appeared, and more low vegetation developed and diversified (Fig. 3.14b).

The Salt River Habitats

This designation included (until the January 1978 floods) three relatively different plant communities: a broad, sparsely vegetated gravel riverbed, an effluent channel lined with cattails and young willows, and cottonwood-willow groves.

In modern times the lower Salt River (forming the northern border of the reservation from 83rd to 115th avenues) consisted of a broad gravelly channel, choked in places with saltcedars. Housholder, Dickerman, Werner, and Simpson collected in a marsh at 107th Avenue. It was still a large cattail marsh in 1943 (Housholder, personal communication to Simpson). But in the early 1950s when Simpson began collecting here it was an open pond 32 to 40 feet (10 to 12 m) across with cattails and a little open water in the middle. A channel, usually dry, disappeared into the surrounding saltcedars. It dried in 1957 (Simpson, personal communication).

In November 1958 the 91st Avenue sewage disposal plant went into operation, discharging effluent into the old bed of the Salt River. Within a few years cattails (but not *Scirpus*) began to line the effluent channel. The Christmas–New Year's flood of 1965–1966 washed out most of the channel growth. Gradually young willow and cottonwood stands developed (1970–1971) along some sections between 91st and 107th avenues.

At the Salt–Gila confluence above 115th Avenue the Arizona Game and Fish Commission constructed a 35-acre (14-ha) pond in 1966. This was the Confluence or Base Meridian Pond. Along most of the edges were stabilized cattail stands. The maximum depth of the water was 5 feet (1.5 m). No spillway was provided, the water level being regulated at the east end by an inlet and a runoff ditch running south outside the dike. A grove of cottonwoods and willows developed in the seepage below the dam. The winter of 1972–1973 was exceptionally wet. The dams on the Salt and Verde rivers filled and on 21 February 1973 the Salt River Project began releasing water. A total of 1.2 million acre-feet (1480 million m^3) flowed down the Salt channel into the Gila during the next ninety-four days. The confluence dam and pond were washed out.

I first visited the Salt–Gila confluence 17 June 1973. There was little evidence of the former pond except for the runoff ditch which had become a shallow marsh filled with cattails and lined with willow saplings about 10 to 17 feet (three to five m) high. This dried in the summer of 1975 because of erosion of the riverbed at the head of the canal. The floods thinned the exotic saltcedars. A good-sized grove of Goodding willow (*Salix gooddingii*) and cottonwood trees (*Populus fremontii*), growing in clumps, persisted in the area that was immediately below the former dam. Some willows measured up to 6.5 inches (17 cm) in trunk diameter, though most were in the 2.5–3 inch (6–7 cm) range. Apparently the smaller saplings had been washed out. This stand continued to mature from 1974 to 1978 (see Fig. 3.15), but the canopy became more open due to deliberate burning (winter of 1973–1974) and vandalism. Some of the tree mortality Simpson (personal communication) attributes to the several feet (approximately a meter) of sand and silt deposited above the root crowns of the trees. The major tall vegetation skirting the grove consists of: *Tamarix ramosissima,* giant reed (*Arundo donax*), arrowweed (*Pluchea sericea*), tumbleweed (*Salsola iberica*), sunflower (*Helianthus annuus*), tree tobacco (*Nicotiana glauca*), four-wing saltbush (*Atriplex canescens*), quail-brush (*A. lentiformis*). Streamside and emergent vegetation of importance to both wintering and breeding birds included: *Sonchus* sp., *Polygonum* sp., *Sisymbrium* sp., *Rumex* sp., *Salix gooddingii,* and *Typha domingensis* (Fig. 3.16).

Except for the effluent channel, most of the bed of the Salt River from 83rd Avenue to about 107th Avenue is exposed sand and gravel with sparse xerophytic bushes such as sandpaper plant (*Petalonyx thurberi*), burrobush (*Hymenoclea monogyra*), yerba del venado (*Porophyllum gracile*), and saltbushes such as *Atriplex linearis.* This apparent wasteland is important to breeding birds such as Lesser Nighthawks (*Chordeiles acutipennis*) and (when near shallow water) Killdeer (*Charadrius vociferus*) and Black-necked Stilts (*Himantopus mexicanus*).

Fig. 3.15. Salt-Gila confluence grove.

Gooding willows, interspersed with cottonwoods, formed an almost continuous canopy grove between 115th Avenue and the game department's Base Meridian Pond. The trees were ten years old when photographed. Sewage effluent supported this plant community. Vagrants, migrants and wintering birds were attracted in numbers. The grove was colonized by such breeding birds as the Yellow-billed Cuckoo, Green Heron, Summer Tanager, and Lesser Goldfinch, all species favoring a broad-leafed overstory. Several warbler species, previously unknown in winter, began overwintering along the lower Salt. (Photographed by Wade C. Sherbrooke, March 1976.)

Fig. 3.16. Channel above the Salt-Gila confluence with cattails and willows.

This photograph was taken east of 115th Avenue and upstream from the main cottonwood-willow grove. The willows shown are three years old. The water is effluent from the Phoenix sewage treatment plants at 35th and 91st avenues. (Photo taken 28 March 1976, by Wade C. Sherbrooke.)

Fig. 3.17. Rapid regeneration of willows: Salt River channel at 91st Avenue, looking west.

The January 1979 floods completely scoured the vegetation from this area (middle channel). In twelve months the new willow thickets grew to about 10 feet (3 m). (The standing person is 1.7 m tall.) Yellow-rumped Warblers were abundant in these new thickets in December. (Photo taken by A. M. Rea, 24 December 1979.)

Fig. 3.18. Chandler boundary cottonwoods.

The cottonwoods are mature, but there is no understory left along the canal banks. The foreground is an alfalfa field. By 1976 only Horned Larks, Common Starlings, House Sparrows, and a few Red-winged Blackbirds were found nesting. Formerly several additional native species bred here. Even the row of mature cottonwoods in the middle had been removed by December 1977. (Photo taken March 1976 by Wade C. Sherbrooke.)

Starting in January 1978 there were severe floods for three winters, which scoured the main channel vegetation of the lower Salt River. Many of the saltcedar thickets were washed away, permitting rapid replacement by Gooding willow. At the 91st Avenue crossing willow thickets grew to well over 17 feet (5 m) in a single season (Fig. 3.17). Thousands of migrating Barn Swallows were feeding just over these willows in mid-October 1979. In late December 1979 myriads of Yellow-rumped Warblers and numerous Orange-crowned Warblers were wintering in these still green thickets. Most of these groves west of 91st Avenue, including the older ones along parallel channels that had survived the floods, had been bulldozed by December 1980 as a "flood control" measure.

The Chandler Boundary Cottonwoods

I have included in this study area the fields bordering the reservation south of Chandler and immediately west of State Highway 87 (Fig. 3.18). Extensive fields here are usually planted in alfalfa while others are rotated. Mature cottonwood trees grew along the ditch banks until 1977. The area is always thoroughly manicured and there is no understory or undergrowth except for the cultivated crops. Starting in 1973, I surveyed about a mile (1.6 km) of these cottonwoods along the reservation boundary throughout the seasons. These were the only mature cottonwoods in this study area except for those in the Upper and Lower Santan area. Numerous cavities, both natural and woodpecker excavated, afforded nest and roost sites for various sized birds.

CHAPTER 4

Riparian Analogs

It is impossible to know fully what bird life on the Gila River was like aboriginally because European contact brought about a series of interrelated changes, both cultural and biological. A search for comparable habitats surviving today in the Salt – Gila Valley has revealed only two areas probably indicative of the riparian conditions once found on the reservation: Blue Point Cottonwoods, about 24 miles (39 km) northeast of the Gila River Indian Reservation, and "The Buttes" area at Cochran ghost town, approximately 22 miles (35 km) up the Gila from the reservation. Both the vegetation and the bird life of the first area have been studied rather thoroughly, but the second area has been visited only sporadically by ornithologists. These two analogs provide an insight into what the entire middle Gila must once have been like.

Blue Point Cottonwoods

The approximately 40-mile (64-km) reach of the Salt River from its confluence with the Verde to its junction with the Gila was known historically to the Spaniards as Río de la Asunción. Just above the Salt-Verde confluence is an extant riparian community, appearing on maps as Blue Point Cottonwoods. The elevation is 1,340 feet (408.5 m), comparable to that at Blackwater Pima village, 1,362 feet (415 m). Starting in the 1930s the Blue Point bird life was studied intermittently by L. D. Yaeger, L. L. Hargrave and A. R. Phillips. From 1969 to 1975 it was worked intensively by Johnson and Simpson (MS). Johnson and Simpson (1971:379) described the habitat in a preliminary report:

> Blue Point Cottonwoods, located approximately 2 miles [3.2 km] upstream from the confluence of the Verde with the Salt River, is the only large stand of mature

Table 1. Comparison of Breeding Birds of Blue Point Cottonwoods and the Gila River Indian Reservation

Only those species with important differences are included. Symbols: 0 = not present; Ex = historically present but extirpated; 1970 (e.g.) = year of first or last appearance; pr = pair(s); RR = recent recolonization.

Species	Modern Blue Point[a]	Modern Reservation[b]	
		Gila & Santa Cruz	Lower Salt[c]
Green Heron, *Ardeola virescens*	3 pr	Ex	RR 1970 (1 pr)
Great Blue Heron, *Ardea herodias*	colony nearby	Ex	Ex
Black-crowned Night Heron, *Nycticorax nycticorax*	0	Ex	(?) RR 1969
Least Bittern, *Ixobrychus exilis*	2 pr	Ex	Ex (1969–71 only)
Cooper's Hawk, *Accipiter cooperii*	1 pr	0	0
Mexican Black Hawk, *Buteogallus anthracinus*	(1934–51)	0	0
Harris' (Bay-winged) Hawk, *Parabuteo unicinctus*	1955–71 Ex	Ex (after 1915)	0
Bald Eagle, *Haliaeetus leucocephalus*	1971–72 Ex	0	0
Virginia Rail, *Rallus limicola*	0	Ex (after 1901)[d]	Ex (after 1956)
Common Gallinule, *Gallinula chloropus*	0	Ex (after 1901)[d]	RR after 1955
American Coot, *Fulica americana*	6 pr	Ex (after 1901)	RR ca. 1969
Yellow-billed Cuckoo, *Coccyzus americanus*	3 pr	local (1970)[d]	RR 1956, 1972–78 (2+ pr)
Ferruginous Pygmy-Owl, *Glaucidium brasilianum*	Ex (1951)	Ex (after 1915)	Ex (after 1898)
Elf Owl, *Micrathene whitneyi*	7 pr	Ex (after 1915)	Ex
Black-chinned Hummingbird, *Archilochus alexandri*	15–20 pr	Ex	? RR
Brown-crested (Wied's) Flycatcher, *Myiarchus tyrannulus*	15 pr	Ex (after 1915)	0 (Ex ?)
Say's Phoebe, *Sayornis saya*	0	Widespread	0
Vermilion Flycatcher, *Pyrocephalus rubinus*	Ex (1969)	Ex (after 1915)	0 (Ex ?)
Rough-winged Swallow, *Stelgidopteryx ruficollis*	7 pr	? local	RR 1978
Barn Swallow, *Hirundo rustica*	Ex (after 1940)	Ex (ca. 1920)	0
Common Raven, *Corvus corax*	1 pr	0	0
Northern Mockingbird, *Mimus polyglottis*	0	Widespread	Widespread
Loggerhead Shrike, *Lanius ludovicianus*	0	Widespread	Widespread
Bell's Vireo, *Vireo bellii*	14 pr	local (few pr)	local (few pr)
Lucy's Warbler, *Vermivora luciae*	750–800 pr	local (few pr)	local (few pr)
Yellow Warbler, *Dendroica petechia*	2–4 pr	0 (Ex ?)	0

Table 1. (Continued)

Species	Modern Blue Point[a]	Modern Reservation[b] Gila & Santa Cruz	Lower Salt[c]
Yellow-breasted Chat, *Icteria virens*	4–5 pr	Ex (before 1907)[d]	RR 1971 (few pr)
Hooded Oriole, *Icterus cucullatus*	Ex (1969)	Ex[d]	0
Summer Tanager, *Piranga rubra*	4 pr	Ex (ca. 1930s/40s)	RR 1974 (1 pr)
Blue Grosbeak, *Passerina caerulea*	3 pr	0 (?)	0 (?)
Lesser Goldfinch, *Carduelis psaltria*	5–29 pr	0	RR 1978 (1 pr)

[a]Mostly 1969–1975, but some earlier data included.

[b]This study; 1963–1982; see species accounts for details.

[c]As of 1977. Destruction of riparian habitats beginning with January 1978 floods, completed by "flood control" clearing in 1980–1981 have eliminated breeding habitat for virtually all these birds.

[d]Occurs locally in favored habitats away from Santa Cruz or Gila channels, as at Barehand Lane.

> cottonwoods (*Populus fremontii*), with its attendant mesquite (*Prosopis juliflora* [=*velutina*]) understory, left of the entire Salt River. These cottonwoods will be covered by Orme Lake [or so it was thought at this writing], thus virtually completing the extirpation of native riparian groves along the Salt River. This virgin grove extends for nearly a mile [1.6 km] along the north side of the river. A narrow marsh of open water approximately one-half mile long, edged with arrowweed (*Pluchea sericea*) and saltcedar (*Tamarix* sp.), and containing several dense stands of cattail (*Typha domingensis*) contributes to the diversity of the habitat. *Azolla* sp., a water fern unrecorded for central Arizona, grows profusely with duckweed (*Lemna* sp.) in the marsh.

The study area they delimited consists of 237 acres (96.5 ha), broken down approximately as follows:

100 acres (40.7 ha) mature mesquite bosque (including 30 acres [12.2 ha] with cottonwood overstory)
70 acres (28.5 ha) open mesquite
30 acres (12.2 ha) old riverbed
7 acres (2.8 ha) marsh
30 acres (12.2 ha) ridge (saguaro-paloverde)

Judging from the descriptions of early explorers (see Chapter 1), conditions found aboriginally along the Gila above the Salt–Gila confluence (the modern reservation), must have closely approximated the conditions found today at the remnant riparian community of Blue Point (Fig. 4.1). The only major difference is that there are no willow thickets at Blue Point, just four individual trees. Table 1 compares some significant breeding birds of the two areas. The eleven species found breeding at Blue Point Cottonwoods but no longer breeding on the reservation are: Least Bittern (*Ixobrychus exilis*), Cooper's Hawk (*Accipiter cooperii*), Elf Owl (*Micrathene whitneyi*), Black-chinned Hummingbird (*Archilochus alexandri*), Wied's Crested Flycatcher (*Myiarchus tyrannulus*), Vermilion Flycatcher (*Pyrocephalus rubinus*),

Fig. 4.1. Blue Point Cottonwoods.

One of the last natural riparian remnants on the Salt, Blue Point served as an analog of former conditions on the middle Gila. This view looks down the Salt River and across the cottonwood grove and mesquite bosque, which are on an ancient oxbow in the mid-ground. The large cottonwoods near the river enclose a cattail marsh. In the foreground is a saguaro-paloverde-bursage-cholla community. (Photographed by R. Roy Johnson, June 1982.)

Rough-winged Swallow (*Stelgidopteryx ruficollis;* except once, 1978), Common Raven (*Corvus corax*), Yellow Warbler (*Dendroica petechia*), Blue Grosbeak (*Passerina caerulea*), Lesser Goldfinch (*Carduelis psaltria;* except once, 1978). (Several of these species may yet be demonstrated to breed locally on the reservation.) Under the remnant but rapidly deteriorating riparian conditions found in Gilman's day, four of these species were still breeding: Elf Owl, Wied's Crested Flycatcher, Vermilion Flycatcher, and Rough-winged Swallow. Breeding Least Bitterns are dependent on well-developed marsh habitat, not found on the reservation since 1971. Four species found at Blue Point Cottonwoods but not on the Gila River Reservation are correlated with the presence of tall riparian with developed understory: Cooper's Hawk, Wied's Crested Flycatcher, Yellow Warbler, and Lesser Goldfinch. If the cottonwood-willow stands on the reservation from 91st to 115th avenues had continued to mature, conceivably these four species as well as the Mississippi Kite (*Ictinia mississippiensis*) would have colonized (or recolonized) the area.

Fig. 4.2. The Gila River at The Buttes-Cochran area.

This view, looking across the living Gila toward part of South Butte, shows the willow thickets that line much of the river's edge. The non-native saltcedar (at the middle in this view) is not dominant here, as it is on the reservation channel. Extensive mesquite bosques fill the numerous oxbows. Between Cochran and The Buttes there are high breeding densities of Bewick's Wrens, Bell's Vireos, Lucy's Warblers, and Yellow-breasted Chats. This is an important stronghold for the endemic desert races of Yellow Warbler and Summer Tanager. (Photo by A. M. Rea, May 1982.)

Three floodplain-bajada species are found today on the reservation but not at Blue Point: Say's Phoebe (*Sayornis saya*), Northern Mockingbird (*Mimus polyglottis*), and Loggerhead Shrike (*Lanius ludovicianus*). Johnson and Simpson (personal communication) would add the Black-crowned Night Heron (*Nycticorax nycticorax*) to the reservation list. They believe it nested in the saltcedars or the willow groves west of Maricopa Colony before the 1978 and 1979 floods, but this has not been demonstrated conclusively.

Certain additional breeding birds from Blue Point Cottonwoods require comment. Three species are known to have bred at least once there, but not on the reservation: Bald Eagle (*Haliaeetus leucocephalus*) in 1972; Bushtit (*Aegithalos minimus*) in 1972; Bewick's Wren (*Troglodytes bewickii*) in 1951. The eagles took over and enlarged a nest of Harris' Hawks, which never returned. The eagles were subadult, and their eggs never hatched. Yaeger, Phillips, and Hargrave all found the Ferruginous Pygmy-Owl (*Glaucidium brasilianum*) regularly in their work, but the species has not been seen there since 1951. The date for this owl's disappearance from the reservation is unknown, but it was after Gilman's time. Phillips considered the Sora (*Porzana carolina*) a summer resident at Blue Point in 1951, but it has not been found there since. In 1940 Hargrave recorded the Hooded Oriole (*Icterus cucullatus*) as common but found no Northern (Bullock's) Orioles (*Icterus galbula*). In 1951 Phillips found twelve of each species. The last pair of *cucullatus* that Johnson and Simpson found breeding there was in 1969, but twelve pairs of *galbula* bred each season. At first sight this seems a simple matter of displacement, as there have been no obvious structural changes at Blue Point during this period, but differential cowbird parasitism might be involved. There are about twenty breeding female Brown-headed Cowbirds (*Molothrus ater*), and two pairs of Bronzed Cowbirds (*Molothrus aeneus*). On the reservation Gilman found both oriole species common. I have found but one pair of *cucullatus,* while *galbula* is still common.

The Common Black Hawk (*Buteogallus anthracinus*) nested at Blue Point Cottonwoods at least from 1934 to 1951 but was not found by Simpson on his first visit in 1955. Though this sedate stream-loving hawk must have been part of the aboriginal avifauna of the reservation, Gilman did not find it nor did any of my Pima consultants recognize Black Hawk skins.

The Harris' (Bay-winged) Hawk (*Parabuteo unicinctus*) colonized Blue Point some time between 1951 and 1955, nesting through the summer of 1971 until their nest was appropriated by eagles.

There are likewise clear-cut differences between the species wintering at Blue Point Cottonwoods and on the reservation. All the species shown in Table 2 are found more frequently or more abundantly at Blue Point. Six species are known to winter only at Blue Point: Common Crow (*Corvus brachyrhynchos*), Bridled Titmouse (*Parus wollweberi*), Bushtit (*Aegithalos minimus*), White-breasted Nuthatch (*Sitta carolinensis*), Blue-gray Gnatcatcher (*Polioptila caerulea*), Solitary Vireo (*Vireo solitarius*). (There is one modern record for vagrant Bushtits on the reservation; I do not consider it a "wintering" species.) There is a modern winter crow roost north of Fort McDowell village, and some of these birds forage by day at Blue Point. No specimen has been secured to determine the breeding origin of these crows. Earlier in this century flocks of "crows" appeared also on the Gila River Reservation (ethnographic data; Russell, 1908:55).

I attribute the distinct difference (fourteen species) between the winter bird composition of the two areas to two factors: First, there is suitable riparian habitat at Blue Point to attract and keep the birds there. Second, there are essentially riparian avenues along the Salt and Verde for local species breeding at higher elevations upstream to follow down to the lower elevations for wintering. Riparian avenues no longer exist to bring the birds to the Gila River Reservation. The Phoenix metropolitan area with some 40 miles (64 km) of unsuitable habitat (completely eradicated riparian community), is an effective barrier at the northwest end of the reservation. Thus, some species simply do not reach the newly developed riparian areas of the reservation from 91st to 115th avenues. And the break in riparian growth up the Gila River from the reservation is just as great. The species that I think follow these natural avenues along drainages in winter are: Scrub Jay (*Aphelocoma coerulescens*),

Table 2. Comparison of Wintering Birds of Blue Point Cottonwoods and the Gila River Indian Reservation

Only those species with important differences are included.

Species	Modern Blue Point	Modern Reservation
Green Heron, *Ardeola virescens*	0	regularly (lower Salt River only)
Scrub Jay, *Aphelocoma coerulescens*	2 out of 4 winters	3 out of 13 winters
Common Crow, *Corvus brachyrhynchos*	annually	0
Mountain Chickadee, *Parus gambeli*	2 out of 4	0
Bridled Titmouse, *Parus wollweberi*	annually	0
Bushtit, *Aegithalos minimus*	2 out of 6	1 record (1975)
White-breasted Nuthatch, *Sitta carolinensis*	annually	0
Brown Creeper, *Certhia familiaris*	annually (few)	5 out of 13 winters (max. 2 indiv.)
Bewick's Wren, *Troglodytes bewickii*	regularly	3 out of 13
Northern Robin, *Turdus migratorius*	annually	9 out of 13
Western Bluebird, *Sialia mexicana*	annually (fair numbers)	3 out of 13 (few)
Blue-gray Gnatcatcher, *Polioptila caerulea*	probably regularly[a]	0
Hutton's Vireo, *Vireo huttoni*	regularly	0 (transient only)
Solitary Vireo, *Vireo solitarius*	irregularly	0 (possibly 1975?)
Orange-crowned Warbler, *Vermivora celata*	irregularly?	recently (local, lower Salt riparian only)

[a]Found by Phillips almost every winter trip; gnatcatcher species usually not checked by Johnson and Simpson (MS), but specimens of *P. caerulea* taken.

Mountain Chickadee (*Parus gambeli*), Bridled Titmouse (*Parus wollweberi*), Bushtit (*Aegithalos minimus*), White-breasted Nuthatch of at least the local race (*Sitta carolinensis*), Blue-gray Gnatcatcher (*Polioptila caerulea*), and perhaps Hutton's Vireo (*Vireo huttoni*). Gilman found the jays every winter and the titmice in three out of seven winters. But there were still some tall trees (Fig. 2.2a) for the birds to follow.

Johnson and Simpson (MS) recorded the Mountain Bluebird (*Sialia currucoides*), only twice at Blue Point Cottonwoods, but during five out of six winters (25–30 individuals each year) at nearby Fort McDowell. I have found this species during only two of nineteen winters on the Gila River Reservation.

The Buttes–Cochran Area

The Gila is a perennial stream only as far west as the Ashurst-Hayden Diversion Dam, 17 miles (27 km) east of the reservation. Approximately 22 miles (35 km) upstream from the reservation is a section of river with healthy riparian vegetation (Fig. 4.2). The western part of this area is known locally as The Buttes, from a mining community established in the 1880s. Here sheer cliffs rising to nearly 3,000 feet (913 m) force the Gila into a number of snakelike bends. Several river miles above is the turn-of-the-century ghost town of Cochran, marked today by two adjacent clumps of ancient tamarisk trees.

I first visited this area with Gordon L. Fritz on 6–7 August 1976. We camped at Cochran and examined about two miles (3.2 km) along the south side of the Gila, making a small collection of birds, particularly juvenals, to document breeding. Alexander Russell, Jr., and I flew into the area briefly on 1 December 1977, but a cold storm had just arrived and we found few birds. On 7–8 May 1978 Fritz and I again censused the area, making collections. Robert Schmalzel and I worked all the major habitats on 5–6 June the following year. In late December 1981, Takashi Ijichi and I camped three days here making a winter census. Sylvester Matthias and I returned on 10–11 May 1982 to make a brief survey of plants and animals.

Unlike the broad, open floodplain found across the reservation downstream, the geomorphic structure along this section of the river restricts the alluvial bottoms to narrow areas (Fig. 4.3). The elevation is approximately 1,630 feet (497 m). At this point the Gila is silty and swift, with a gravel bottom. In early August it averaged about 20 inches (half a meter) deep, though wet silt on the banks indicated that the water level had recently been about 3 feet (almost a meter) higher, following the July rains. The river channel is well defined, about 100 feet (30 m) wide, with thick vegetation growing to the water's edge. The overstory consists of mature mesquites, saltcedars, young willows, and scattered tall cottonwoods. Saltcedars do not dominate here as they do along the reservation channel. Willows form occasional thickets along the north bank, but I found no large, old trees. Nor are there any groves of cottonwoods. A well-developed understory is composed of seepwillow, thickets of arrowweed, graythorn, spiny aster, *Haplopappus,* and other low vegetation. For the most part the mesquite bosques here are sufficiently mature that one can move beneath the canopy with relative ease along numerous cow trails. I found no ash, walnut, sunflower, or emergent marsh vegetation. About a mile (1.6 km) west of the Cochran site are a large pasture and corrals, with fences now lined with tall, second-growth mesquite and graythorn. Immediately across the river from the corrals the riparian growth is over 1,000 feet (300 m) broad and appears to have the best-developed mesquite bosque. Near The Buttes small coves and arroyos just back of the river have dense thickets of catclaw (*Acacia greggii*), desert hackberry (*Celtis pallida*), and mesquite.

The bosque structure seems to be much like that described by Pima consultants (see Chapter 2) for the western villages on the reservation, except that cottonwood groves and marsh habitats are absent. Early mining activities, especially coke production, must have made heavy demands on mesquite and other hardwood trees. Old willows and cottonwoods near Cochran are perhaps periodically destroyed by floods. However, vigorous riparian growth undoubtedly can withstand occasional flooding. In June 1979 I found con-

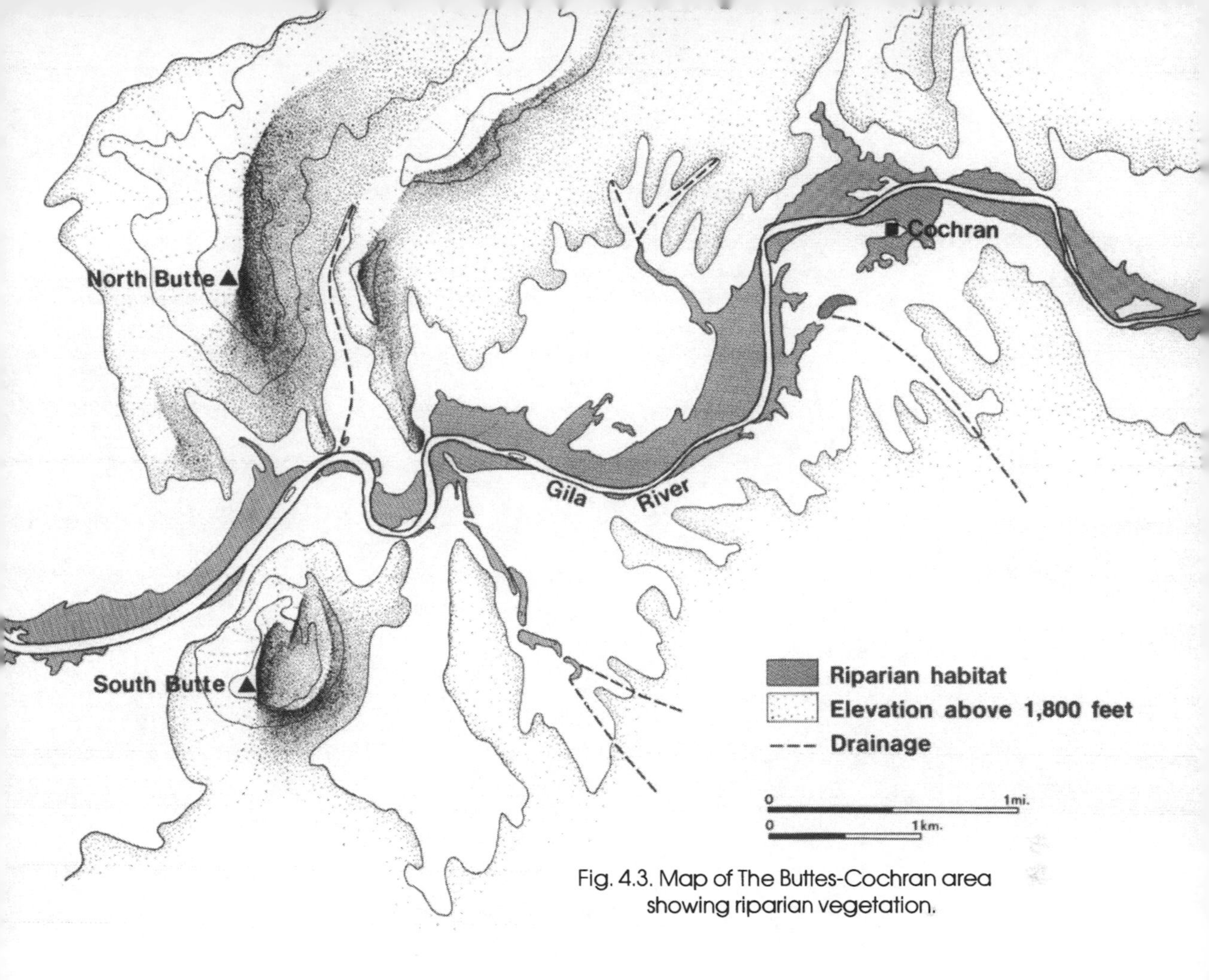

Fig. 4.3. Map of The Buttes-Cochran area showing riparian vegetation.

siderable amounts of flood debris stuck to heights of 8 feet (2.4 m) throughout the bosque. But there was no evidence of damage to any of the riparian vegetation (*Prosopis, Ziziphus, Pluchea, Baccharis, Tamarix,* or *Aster spinosus*). Young willow thickets throughout the surveyed area are evidence that grazing pressure is not so great as to prevent vigorous rejuvenation.

During only brief visits it is not possible to discover all the breeding birds of an area, but a sufficient number of species were conspicuous and vocal in the Cochran bosque to permit some significant comparisons (Table 3) with the reservation portion of the Gila.

Nesting abundantly in the riparian thickets were Bewick's Wrens (*Troglodytes bewickii*), Yellow-breasted Chats (*Icteria virens*), Bell's Vireos (*Vireo bellii*), Lucy's Warblers (*Vermivora luciae*), Lesser Goldfinches (*Carduelis psaltria*), Abert's Towhees (*Pipilo aberti*), and Verdins (*Auriparus flaviceps*). This is the westernmost point on the Gila drainage for breeding Bewick's Wrens. Territories of the wren and chat were densely packed, followed by only slightly fewer vireos and Lucy's Warblers. Also nesting commonly in the thickets (but not confined there) were Mourning Doves (*Zenaida macroura*), White-winged Doves (*Zenaida asiatica*), and Common Screech-Owls (*Otus asio*). In the approximately 2.5 miles (4 km) censused, there were at least two or three pairs each of Cooper's Hawks (*Accipiter cooperii*), Gila Woodpeckers

Table 3. Comparison of Breeding Birds from Two Sections of the Gila River: The Buttes–Cochran Area and the Gila River Reservation

Symbols: + = present (see text for presumed status); 0 = absent (1960s and 1970s); T = territory; N = nest found; J = juvenal (seen/ √collected); pr = pair(s); * = obligate riparian breeding (species column).

Species	The Buttes–Cochran Area				Reservation[1]
	(6–7) Aug. 1976	(7–8) May 1978	(10–11) May 1982	(5–6) June 1979	
Cooper's Hawk, **Accipiter cooperii*	+ J	+ N (2)	(heard only)	+ N	0
Harris' (Bay-winged) Hawk, *Parabuteo unicinctus*	+	+	+ T (1 pr)	?0	0
White-winged Dove, *Zenaida asiatica*	+ N	+	+	+	+
Mourning Dove, *Zenaidu macroura*	+ N	+	+	+	+
Common Screech-Owl, *Otus asio*	+ T	?	+	?	+
Great Horned Owl, *Bubo virginianus*	+ T	+ T	+	?	+
White-throated Swift, *Aeronautes saxatalis*	0	+ (3–4)	+(?)	+ (4)	+
Black-chinned Hummingbird, **Archilochus alexandri*	+ ♂	+ (♂♂, ♀♀, N)	+ (common)	+ N	0
Ladder-backed Woodpecker, *Dendrocopos scalaris*	?	+ ♀ (♂?)	+ (few pr)	+ (2 pr)	+
Gila Woodpecker, *Melanerpes uropygialis*	+ (1 pr)	+ (few)	0	+ (1 pr)	+
Vermilion Flycatcher, **Pyrocephalus rubinus*	+ T (♂, ♀)	0	0	0	0
Wied's Crested Flycatcher, *Myiarchus tyrannulus*	+ T, J √(1 pr)	+ T	+ (1 seen)	+ (1 seen)	0
Bushtit, *Aegithalos minimus*	0	0	+J√	0	0
Ash-throated Flycatcher, *Myiarchus cinerascens*	+ (abundant)	+ (1 pr)	+ (common)	+ (1 pr)	+
Bewick's Wren, **Troglodytes bewickii*	+ T, J √(2 pr)	+ T (common)	+ T, J (common)	+ T (abundant)	0
Black-tailed Gnatcatcher, *Polioptila melanura*	+ T	+ T, J	+ T, J√	+ (♀)	+
Phainopepla, *Phainopepla nitens*	0	+(♂, ♀ along wash)	(4 or 5)	+ (3+3+1)	+
Bell's Vireo, *Vireo bellii*	+ T, J √	+T (abundant)	+ (common)	+ T, J √ (common)	+ (local)

Table 3 (Continued)

Species	The Buttes–Cochran Area (6–7) Aug. 1976	(7–8) May 1978	(10–11) May 1982	(5–6) June 1979	Reservation[1]
Lucy's Warbler, *Vermivora luciae*	+ T, J (common)	+ T, J (abundant)	+ J (common)	+ T (common)	+ (local)
Yellow Warbler, *Dendroica petechia*	+	+ T, J √	+ T (2 pr ±)	+ T (2 pr)	0
Yellow-breasted Chat, *Icteria virens*	+ T (common)	+ T (abundant)	+ (abundant)	+ T (abundant)	0
Brown-headed Cowbird, *Molothrus ater*	+ (6 birds)	+ (♀ + 2 ♂♂)	+ (several)	+ (several)	+
Bronzed Cowbird, *Molothrus aeneus*	+ J (2, separated)	0	0	0	+
Hooded Oriole, **Icterus cucullatus*	+ T (2 pr)	+ (1 pr)	+ (several)	+ T (1 pr)	0
Summer Tanager, **Piranga rubra*	+ T (2 pr)	+ T (1 ♀)	+ T(8–10 pr)	+ T (2 pr)	0
Northern Cardinal, *Cardinalis cardinalis*	+	+ T (1 pr)	+ T (10–12 pr)	+ T (2 pr)	+
Blue Grosbeak, **Passerina caerulea*	+ T (widespread)	+ T, N	+ (1 ♀ only)	+ T (♂)	0 ?
Lazuli Bunting, **Passerina cyanea*	+ (♂ seen)	+ T (singing)	0	0	0
Abert's Towhee, **Pipilo aberti*	+ N	+ T, J (few pr)	+ (common)	+ N (few pr)	+
Song Sparrow, **Passerella melodia*	0	+ T (few)	+ (T?) (2 pr)	+ T (2 pr only)	0
Lesser Goldfinch, **Carduelis psaltria*	0	+ T, J (common)	+ (common)	+ J (common)	0

[1]Exclusive of the lower Salt effluent and other wastewater-supported communities.

(*Melanerpes uropygialis*), Ladder-backed Woodpeckers (*Dendrocopos scalaris*), Yellow Warblers (*Dendroica petechia*), Hooded Orioles (*Icterus cucullatus*), Summer Tanagers (*Piranga rubra*), Northern Cardinals (*Cardinalis cardinalis*), Blue Grosbeaks (*Passerina caerulea*), and Song Sparrows (*Passerella melodia*). One pair of Cooper's Hawks nested in the same tamarisk tree in successive years at Cochran, the other in old mesquites downstream. In August 1976 family groups of Blue Grosbeaks were common and widespread, but in 1978 and 1979 only one pair was found. Perhaps the winter floods had reduced a critical food resource. A pair of Great Horned Owls (*Bubo virginianus*) was calling in the Cochran bosque in May 1978 and in August. On the June trip we camped in the uplands and did not census night birds along the river. In May 1982 I heard only the Barn Owl (*Tyto alba*) at Cochran and saw a Great Horned Owl at North Butte several miles west. Song Sparrows were sparse with widely separated territories in willow thickets on the north side of the river. There were surprisingly few Brown-headed Cowbirds (*Molothrus ater*) considering the great numbers of potential hosts. I saw three in May, heard a small group several times in June, and found only six individuals in August. One juvenal Bronzed Cowbird (*Molothrus aeneus*) was associated with a chat and another with an adult Hooded Oriole in August, but the species was not noticed in May or June. Family groups of both Rock Wren and Cañon Wren were foraging on bluffs and railroad cuts on 11 May 1982. On this day a pair of Black Phoebes (*Sayornis nigricans*) was at a nest where the river skirts South Butte. My only other observation of this water-loving flycatcher was over the river one evening in August. In December 1981 we found a flock of Bushtits (*Aegithalos minimus*) in a hackberry-catclaw-mesquite–filled cove. I found a dozen or so birds in this thicket the following May and took two juvenals and an adult. It is not known whether this normally Upper Sonoran Zone bird breeds here regularly or if the colony will persist.

Seven additional species may breed in The Buttes–Cochran riparian community. A male Lazuli Bunting was present in August and apparently in May (very shy, singing in a mesic hackberry thicket along the railroad), but I found none in June the following year. At dusk on 5 June a Violet-green Swallow (*Tachycineta thalassina*) and a number of Rough-winged Swallows (*Stelgidopteryx ruficollis*) were feeding over the river. I watched a Violet-green again over the Gila the following day. A localized pair of Vermilion Flycatchers (*Pyrocephalus rubinus*) defended the mesquite-lined corral and pasture area against Western Kingbirds (*Tyrannus verticalis*) in August, but they were not found on any subsequent visits. They may nest immediately across the river in a large mesquite bosque, which we have never visited. In a willow thicket on 7 May I found a silent (stray?) Traill's Flycatcher (*Empidonax traillii*) that appeared to be the extremely gray local race, *E. t. extimus.* This form at least formerly bred in the lower San Pedro River not far to the east and might be expected to nest here, too. I did not find any on 11 May 1982, but these dates are early; territorial birds should be sought about six weeks later in the fine willow thickets. The status of Phainopeplas in the riparian habitat is unclear. They were absent in August, and only a male and a female were found along an arroyo in May 1978. Along the river in early June there were two groups of three birds each and a solitary female in prebreeding condition (brood patch, many ova 1.5 to 2 mm, but no yolks or ruptured follicles). In May 1982 I found only four or five individuals along the Gila. Nests should be sought in the mesquite thickets from mid-February to mid-April. Crissal Thrashers (*Toxostoma crissale*) were heard only in August, but they probably nest in the mesquite thickets here, as they do in the elevated country just south.

Evidently breeding in the rugged buttes to the north and west of the study area was a pair of Turkey Vultures (*Cathartes aura*), which was seen each trip, and small numbers of White-throated Swifts (*Aeronautes saxatalis*) which were seen in May and June but not in August. I saw three or four Common Ravens in May 1978 and found an active nest on the railroad trestle in

May 1982. These species use the river and its vegetation only incidentally and cannot be considered riparian species.

Found in the bosque and other riparian growth, but breeding as well out on the surrounding bajadas and along arroyos were Harris' Hawk (*Parabuteo unicinctus*), Gambel's Quail (*Callipepla gambelii*), White-winged and Mourning Doves, Greater Roadrunner (*Geococcyx californianus*), Great Horned and Common Screech-Owls, Gila and Ladder-backed Woodpeckers, Wied's Crested and Ash-throated Flycatchers, Verdins, Cactus Wrens (*Campylorhynchus bruneicapillus*), Black-tailed Gnatcatchers (*Polioptila melanura*), Phainopeplas (*Phainopepla nitens*), Brown-headed Cowbirds, Northern Cardinals (*Cardinalis cardinalis*), and House Finches (*Carpodacus mexicanus*).

Away from the river, in the dissected uplands dominated by *Cereus, Cercidium, Celtis, Acacia, Ambrosia,* and grasses, we found Harris' Hawk, American Kestrel, Elf Owl (abundantly), Lesser Nighthawk (*Chordeiles acutipennis*), Gilded Flicker (*Colaptes auratus*), Purple Martin (*Progne subis;* adult male in May and June), Common Raven, Loggerhead Shrike (*Lanius ludovicianus*), Cañon Wren (*Catherpes mexicanus*), Curve-billed Thrasher (*Toxostoma curvirostre;* abundantly), and Desert Sparrow (*Amphispiza bilineata*).

A number of birds found in early August but not on the subsequent visits may have represented a postbreeding dispersal from elsewhere. Western Kingbirds were in groups of twos and threes all along the river. An Inca Dove (*Scardafella inca*) was heard at the edge of the river. The absence in May and June of Inca Doves and Western Kingbirds, both highly vocal species, suggests that they are probably not part of the breeding avifauna. Lark Sparrows (*Chondestes grammacus*), which were common in the flats, and three Black-headed Grosbeaks (*Pheucticus ludovicianus*) were early fall migrants. A female Great-tailed Grackle (*Quiscalus mexicanus*), seen at dusk on 7 May but not found the following day, was probably a nonbreeding vagrant.

Absent on all visits were fishing birds (ardeids and kingfishers), Killdeers (*Charadrius vociferus*) and other waders, Long-billed Marsh Wrens (*Cistothorus palustris*), Northern Mockingbirds, European Starlings (*Sturnus vulgaris*), Common Yellowthroats (*Geothlypis trichas*), House (English) Sparrows (*Passer domesticus*), Western Meadowlarks (*Sturnella neglecta*), Red-winged Blackbirds (*Agelaius phoeniceus*), and Pyrrhuloxias (*Cardinalis sinuatus*). The Buttes–Cochran area appears suitable for both Yellow-billed Cuckoos (*Coccyzus americanus*) and Common Ground Doves (*Columbina passerina*), but these secretive birds may easily go undetected on brief visits. Only once did I see a wary Say's Phoebe. The area seems suitable for them, but they may have bred and vacated the area by early May, the time of my earliest visits. In early August no Phainopeplas were found in two days' search of bosque and arroyos, indicating a postbreeding exodus. Family groups of Lesser Goldfinches were common all along the river in May and June, but I found none at all in August. Apparently they too vacate the hot lowlands after breeding.

The Bullock's Oriole appears not to be breeding in the riparian habitat. An adult male *I. g. bullockii,* taken by Fritz in the bosque 8 May 1978, still had moderate fat and gonads not yet fully enlarged, suggesting it was not breeding. In June I saw an adult male along an arroyo away from the river, indicating that the species no doubt breeds at least in the surrounding area.

Obvious environmental conditions explain why certain birds are either absent or few in number in The Buttes–Cochran riparian community. In December, when the Gila was clear, I was unable to find any fish. Gordon Fritz, who made an archaeological reconnaissance of this area, assured me that his crew found no fish or frogs in this part of the Gila. Food is evidently unavailable for piscivorous species such as herons, egrets, and kingfishers. Since there are no shallow backwaters with emergent vegetation (*Typha, Scirpus, Carex, Polygonum*), coots, rails, gallinules, yellowthroats, marsh wrens, and blackbirds are absent as breeding birds and Song Sparrows are relatively few. The sparsity of tall cottonwoods and the absence of an overstory from mature groves limits the numbers of Yellow Warblers, Hooded Orioles, and Summer Tanagers, and may account for the apparent lack of cuckoos. At the times of my visits there were very few head of cattle in the riparian areas, which may explain the unusually low numbers of Brown-headed Cowbirds. There are no human habitations along this part of the river, and House Sparrows, Inca Doves, and Great-tailed Grackles are absent as breeding birds.

Starlings were absent on all five visits. They are not dependent on the proximity of humans and often nest far out in deserts where saguaros provide nest holes within a few miles of water.

As would be expected, there are significant differences between the breeding avifauna of The Buttes–Cochran riparian habitat and the reservation portion of the Gila River, exclusive of the Salt–Gila confluence and Barehand Lane Marsh. Eleven species breed at Cochran but not on the Gila within the reservation: Cooper's Hawk, Harris' Hawk, Black-chinned Hummingbird, Wied's Crested Flycatcher, Bewick's Wren, Yellow Warbler, Yellow-breasted Chat, Summer Tanager, Blue Grosbeak, Lazuli Bunting (apparently), Song Sparrow, and Lesser Goldfinch. Additionally, the Bell's Vireo and Lucy's Warbler, common at Cochran, are only very local on the reservation, and the Hooded Oriole in recent decades has been found only in a few palms planted near dwellings, never in the riparian community. Surface water, supporting a dense, floristically diverse understory only partially invaded by saltcedars and at least some cottonwood and willow overstory, accounts for these major differences in bird life. With the possible exception of the Bewick's Wren, all these species must have bred in the reservation riparian habitats before the severe ecological disruption of the past 100 years.

Breeding also in the uplands around Cochran but not on the reservation today are Harris' Hawk, Elf Owl, Common Raven, and apparently Purple Martin. The Elf Owl and raven are also at Blue Point, as was the Harris' Hawk, until displaced.

Several species breeding commonly on the reservation have not yet been found in the Cochran area: Killdeer, Say's Phoebe, Northern Mockingbird, Loggerhead Shrike, Red-winged Blackbird, and Pyrrhuloxia. Apparently, these birds benefit from traditional Pima farming methods, proliferating in irrigated fields with fence rows and wet, weedy borders. As noted earlier, the Say's Phoebe and shrike probably breed at Cochran earlier in the season.

Along the narrow, swift channel at Cochran there is nothing comparable to the two mesic areas with emergent vegetation at the northwest end of the reservation—the Salt River, fed by eutrophic sewage effluent, and Barehand Lane Marsh, fed by irrigation runoff. Breeding at some time in the 1970s at these two places but absent from Cochran were: Cinnamon Teal (*Anas cyanoptera*), Green Heron (*Ardeola virescens*), Virginia Rail (*Rallus limicola*), Common Gallinule (*Gallinula chloropus*), American Coot (*Fulica americana*), Common Yellowthroat, and perhaps Black-crowned Night Heron.

The Buttes–Cochran area is an important analog of former conditions on the Reservation. It should be studied during both spring and fall migrations, but especially during the winter and early in the breeding season, before possible postbreeding exodus of certain species.

Implications of the Analogs

The two riverine analogs, at comparable elevations in the lower Sonoran Desert not far from the reservation, give a strong suggestion of the predeterioration conditions which must have existed on the reservation until a century ago. Birds breeding in the riparian community must have included Least Bittern, Cooper's Hawk, Common Black Hawk, Black-chinned Hummingbird, Wied's Crested Flycatcher, Bell's Vireo, Lucy's Warbler, Yellow Warbler, Yellow-breasted Chat, Blue Grosbeak, Summer Tanager, and Lesser Goldfinch. Changes in the reservation floodplain and bajada habitats caused by excessive tree cutting, overgrazing, and simplification of floristic composition have greatly reduced the numbers of some species which are still breeding commonly to abundantly in the two analogs. On the basis of the Blue Point analog, the reservation must have once supported a much richer wintering avifauna as well.

A final observation from the Cochran analog. In what appears to be a quite stable desert ecosystem, there is considerable annual variation in both the number of species and the number of individuals breeding in the riparian community. For instance, Ash-throated Flycatchers in the early May and early June period were common one year, but only a single pair was seen the others. In one year I found one pair of Northern Cardinals, in another two or three, and in 1982 ten or twelve pairs. Also, as emphasized in the individual species accounts, there is no defined spring breeding season as in most temperate ecosystems. While many birds do breed during the conventional April to June period, others may begin breeding in what is actually late winter, then depart, while still others may wait for the great crops of insects and seeds that are produced by the mid-summer rains. And a few species are opportunists, breeding throughout the entire period if food is available. But no single month, and much less no single year, can be used as typical.

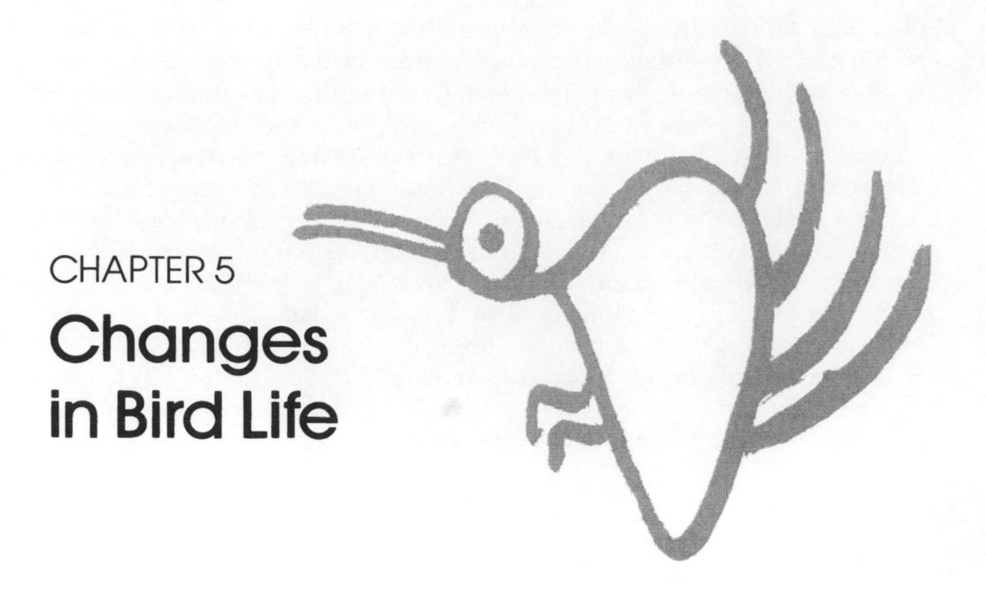

CHAPTER 5

Changes in Bird Life

In the delimited area of the 581 square miles (1,500 km^2) of the Gila River Indian Reservation, with historical data extending back approximately a century, it is possible to examine in some detail changes in bird life. Eight major types of avifaunal change are evident: (1) extirpations; (2) changes in population sizes associated with habitat deterioration; (3) local riparian recovery and recolonization; (4) northward movements; (5) changes in subspecies distributions; (6) changes in wintering status; (7) changes in departure times of winter residents; (8) winter flight years.

The major changes in the bird life of the reservation are attributable to human-induced habitat modifications. The most conspicuous aspect has been the loss of the Gila River as a live stream with its associated riparian community stretching some 65 miles (104 km) across the reservation. By assembling data from a number of baseline sources (archaeological, ethnographic, historic) it is possible to reconstruct the aboriginal riparian avifauna fairly accurately and to document its change. It is not possible to reconstruct the avifaunal composition of the grasslands, pastures, and meadows mentioned by early explorers (Chapter 2). These were grazed out of existence before ornithologists visited them, and there are no modern analogs at similar elevations in Lower Sonoran deserts. The least modified part of the reservation appears to be the bajada and mountain habitats.

In limited areas of the reservation there have been riparian habitat ameliorations and consequent recolonizations by breeding as well as wintering birds. These changes are recent and have been monitored by ornithologists. I have attempted to determine the successional sequence of recolonization for two such areas. These case histories admirably demonstrate the management potential of wastewater.

Some increments to the breeding avifauna are the result of rather widespread northward movements of essentially Neotropical species. Two instances of subspecies range extensions have occurred in south-central Arizona so that formerly isolated populations are now interbreeding.

These initial seven types of changes are apparently long-term and for the most part appear to be correlated with human modifications of local habitats. The final type of change is an exception. Annual fluctuations in species compositions of wintering birds are attributable in part to phenomena called "flight years" or irregular irruptions of montane birds reaching the low desert.

Extirpations

There is unfortunately no single set of baseline data from the aboriginal avifauna against which to compare the modern bird life of the reservation. The basic reason is that no ornithologist visited and recorded the bird life *before* the disintegration of the riparian community began. Breninger's brief survey of 1901 was near the end of a prolonged drought and after more than two decades of heavy grazing of the upper Gila watershed by Anglo ranchers. That some integrity of the riparian community remained is indicated by Breninger's finding, in evident numbers, the Black-crowned Night Heron (*Nycticorax nycticorax*), Virginia Rails (*Rallus limicola*), Common Gallinules (*Gallinula chloropus*), American Coots (*Fulica americana*), Belted Kingfishers (*Megaceryle alcyon*), Song Sparrows (*Passerella melodia*), and Common Yellowthroats (*Geothlypis trichas*). As his visits were in the latter half of September, some of these species may have been only transients. He observed no Long-billed Marsh Wrens (*Cistothorus palustris*), Least Bitterns (*Ixobrychus exilis*), Soras (*Porzana carolina*), or Yellow-breasted Chats (*Icteria virens*). At least some of these would have been relatively silent and secretive at that season. But the more vocal and curious marsh wren seemingly would not have been overlooked had it been present. In the light of Russell's (1908) remarks on the river conditions in 1901, it is surprising that Breninger found as many aquatic associated species as he did.

Gilman's period of residence (1907–1915) began two years after another set of damaging floods began sweeping down the Gila (Burkham, 1972; Dobyns, 1981). The channel became an ill-defined sandy expanse (Fig. 2.2a). The gallery forest was sufficiently weakened that Gilman did not find breeding Great Blue Herons (*Ardea herodias*) or Black-crowned Night Herons. Both Yellow Warblers (*Dendroica petechia*) and Summer Tanagers (*Piranga rubra*) were scarce, if breeding at all. Fishing birds such as the Osprey (*Pandion haliaetus*) and Double-crested Cormorant (*Phalacrocorax auritus*) occurred only as rare transients (one record each). Most conspicuously absent from Gilman's notes are references to such marsh-breeding species as Pied-billed Grebe (*Podilymbus podiceps*), Virginia Rail, Common Gallinule, Common Yellowthroat, and Yellow-breasted Chat. He found the coot, marsh wren, and Lincoln's Sparrow (*Passerella lincolni*) only as migrants. One would expect the wren and sparrow to have been common winter residents had emergent aquatic vegetation been left on the Gila. Yet in his rather extensive writing, Gilman never mentioned cattails (*Typha*) and made only a single reference to "tule" (presumably *Scirpus*).

Definitely extirpated from the middle Gila (Sacaton area) between 1901 and 1907 was the Common Gallinule and probably the Black-crowned Night Heron, Virginia Rail, and Common Yellowthroat.

It is evident both from the history of the founding of the western villages (see Chapter 2) and from interviews with older Pima consultants and Franciscan missionaries that the Gila River persisted longest as a living stream at the northwestern end of the reservation, about the villages of Santa Cruz, Gila Crossing, Komatke, and Co-op Colony. For this reason the ethnographic accounts from this area give perhaps the best picture of the original bird life on the middle Gila River. Such data are inherently limited by the ethnotaxonomic distinctions made by the Pima. For example, because all the anseriforms are subsumed under a single generic taxon, *vápkuch,* an investigator cannot extract data for a particular Linnaean species such as Mallard or Shoveler. Life history data are obtainable only for those species which were distinguished in the Pima lexicon. At least seventy species are so distinguished, approximately sixty-three of which have a one-to-one correspondence with Linnaean taxa. Fortunately, among those taxa are a number whose status in the early twentieth century had been in question. Details are given in each species account.

The ethnographic record provides valuable data on nesting colonies of Great Blue Herons and Black-crowned Night Herons and breeding localities for the Golden Eagle (*Aquila chrysaetos*) and Belted Kingfisher. Data on the former occurrence of the Crested Caracara (*Caracara cheriway*), Sandhill Crane (*Grus canadensis*), and Summer Tanager agree with observations of contemporaneous field ornithologists who were working in nearby areas (Gilman, Housholder, Yaeger). A second species of corvid, presumably the Common Crow (*Corvus brachyrhynchus*), mentioned by Russell (1908:92), is confirmed in the ethnographic record and indicated to have nested. Hibernation of the Poor-will (*Phalaenoptilus nuttallii*) and egg movement by the Lesser Nighthawk (*Chordeiles acutipennis*) were known to the Pima far in advance of the formal discovery of these phenomena by "science" (Evans, 1967; Ferguson, 1967; Jaeger, 1948, 1949; Phillips et al. 1964:57). The record reveals the former nesting of the Barn Swallow (*Hirundo rustica*) in shallow wells at Komatke and Santa Cruz villages, the only known breeding on the Gila River (see map in Phillips et al., 1964:98). The kingfisher is confirmed as a breeding species for southern Arizona (see discussion by Phillips, 1968).

In addition to suitable locations for nest-building, an important factor limiting the breeding of most herons and kingfishers is the availability of fish. When Russell conducted his fieldwork in the Sacaton area (1901—1902), he reported that the Pima there had but the single life-form term, *vatop,* for 'fish.' He concluded (Russell, 1908:83), "Either the long series of dry years and the absence of fish have caused the people to forget former classifications, or else they never distinguished one species from another." However, my consultants living in the northwestern villages where the Santa Cruz and Gila formed living streams until recent decades, provided me with six or seven specific fish ethnotaxa (folk generics), in addition to the life form *vatop.* The Pima living here regularly utilized fish until the 1930s (Rea, MS). This part of the reservation was the last refugium for breeding kingfishers, Great Blue Herons, night herons, and perhaps the Osprey. The Green Heron, which Gilman (MS) found "nesting annually in the yard" at Sacaton, can subsist without fish (Bent, 1926).

The archaeological record, consisting of bird bones from four Hohokam sites, is rather meager. The most interesting assemblage is from Snaketown on the Gila River (McKusick *in* Haury, 1976). Of some sixty-eight identifiable avian bones from Snaketown, twenty-four or 35 percent are aquatic or marsh species. These include a heron, geese, ducks, crane, avocet, and marsh icterids. Though the sample is small, it is at least noteworthy that none of these aquatic species is associated with the late phases of Snaketown (Santa Cruz–Sacaton, A.D. 700–1100). Haury (personal communication) attributes this apparent difference to sampling error. The wintering or transient Canada Goose (*Branta canadensis;* one recovery), White-fronted Goose (*Anser albifrons;* four individuals), Snow Goose (*Anser caerulescens;* four), and Pintail (*Anas acuta;* one), are indicators of more extensive bodies of open water on the Gila during Hohokam occupation. These large anseriforms no longer occur on the reservation.

Table 4 summarizes the species that have been extirpated from the Gila River portion of the reservation, based on the four major sources of baseline data (archaeological, ethnographic, historic, modern). A few of these species do occur today as rare transients or vagrants. Two of the species (Elf Owl, *Micrathene whitneyi,* and Rough-winged Swallow, *Stelgidopteryx ruficollis*) might yet be found breeding somewhere on the reservation.

Twenty-five of the twenty-nine species on this list are directly associated with the riparian woods and/or open water of the former river. The Elf Owl would appear to be independent of the riparian forest, as Gilman (1909b) found it in saguaros as well as cottonwoods. However, both Breninger (1898) and Gilman (1909b) found the Ferruginous Pygmy Owl (*Glaucidium brasilianum*) associated with river growth, as did Phillips, Yaeger, and Hargrave on their visits to the Salt-Verde confluence.

Although there are no ethnographic or historical references to them, I suspect on geographical grounds that the following species probably bred on the reservation under aboriginal riparian conditions:

Common Black Hawk, *Buteogallus anthracinus*

Osprey, *Pandion haliaetus*

Willow Flycatcher, *Empidonax traillii*

Long-billed Marsh Wren, *Cistothorus palustris*

Yellow Warbler, *Dendroica petechia*

Yellow-headed Blackbird, *Xanthocephalus xanthocephalus*

Blue Grosbeak, *Passerina caerulea*

Of these seven species, the Common Black Hawk, Yellow Warbler, and Blue Grosbeak have bred in modern times at Blue Point Cottonwoods. Both the

Table 4. Species Extirpated from Gila River Reservation

Barehand Lane and recent riparian development on Salt River (91st to 115th avenues) not included. Symbols: Probable Former Status, B = breeding, W = wintering (transient status not considered); Baseline Source, A = archaeological, E = ethnographic, H = historic, M = modern (primarily from Salt-Gila confluence and Barehand Lane); Modern Analog, BP = Blue Point Cottonwoods, C = Cochran-Buttes area.

Species	Probable Former Status	Baseline Source	Modern Analog
Pied-billed Grebe, *Podilymbus podiceps*	B	M	
Great Blue Heron, *Ardea herodias*	B, W	A, E, H	BP
Green Heron, *Ardeola virescens*	B	H, M	BP
Black-crowned Night Heron, *Nycticorax nycticorax*	B, W	E, H	
Least Bittern, *Ixobrychus exilis*	B	M	BP
Whistling Swan, *Olor columbianus*	W	A, H	
Canada Goose, *Branta canadensis*	W	A	
White-fronted Goose, *Anser albifrons*	W	A, H	
Snow Goose, *Anser caerulescens*	W	A, E	
Pintail, *Anas acuta*	W	A, H, M	
Common Merganser, *Mergus merganser*	W	H	
Harris' (Bay-winged) Hawk, *Parabuteo unicinctus*	B, W	E, H	BP, C
Golden Eagle, *Aquila chrysaetos*	B, W	E	
Sandhill Crane, *Grus canadensis*	M, W	E, H	
American Coot, *Fulica americana*	B, W	E, M	BP
Yellow-billed Cuckoo, *Coccyzus americanus*	B	M	
Ferruginous Pygmy-Owl, *Glaucidium brasilianum*	B, W	H	BP
Elf Owl, *Micrathene whitneyi*	B	H	BP, C
Belted Kingfisher, *Megaceryle alcyon*	B	E	
Brown-crested Flycatcher, *Myiarchus tyrannulus*	B	H	B, C
Vermilion Flycatcher, *Pyrocephalus rubinus*	B	H	BP, C
Barn Swallow, *Hirundo rustica*	B	E	
Corvid, sp? *Corvus* (*brachyrhynchos?*)	B, W	E, H	BP
Bridled Titmouse, *Parus wollweberi*	W	H	BP
Long-billed Marsh Wren, *Cistothorus palustris*	B, W	M (nearby)	
Yellow Warbler, *Dendroica petechia*	B	M	C
Common Yellowthroat, *Geothlypis trichas*[a]	B	M	BP
Yellow-breasted Chat, *Icteria virens*	B	M	BP, C
Summer Tanager, *Piranga rubra*	B	E, H, M	BP, C

[a]Possible local recent recolonization, see Species Accounts.

Yellow-headed Blackbird and the marsh wren are dependent on extensive *Scirpus* marshes for breeding. A relict population of marsh wrens bred in a former marsh about 25 miles (40 km) downstream from the Salt–Gila confluence (Simpson and Werner, 1958; Johnson et al., MS). Earlier they must have bred throughout the great marsh formed by the confluences of the Santa Cruz, Salt, and Agua Fria with the Gila River. Possibly a few grosbeaks still breed locally. Yellow-headed Blackbirds breed in still hotter areas at lower altitudes as on the Gila in Yuma County, Arizona (personal observation). Phillips (1968:135) remarked of the black hawk, "It may be significant that none of the early explorers identified this striking hawk along the Gila River. It ranges up this river well into New Mexico, but I know of no definite record prior to 1918."

Three falconiform species have been extirpated from the reservation: Harris' (Bay-winged) Hawk (*Parabuteo unicinctus;* found nesting by Gilman and in ethnographic record); Gold Eagle (*Aquila chrysaetos;* ethnographic record of nest locations); and Crested Caracara (ethnographic and historic records; breeding unknown). Additionally, the Prairie Falcon (*Falco mexicanus;* breeding, both ethnographic record and Gilman) has retreated to the most remote parts of the reservation and is now found nesting in the Sierra Estrella. The Harris' Hawk still breeds in numbers about Florence, not far east of the reservation (Wayne Whaley and William Mader, personal communication) and between Florence and Cochran (personal observation). I suspect that these larger species of raptors simply may have been shot out. The Bald Eagle (*Haliaeetus leucocephalus*) probably bred along the middle Gila, as it does on undisturbed reaches of the Salt and Verde (Rubink and Podborny, 1976); but conditions on the Gila have long been unsuitable for this fishing species and there is no ethnographic record of this eagle. There was a breeding attempt at Blue Point Cottonwoods (see Chapter 4).

Changes in Population Sizes Associated With Habitat Deterioration

In addition to the approximately twenty-nine species completely extirpated from the middle Gila and lower Santa Cruz rivers, exclusive of the lower Salt, as either breeding or wintering species (see Extirpations, above, and Table 4), several avian species have noticeably declined in population size because of habitat deterioration. Again, the baseline data sources are variable; former status and relative abundances are derived from archaeological evidence, from direct historical or ethnographic accounts, or from fieldwork by myself and others (Table 12) in the 1960s and 1970s. While this list (Table 5) is probably quite conservative, at least twenty-five species have conspicuously declined in numbers. Some species such as the Great Blue Heron (*Ardea herodias*), Green Heron (*Ardeola virescens*), Black-crowned Night Heron (*Nycticorax nycticorax*), American Coot (*Fulica americana*), and Belted Kingfisher (*Megaceryle alcyon*) have been extirpated as *breeding* species but are included here too because the numbers occurring as transients and as wintering birds have been reduced also.

Table 5. Population Declines Associated with Habitat Deterioration

Symbols: Status B = breeding, T = transient, W = wintering; Baseline Source, A = archaeological, H = historic, E = ethnographic, M = modern.

Species	Former Status	Baseline Source	Primary Habitat
Great Blue Heron, *Ardea herodias*	*B, W	E, H	riparian + open water
Snowy Egret, *Ardea thula*	*B, W	E	riparian, open marsh, fields
Black-crowned Night Heron, *Nycticorax nycticorax*	*B, W	E	riparian + marsh
Cooper's Hawk, *Accipiter cooperii*	W	H	widespread
Crested Caracara, *Caracara cheriway*	W	E, H	widespread
American Kestrel, *Falco sparverius*	B, W	H	riparian, etc.
Virginia Rail, *Rallus limicola*	B, W	H, M	emergent vegetation
Sora, *Porzana carolina*	B ?, W	H, M	emergent vegetation
Common Gallinule, *Gallinula chloropus*	B, W	H, M	marsh, riparian
American Coot, *Fulica americana*	B, W	E, H, M	riparian, marsh, open water
Killdeer, *Charadrius vociferus*	B, W	E, H	fields, open riverbeds
Common Snipe, *Capella gallinago*	W, T	M	low emergent vegetation
Yellow-billed Cuckoo, *Coccyzus americanus*	B	E, H, M	riparian
Belted Kingfisher, *Megaceryle alcyon*	*B, T, W	E	streams and banks
Western Kingbird, *Tyrannus verticalis*	B, T	H, M	farmlands, cottonwoods
Black Phoebe, *Sayornis nigricans*	*B, W	H	open water with cliffs or bridges
Rough-winged Swallow, *Stelgidopteryx ruficollis*	B	H	open water + dirt cliffs
Scrub Jay, *Aphelocoma coerulescens*	W	H	riparian + fields
Common Raven, *Corvus corax*	*B ?, W	A, E	widespread
Long-billed Marsh Wren, *Cistothorus palustris*	*B ?, W	M	emergent vegetation
Lucy's Warbler, *Vermivora luciae*	B	H	mesquite bosques
Common Yellowthroat, *Geothlypis trichas*	B	M	emergent vegetation
Hooded Oriole, *Icterus cucullatus*	B	H	riparian + palms
Song Sparrow, *Passerella melodia*	B, W	H	riparian + marsh

*Extirpated as breeding bird but still occurring in diminished numbers in another status.

Within the confines of the reservation, over 65 miles (100 km) of the Gila River and over 16 miles (25 km) of the lower Santa Cruz River are essentially dead as far as surface water, marsh, and riparian communities are concerned. On these channels the native broad-leafed riparian vegetation has been completely replaced by saltcedars and velvet mesquite. The effects on bird life are considerable. Seven species have declined because of loss of the riparian timber, five because of loss of open water to fish in or forage over, and seven because of the almost total loss of emergent vegetation, particularly cattail and bulrush. Probably several species listed here (for example, the rails and cuckoo) really belong, since 1980, on the extirpated list. If they occur at all, except as transients, it would be only along the lower Salt, but that habitat too has suffered serious deterioration.

Excessive mesquite woodcutting and the loss of mature bosques caused by dropping water tables has clearly affected at least the Lucy's Warbler (*Vermivora luciae*), which Gilman found a common breeding species early in this century. It is still common at Cochran, along the Verde River on the Fort McDowell Reservation (personal observation) and at Blue Point Cottonwoods—places with surface water and mature bosques. Undoubtedly several thousand pairs of Lucy's Warblers once bred on the reservation, whereas in the late 1970s and early 1980s there were only a few scattered, apparently recolonizing pairs. Though baseline data are unavailable to demonstrate it, bosque deterioration has also reduced the numbers of breeding orioles, chats, vireos, towhees, and owls.

Historical and ethnographic accounts indicate that the decline of the traditional Pima rancherias with their small fields, charcos, weedy ditches, fence rows, and tamarisk or cottonwood groves shading the houses, have caused a reduction in at least four species: Snowy Egrets (*Ardea thula*), Killdeer (*Charadrius vociferus*), Gila Woodpecker (*Melanerpes uropygialis*), Western Kingbird (*Tyrannus verticalis*), Scrub Jay (*Aphelocoma coerulescens*), and both species of orioles (*Icterus galbula* and *I. cucullatus*). But probably the other typically rancheria-loving species such as the Gambel's Quail (*Callipepla gambelii*), Northern Mockingbird (*Mimus polyglottos*), Bendire's Thrasher (*Toxostoma bendirei*), and Abert's Towhee (*Pipilo aberti*) have declined considerably and will continue to diminish with the conversion of more land to mechanized farming.

The reasons for the declines in some other species may be less evident. The American Kestrel (*Falco sparverius*) was formerly a common breeding species, nesting in cottonwoods, but now only a few pairs remain in saguaro stands. Widespread species such as the Cooper's Hawk (*Accipiter cooperii*), Harris' Hawk (*Parabuteo unicinctus*), Crested Caracara (*Caracara cheriway*), and Common Raven (*Corvus corax*) have declined. Such food resources as insects, fish, rodents, and carrion were greater when the habitats were more complex. A drastic simplification of the plant community results in a simplification of both numbers and species of all other classes of organisms. Complex food chains and species interactions are broken with the loss of surface water and hydrophylic vegetation. The summer resident Turkey Vulture (*Cathartes aura*) population on the reservation has declined in the past two decades. While improved sanitation may be one contributing factor, poisoning by pesticides, shooting, and removal of the cottonwood rows along ditches, used for roosting, have all added their toll to diminishing these slowly reproducing scavengers. Although DDT was banned in the United States in 1971, migratory birds are still contaminated during their sojourn in Latin America. A first-year Turkey Vulture that died at Komatke on 7 May 1975 carried 3,000 parts per million (lipid weight) DDE (the metabolite of DDT), as well as other contaminants.

In spite of continuing habitat deterioration, a few species have increased in numbers during this century, but most of these probably invaded during this period: Inca Dove (*Scardafella inca;* starting on the reservation about 1905), Horned Lark (*Eremophila alpestris;* by 1972), European Starling (*Sturnus vulgaris;* between 1954 and 1963), House Sparrow (*Passer domesticus;* spring 1908), Bronzed Cowbird (*Molothrus aeneus;* summer 1909), and Pyrrhuloxia (*Cardinalis sinuatus;* between 1947 and 1964). Great-tailed Grackles (*Quiscalus mexicanus*) are primarily transients; they have not been very successful in colonizing the reservation because it is so dry. The dove, cowbird, House Sparrow, and Pyrrhuloxia are found primarily about rancheria-type settlements, and all but the last adjust to urbanization. Only the starling has colonized the upland bajadas away from the disturbed floodplains.

Local Riparian Recovery

Avifaunal Succession at Barehand Lane

As noted in Chapter 3, Barehand Lane underwent a rather continuous vegetational development consisting of an approximate annual doubling in area of *Typha* and *Scirpus,* and a thickening of tree growth, particularly of saltcedar. More soil became waterlogged and low aquatic vegetation (*Rumex, Ranunculus, Veronica*) spread.

When I found the swamp in 1966, it already supported breeding populations of Song Sparrows and Redwing Blackbirds (*Agelaius phoeniceus*). Also breeding here were such species as Northern (Bullock's) Orioles (*Icterus galbula*), Verdins (*Auriparus flaviceps*), Northern Cardinals (*Cardinalis cardinalis*), Brown-headed Cowbirds (*Molothrus ater*), thrashers, and other birds, but these are independent of the marsh habitat per se. In the winter starlings and icterids, particularly Red-wings, roosted here by the thousands. The only common charadriiform was wintering Common Snipes (*Capella gallinago*) found mostly in the open marshy meadow at the west end.

In the summer of 1967 the only new breeding bird was the Bell's Vireo (*Vireo bellii*). No additional avian species colonized in 1968, 1969, or 1970. During the summer of 1971 young Phainopeplas (*Phainopepla nitens*) were being raised in the marsh, though possibly a few bred there in previous years. The maturation and fruiting of elderberry trees (*Sambucus mexicana*) were the important vegetational development supporting the early summer breeding Phainopeplas. This development also augmented the numbers of breeding Northern Mockingbirds. The Lucy's Warbler (*Vermivora luciae*) and Common Gallinule colonized the marsh as breeding birds in 1972. Probably breeding that year also was the Sora (seen 1 August), but this was not confirmed by finding nests or juvenals.

In 1973 the first Common Yellowthroats (*Geothlypis trichas*) bred, and one Yellow-breasted Chat (*Icteria virens*) was territorial. That summer gallinules were found throughout the swamp rather than confined to the eastern open-water marsh.

Though Barehand Lane was ditched and the emergent vegetation, then at its maximum extent, began drying in the summer of 1974, still a number of marsh species continued to breed. Yellowthroats and Lucy's Warblers produced young, and three chats held territories. The Virginia Rail (*Rallus limicola*) apparently bred successfully that summer (one young seen).

In 1975, following the complete destruction of growth and marsh at Barehand, the only birds found breeding were Phainopeplas, Common Ground Doves (*Columbina passerina*), and perhaps the Northern Mockingbird (*Mimus polyglottos*). *Sambucus,* growing up from burnt trunks, and peripheral *Prosopis,* which had escaped burning, were the only taller vegetation at that time (see Fig. 3.14a).

The avifaunal succession at Barehand Lane was fairly straightforward. (Arrows indicate cumulative annual succession.) It was colonized by Red-wings, Song Sparrows > Bell's Vireo > Phainopepla > Common Gallinule, Lucy's Warbler (and perhaps Sora) > Yellowthroat and Yellow-breasted Chat > Virginia Rail. Of the first pair of species to colonize, I think Red-wings would have preceded Song Sparrows. The blackbird is more flexible in its requirements and will breed even in saltcedars and in rank annual growth such as mustards and sunflowers. The Song Sparrow, on the other hand, requires more mature perennial growth such as *Pluchea* and more stabilized aquatic conditions.

Avifaunal Succession on the Salt River

Unlike the relatively uncomplicated vegetational and avifaunal succession at Barehand Lane, changes on the Salt River are complex. At least three distinct aquatic communities have occurred between the 91st Avenue Phoenix Sewage Plant and the lower Salt–Gila confluence at 115th Avenue: the channel itself, the Base Meridian or Confluence Pond, and the confluence grove. These habitats, with their histories and floristic compositions, are described in Chapter 3.

The discharge of effluent from the sewage plant began the winter of 1958–1959. The year of first breeding in the resulting riparian community is not known for Redwing Blackbirds, Song Sparrows, Common Gallinules, American Coots (*Fulica americana*), and Common Yellowthroats. The first recolonizations were probably by the Red-wings and Song Sparrows, as for other areas under study. By 1969, Johnson and Simpson (MS) found gallinules, coots and yellowthroats breeding in numbers along the channel between 91st and 107th avenues. Some clumps of willows were present. At this time there was one pair each of Yellow-billed Cuckoos (*Coccyzus americanus*) and Bell's Vireos nesting. By the spring of 1971 a pair of Green Herons (*Ardeola virescens*) and three pairs of chats were found. In the summers of 1974 and 1975 night herons were present and may have bred, but this was not demonstrated conclusively.

The confluence pond, constructed in 1966, had Ruddy Ducks (*Oxyura jamaicensis*) present during the summer of 1969, but their breeding status was not ascertained. In 1970 Johnson and Simpson (MS) found nesting Pied-billed

Grebes (*Podilymbus podiceps*), Least Bitterns (*Ixobrychus exilus*), and Green Herons. The following summer yellowthroats colonized the cattails, and Ruddy Ducks were breeding (dying young collected). The summer of 1972 was the last breeding season the pond was in existence. No new species were found to have colonized. At least the yellowthroats, Ruddy Ducks, and Green Herons were present, though the herons were probably dependent on the nearby cottonwood-willow grove rather than on the emergent vegetation of the pond itself. Significantly, neither *Scirpus* nor Long-billed Marsh Wrens colonized the pond, though the wrens were present as winter residents in all suitably large *Typha* clumps along the Salt. The floods of the following winter washed out the pond, leaving a dry, sandy expanse that grew up in *Salsola, Polygonum, Helianthus, Tamarix,* and *Salix,* and along the channel's edge, *Typha* (see Fig. 3.16).

The confluence grove, just east of 115th Avenue and west of the confluence pond, probably began growth about the time the pond was constructed in 1966 (see Fig. 3.15). The grove was the probable nesting place of the Green Herons mentioned above (1970, 1971). Johnson and Simpson (MS) found a territorial chat here in 1971 and a pair of cuckoos in 1972. My first visit to the grove was in June of 1973, after the extensive floodwaters earlier in the year had scoured out the emergent vegetation, very young willows, and much of the saltcedars. The following species were present, nesting in the taller overstory: cuckoo, White-winged Dove (*Zenaida asiatica*), Mourning Dove (*Zenaida macroura*), Common Ground Dove (*Columbina passerina*). A Black-chinned Hummingbird (*Archilochus alexandri*), with an ovary in post-breeding condition, may have also bred in the grove. Gallinules were nesting in emergent annuals (*Polygonum*), but only one territorial Song Sparrow (*Passerella melodia*) and no chats or yellowthroats were found. These apparently require some maturity of emergent vegetation and density of understory not found immediately following the flooding.

However, by the summer of 1974 low growth, both annual and perennial, had sufficiently recovered for at least one territorial Bell's Vireo and abundant Lucy's Warblers to breed. Abundant also were Song Sparrows, yellowthroats, and Abert's Towhees (*Pipilo aberti*). One pair of Summer Tanagers (*Piranga rubra*) bred in the grove, the first known recolonization on the reservation. In addition to the three dove species, a number of other species not necessarily dependent on riparian growth bred abundantly in the grove. These included Bullock's Orioles, Ash-throated Flycatchers (*Myiarchus cinerascens*), Abert's Towhees, and Verdins. Blue Grosbeaks (*Passerina caerulea*) and two singing male Indigo Buntings (*Passerina cyanea*) were also present but were not demonstrated to be breeding. These were apparently attracted by a peripheral growth of *Helianthus.*

The following summer (1975) Green Herons, gallinules, cuckoos, chats, Summer Tanagers, and Bell's Vireos were present in about the same numbers as the previous summer, but Song Sparrows had declined in number and no Lucy's Warblers were found. Much of the area had dried, except for the three main channels flowing through the grove. At least by 1978 the Ladder-backed

Woodpecker (*Dendrocopos scalaris*) and Lesser Goldfinch (*Carduelis psaltria*) were nesting in the grove.

Because of the interplay of several major ecological components in this grove (overstory, understory, emergent vegetation), it is possible to draw only a rough temporal sequence of recolonization related to the maturation of the *Salix-Populus* grove itself: Green Heron + Yellow-billed Cuckoo > (Black-chinned Hummingbird?) > Bell's Vireo + Summer Tanager > Lesser Goldfinch + Ladder-backed Woodpecker. (This annual sequence is cumulative.) The next species I would expect to colonize the grove, were it allowed to mature, is the Yellow Warbler (*Dendroica petechia*), not yet found breeding on the reservation. Gilded Flickers (*Colaptes a. mearnsi*) and Gila Woodpeckers (*Melanerpes uropygialis*) forage throughout the year in the grove, but apparently they nest only in adjacent bajada and mountain areas (*Cereus-Cercidium* association).

Northward Movements

Since European settlement of the United States, various Neotropical bird species are known to be extending their breeding ranges northward, successfully colonizing rather extensive new areas. In the Southwest at least nineteen avian species (Tables 6 and 7) have demonstrated remarkable changes within the past hundred years. Ten of these (Table 6) were unknown in Arizona before this century. And it must be kept in mind that during the latter half of the nineteenth century and the early twentieth century extensive fieldwork and collecting were conducted in southeastern Arizona by Stephens, H. Brown, Bendire, Scott, Nelson, Henshaw, A. P. Smith, Cahoon and the Biological Survey parties, Law, Mearns, Holzner, Swarth, and Kimball; ten species new to Arizona first colonized the well-worked southeastern section of the state.

During this period four southern species invaded the reservation: the Black Vulture (*Coragyps atratus*), Great-tailed Grackle (*Quiscalus mexicanus*), Bronzed Cowbird (*Molothrus aeneus*), and Pyrrhuloxia (*Cardinalis sinuatus*).

The Black Vulture, first observed in Arizona in 1920 by Gilman near Sells Papago Agency, appeared on the reservation by 1944 (Phillips, 1946). There is no indication that it is permanently established on the reservation or in the Phoenix region, though in recent years the species has been reported with increasing frequency along the northern section of the Sierra Estrella. Nevertheless, I suspect that the Black Vulture ranges widely at the periphery of its range, dependent as it is on fortuitous food supplies.

Range extension of the Great-tailed Grackle into the Southwest began in the late 1930s (Phillips, 1950a). I have documented in detail (Rea, 1969) the movements and interrelationships of the two populations involved (see next section). Local human modifications of the environment (irrigation, grain agriculture, cattle feeding, planting of exotic trees) facilitate local colonization

Table 6. Species Colonizing Arizona During the Twentieth Century

Records based primarily on Phillips et al. (1964) and Phillips (1968). Symbols: + = readily demonstrable association with human settlement or man-modified habitats; 0 = no readily apparent association with man-modified habitats.

Species	Primary Habitat	Human Facilitation
Black Vulture, *Coragyps atratus*	Lower Sonoran	+
Mississippi Kite, *Ictinia mississippiensis*	riparian	+
Anna's Hummingbird, *Calypte anna*	riparian & urban	+
Violet-crowned Hummingbird, *Amazilia violiceps*	riparian	0
Rose-throated Becard, *Pachyryamphus aglaiae*	riparian	0
Thick-billed Kingbird, *Tyrannus crassirostris*	riparian	0
Tropical Kingbird, *Tyrannus melancholicus*	riparian	0
Great-tailed Grackle, *Quiscalus mexicanus*	agric. & urban	+
Bronzed Cowbird, *Molothrus aeneus*	Lower Sonoran	+
Black-capped Gnatcatcher,[a] *Polioptila nigriceps*	high Lower Sonoran	0
Five-striped Sparrow,[b] *Aimophila quinquestriata*	high Lower Sonoran	0

[a]Based on Phillips, Speich, and Harrison (1973).

[b]Based on G. Monson and S. Russell (personal communication).

Table 7. Species with Range Extensions Within the Southwest During Historic Times

Based primarily on Phillips et al. (1964) and Phillips (1968). Symbols as for Table 6.

Species	Primary Habitat	Human Facilitation
Common Black Hawk, *Buteogallus anthracinus*	riparian	0
White-winged Dove, *Zenaida asiatica*	Lower Sonoran	0
Inca Dove, *Scardafella inca*	Lower Sonoran	+
Gila Woodpecker, *Centurus uropygialis*	Lower Sonoran	0
Olive Warbler, *Peucedramus taeniatus*[a]	montane	0
Red-faced Warbler, *Cardellina rubrifrons*[a]	montane	0
Hooded Oriole, *Icterus cucullatus*	Sonoran Zones	?
Northern Cardinal, *Cardinalis cardinalis*	Lower Sonoran	0

[a]Gale Monson, personal communication.

by this essentially marsh icterid. Most of the reservation (except about the Sacaton Agency) is too dry for grackle nesting. But conceivably the species might have colonized either Hohokam or aboriginal Riverine Pima settlements, had the initial range expansion occurred in an earlier century.

The Bronzed or Red-eyed Cowbird was discovered in the Southwest in 1909, simultaneously at Tucson and Sacaton. It colonized successfully and has become widespread at villages and rancherias where horses and cattle are maintained. It now far outnumbers its principal host, the Hooded Oriole. At least on the reservation the cowbird is completely migratory, departing by August.

The Pyrrhuloxia's massive colonization of the agricultural areas of the reservation occurred after 1915 and almost certainly after 1947. Unlike the preceding three species, which were unknown in Arizona at the turn of the century, the Pyrrhuloxia appears to have had a stable range in southern Arizona. On the Gila River it occurs in sorghum and other fields with fence rows. The reservation portion of the Gila has been farmed for centuries, with wheat and other grains being principal crops since Kino's day, but the species has invaded only recently.

Causality for widespread range extensions is difficult to demonstrate. Except the Mississippi Kite and Anna's Hummingbird, all the species in Table 6 colonizing the Southwest originated directly from the south. Of the nineteen species with known extensions in the Southwest, twelve occur in habitats apparently unmodified by man, six are definitely associated with human modifications, and one (Hooded Oriole) is equivocal. Seven of these species breed in riparian habitats: Mississippi Kite, Common Black Hawk, Anna's Hummingbird, Violet-crowned Hummingbird, Rose-throated Becard, Thick-billed Kingbird, Tropical Kingbird. Their colonization has coincided with the greatest loss of riparian habitat known in the Holocene. In these cases human modifications seem clearly to be negative rather than facilitating factors.

The kite is the noteworthy exception to the direct northward invasion. Interestingly, the only other species in the genus *Ictinia* is largely Neotropical in overall distribution, and Sutton (1944) has presented strong evidence suggesting that the two taxa are conspecific. This riparian species was confined until recently to the southeastern part of the United States. Oberholser (1974:211) discussed its habitat shift since 1957 from the mesic eastern parts of Texas to the arid western portions of the state. Rich Glinski (personal communication) believes the kite's successful colonization and continued expansion in the Southwest is attributable to habitat modifications: there is an enormous buildup of cicada populations in the saltcedar understory along the riparian habitats. This insect appears to be a principal food for kites. I suspect kites would colonize the Salt River portion of the reservation if the willow and cottonwood thickets were permitted to develop.

The Anna's Hummingbird has invaded the Southwest from the opposite direction. Originally a bird of the Pacific coast, the species appeared first as a winter visitant across southern Arizona (Phillips, 1947), then began nesting in the 1960s. Exotic plantings, at least in part, have facilitated its establishment in urban areas. The tree tobacco, naturalized from South America, is an important winter food in the Southwest. Zimmerman (1973) documented its continued eastward spread even into New Mexico and Texas. Within the species' original coastal range, habitat modifications and artificial feeding apparently produced a population buildup.

I think the invasions (Tables 6 and 7) in themselves are usually part of widespread range extensions in the species, independent of, but sometimes facilitated by, local habitat modifications. I would attribute most such broad-fronted expansions to a continuing overall postglacial (Holocene) movement of fauna and flora (Van Devender, 1973, 1977) following the Pleistocene.

Nowak (1975) has analyzed range extensions of twenty-eight animal species (vertebrates and invertebrates) in Europe. Of the eight avian species that have recently colonized, only two (both Palearctic ducks) appear to have been facilitated by human modifications of the environment. Six species appear to be naturally dispersed into habitats that seem to show no obvious changes in kind or degree of modification in recent centuries. (Quite the opposite situation holds for the New World.) Nowak (1975:112) concluded: "The extent of contemporary processes of spreading of animals is very great and confirm[s] the thesis that the European fauna has still not overcome its reduction during the last glaciation. The ease with which many species of animals spread points to the existence of many free niches in the European biocenoses."

Changes in Subspecies Distributions

It is generally believed that subspecies have been rather stable entities in the Holocene, presumably without changes in historic times. Undoubtedly lengthy periods of local stability (at least during the breeding season) are necessary for natural selection to produce recognizable subspecies. In south-central Arizona two instances of racial instability and phenotypic fluctuation affect the reservation avifauna: Horned Larks (*Eremophila alpestris*) and Great-tailed Grackles. The larks are a case of two peripheral range extensions from within the Sonoran Desert, the grackles a case of dual northward invasions. (See distribution maps, pp. 192, 193, and 249.)

I have detailed the history and phenotypic outcome of the grackles (Rea, 1969). Briefly, two distinctive races, differing in both color and size, invaded the Southwest from Mexico: *Q. m. monsoni* from the plateau and *Q. m. nelsoni* from the Pacific coastal plain. The resulting secondary intergrades were variable but now appear to have stabilized in both size and color nearest *monsoni*, the race that is larger and darker, though more distant in origin. The intergrades (rather than the original parental forms) are responsible for recent range extensions into the lower Gila Valley, the Verde Valley, and even southern Nevada (AMR, MNA, UNLV; for abbreviations of collections, see Table 13).

Hurley and Franks (1976) have shown that the eastern race of Horned Lark (*E. a. praticola* (Henshaw)), has significantly extended its range in historic times as a result of habitat modification. But the situation with the Arizona Horned Larks is perhaps less expected because both races involved are believed to be sedentary (Phillips, 1946) and to occupy different habitats. Both desert taxa are well defined, differing conspicuously in color as do the grackles, but having no significant size differences.

Presumably Horned Larks were entirely absent formerly from the Phoenix region. The Yuma race, *E. a. leucansiptila,* originally was known only from the lower Colorado River Valley. Starting in 1939 Phillips, Hargrave, L. Miller, and van Rossem (Monson and Phillips, 1941; van Rossem, 1947) located probable or actual breeding colonies in western Maricopa County (Gila Bend,

Hassayampa, Aguila). This race reached the immediate Phoenix region and bred at least by 1958 (see species account for details). Invasion of the reservation followed the wet winter of 1972–1973. Data documenting the spread of the southeastern Arizona grassland race, *E. a. adusta,* are less precise. Monson's breeding record (GM, ARIZ) from near Oracle Junction, north of the Santa Catalina Mountains, and the presence of some phenotypic *adusta* at Ventana Ranch, demonstrate that this race also has a potential for colonizing new areas outside of its normal range. *Adusta* reached the Phoenix region at least by 1966 and probably by 1959. The phenotypic outcome of these two range extensions is not yet known. Further collecting of adequate samples of breeding Horned Larks throughout the newly cultivated areas of southern Arizona is necessary to determine the genetic status of these populations.

Changes in Wintering Status

Several warblers (family Parulidae) began wintering at the northwest end of the reservation during my study (see respective species accounts for details). Starting in 1972, I found Orange-crowned Warblers (both Pacific *V. c. lutescens* and interior *V. c. orestera*) wintering rather commonly in saltcedars mixed with other deciduous vegetation at Barehand Lane on the lower Salt. Three additional warblers were found wintering only on the lower Salt. Beginning in 1973, Black-throated Gray Warblers wintered in willow thickets in increasing numbers each year, becoming common by 1976 and 1977. In December 1974 I found one or two Townsend's Warblers at the confluence. These, too, increased in numbers during succeeding winters, becoming actually common by late December 1977. Finally, between late November and late December 1977 I found Hermit Warblers in small numbers and took three substantiating specimens. On 28 November 1977 I noted that the Audubon's Warbler was the most abundant warbler at the confluence, followed by Black-throated Gray, then Townsend's, with Orange-crowned Warblers being less common than these three. None of the four species (Orange-crowned, Black-throated Gray, Townsend's, and Hermit Warblers) was previously known to be a wintering bird in southern Arizona.

An additional species must be mentioned as a possible change in wintering status. Today the Audubon's (Yellow-rumped) Warbler is the most common and widespread wintering insectivore on the reservation. Where willow thickets retain their leaves through December and much of January, this warbler winters in considerable numbers. For instance, in late 1979 there were literally hundreds on the lower Salt, in spite of flood damage. Surprisingly, Gilman (MS), after spending seven winters on the reservation, stated pointedly "none seen in winter," which suggests that there has been a change in wintering with these species as well. During my first brief mid-winter visit to The Buttes-Cochran area, I found no Audubon's Warblers along the river. There were a few in December 1982. But the species normally winters commonly throughout southern Arizona (Phillips, 1946; Phillips et al., 1964).

I think the wintering of these warblers (except the Audubon's and perhaps the Orange-crowned) is a direct response to the development of deciduous riparian growth along wastewater channels. Broad-leafed communities with understory exist nowhere else on the reservation today. When these lower Salt communities were damaged by floods, starting in early 1978, followed by bulldozing much of the channel vegetation for flood control, the warblers ceased wintering in any numbers. I found no Townsend's or Hermit Warblers and only one Black-throated Gray in December 1978 and none of these at all in 1980.

Changes in Departure Times of Winter Residents

A comparison of Gilman's departure dates of winter residents with those obtained during my own field work (Table 8) indicates at least thirteen species were wintering longer in the lowlands during the early part of the century than they do today. If anything, my data are biased in favor of late dates: because of my special interest in migration and taxonomy, I have consistently attempted to secure early and late records, particularly of polytypic species. Probably a few additional species might be added to the list, but Gilman does not give precise departure dates for some, such as the Sharp-shinned Hawk. Gilman called the Cooper's Hawk, another wintering species, numerous in spring. By contrast my latest dates for Cooper's Hawk are 7 January and 21 February (one each) with none in spring.

From the wording of Gilman's account, it is not certain that Hermit Thrushes were wintering. My late dates for migrating races of Hermit Thrushes (see species account) are 16 and 25 April and 5 May, whereas Gilman recorded the species only to 12 April. My latest date for *Catharus g. guttatus,* the race known to winter on the reservation, is 14 March. My late dates for Chipping Sparrows are 25 and 30 April. Gilman recorded them as wintering "until March." But as in the case of the thrushes, my late *Spizella* are clearly transients, migrating with parulids.

In spite of several ambiguous cases, it is evident that the northern species actually wintering on the reservation now depart appreciably earlier than they did formerly. In fact, in the 1960s and 1970s during late winter (most of January through early March) many wintering species occurred in very low numbers or not at all in the floodplains. This exodus has been noted also in the Salt River Valley by Johnson and Simpson (personal communication). I attribute this change to habitat deterioration in the lowlands. With the disappearance of the riparian community, which formerly supplied more favorable habitat (shelter, food, and cooler temperatures as the desert becomes hot in spring), wintering birds presumably must seek more suitable habitats elsewhere, perhaps by moving to somewhat higher elevations.

When I camped atop the Sierra Estrella on 5–6 May 1975, I was surprised to find House Wrens and abundant Green-tailed Towhees and Hermit Thrushes in the live oak-gooseberry thicket (above 4,000 feet or 1,220 m). These species had already deserted the lowlands that year.

Table 8. Species Formerly Wintering Later on the Gila River

Species	Gilman (1907–1915)	Rea (1963–1976)
Red-shafted Flicker, *Colaptes auratus* (*cafer* group)	15 April	10–17 March
Yellow-bellied Sapsucker, *Sphyrapicus varius*	17 April	10–12 March
Steller's Jay, *Cyanocitta stelleri*	1 March	23 February
Scrub Jay, *Aphelocoma coerulescens*	21 April	15 February
Common Raven, *Corvus corax*	27 May	14 March, 11 April
Northern Robin, *Turdus migratorius*	12 April	5–14 March
Ruby-crowned Kinglet, *Regulus calendula*	15 May	8 April
Green-tailed Towhee, *Pipilo chlorurus*	regularly to 20 May	18–29 April (6 May, Mountain)
Vesper Sparrow, *Pooecetes gramineus*	"leaving in April"	13–26 March
Red-backed Junco, *Junco hyemalis* (*caniceps* group)	22 May	12 March to 6 April
White-crowned Sparrow, *Zonotrichia leucophrys*		
(dark lored)	20 May	29 April–12 May
(pale lored)	1 May	15 April
Brewer's Blackbird, *Euphagus cyanocephalus*	mid to 30 April	17–30 March
Lawrence's Goldfinch, *Carduelis lawrencei*	12–20 April	11 January, 24 February

Winter Flight Years

Periodic winter invasions of birds into the lowlands are termed "flight years." These movements lack the regularity of true migrations. Such "irruptions" in the Northern Hemisphere are common to both the Old and New World. This phenomenon has been discussed by Cornwallis (1964), Dorst (1962), Lack (1970), Wetmore (1926), and others. Invasions into Arizona deserts, particularly by corvids and fringillids, are discussed by Phillips et al. (1964). At least in part these movements are related to failure of major prey species or crop species (see, for example, Davis and Williams [1957] and Formosov [1933] for Old and New World *Nucifraga* species).

Though causality of such flights is due to complex factors in the breeding ranges of the species involved, these irregular invasions do affect the composition of the avifauna wintering on the Gila River Reservation. Breninger's (1901:45) fieldwork was during a flight year, as is indicated by his observation of Piñon Jays (*Gymnorhinus cyanocephalus*) and his comment on Scrub Jays (*Aphelocoma coerulescens*): "Have never seen so many in this valley before." Earlier that year Steller's Jays (*Cyanocitta stelleri*) were collected on the San Pedro River (Phillips et al., 1964). The winter of 1910–1911 was another flight year, when Gilman recorded Clark's Nutcrackers (*Nucifraga columbiana*), Steller's Jays, Red-breasted Nuthatches (*Sitta canadensis*), and Townsend's Solitaires (*Myadestes townsendi*). He found Piñon Jays also in 1914 and in another year not specified.

During the period of my consistent fieldwork (1963–1976), the winter of 1972–1973 was a major flight year (Table 9). Nine taxa were seen or collected only that winter: *Asyndesmus lewis, Zoothera naevia, Agelaius phoeniceus arctolegus, A. p. neutralis, Carpodacus purpureus, Carpodacus c. cassinii, C. c. vinifer, Carduelis t. tristis, Pipilo erythrophthalmus curtatus.* Five other taxa were seen during some other winters of my fieldwork but appeared in numbers only in the winter of 1972–1973; *Turdus migratorius, Sialia currucoides, Pipilo e. montanus, Carduelis pinus, Carduelis tristis pallidus.* Table 10 shows the taxa apparently involved in this flight year and their breeding ranges.

A number of observations can be made on the 1972–1973 flight year. None of the usual corvids associated with coniferous crops (Piñon Jay, Clark's Nutcracker, Steller's Jay) appeared that year on the reservation. The first two of these are seed-crop specialists and would be expected in the lowlands. Also, some of the more obligatory coniferous species such as Red Crossbill (*Loxia curvirostra* Linnaeus) and Evening Grosbeak (*Coccothraustes vespertinus*) were not discovered on the reservation that winter. (But a grosbeak was taken that winter in the nearby Phoenix region: AMR 4324, 28 October, Youngtown; immature male, nominate race, if bill grown.) Other cardueline finches appeared: Purple Finch, Cassin's Finch (two races), American Goldfinch (two races), Pine Siskin. But as Table 9 demonstrates, a little less than half the taxa were strongly or exclusively seed-eating species. Except for the two races of towhees, no unusual Emberizinae, such as Fox Sparrow (*Passerella iliaca*) or unexpected Song Sparrow races (*P. melodia*), were detected.

As usually conceived, flight years involve birds leaving cold northern coniferous forests for lower latitudes, but this is not always the case. Formerly the Thick-billed Parrot (*Rhynchopsitta pachyrhyncha* (Swainson)), a food specialist, irrupted northward into Arizona (Cornwallis, 1964). Three other taxa of the 1972–1973 flight year originated from the more mild-wintered, maritime climates west of the Cascade-Sierran Axis: *Sturnella neglecta confluenta, Agelaius phoeniceus neutralis,* and *Carpodacus purpureus californicus.*

These peculiarities of this flight year should at least discourage simple association of irregular invasions into the lowlands with seed-crop failures in northern latitudes of the continent.

The 1977–1978 winter was still more unusual in that all the casual or accidental species occurring belonged to a single insectivorous family, the Parulidae (Table 11). Most were found in late December during a brief period of fieldwork on the lower Salt, an area I had worked intensively for a number of years previously. By this time the cattails and broad-leafed vegetation had become very dense along several sewage and other wastewater channels in the riverbed between 91st and 115th avenues. The effluent was warm, no doubt maintaining the entire riparian community at slightly higher than normal temperatures.

Table 9. Years of Appearance of Some Erratic Winter Visitants

Symbols: numeral = actual number seen; • = more than 10; ▲ = actually common.

Species	Winter Starting												
	1963	1964	1965	1966	1967	1968	1969	1970	1971	1972	1973	1974	1975
Acorn Woodpecker, *Melanerpes formicivorous*		1											
Lewis Woodpecker, *Asyndesmus lewis*										2			
Williamson's Sapsucker, *Sphyrapicus thyroideus*								1					
Eastern Phoebe, *Sayornis phoebe*	1												
Steller's Jay, *Cyanocitta stelleri*				1					1				
Scrub Jay, *Aphelocoma coerulescens*								5		1			
Red-breasted Nuthatch, *Sitta canadensis*	2									2	1		1
Bushtit, *Aegithalos minimus*													5+
Brown Creeper, *Certhia familiaris americana*	1		1										1
Brown Creeper, *C. f. montana*	1(±)									2		1	1
Northern Robin, *Turdus migratorius*	•	•			•					▲			
Varied Thrush, *Zoothera naevia*										1			
(Eastern) Hermit Thrush, *Catharus g. "faxoni"*													1
Swainson Thrush, *Catharus ustulatus*						7							1
Western Bluebird, *Sialia mexicana*	▲			•						•			
Mountain Bluebird, *Sialia currucoides*										▲		•	
Townsend's Solitaire, *Myadestes townsendi*	1					2				2			
Cedar Waxwing, *Bombycilla cedrorum*	3–4		2	1	2					2			1
Rufous-sided Towhee, *Pipilo erythrophthalmus curtatus*										1			
Rufous-sided Towhee, *P. e. montanus*		1			1	1	1	2		▲			1–2
Evening Grosbeak, *Coccothraustes vespertinus*	1(+?)	1											
Cassin's Finch, *Carpodacus cassinii*										2–3			
Purple Finch, *Carpodacus purpureus*										1			
Pine Siskin, *Carduelis pinus*							▲			▲			
American Goldfinch, *Carduelis tristis*									5	▲			
Lawrence's Goldfinch, *Carduelis lawrencei*	2–3			2–3		1				3			

Table 10. Taxa Involved in the 1972–1973 Flight Year

Abbreviations: I = essentially insects; S = essentially seeds; I/F = insects with fruits, berries; I/S = insects in summer, seeds in winter.

Species	Origin (Breeding Range)	Food
Lewis Woodpecker, *Asyndesmus lewis*	Western U.S. (Pacific to Rocky Mts.)	I
Red-breasted Nuthatch, *Sitta canadensis*	Canadian Zone, North America	I
Brown Creeper, *Certhia familiaris montana*	Rocky Mts.	I
Northern Robin, *Turdus migratorius propinquus*	Interior West & Rocky Mts.	I/F
Varied Thrush, *Zoothera naevia meruloides*	Interior Northwest	I/F
Mountain Bluebird, *Sialia currucoides*	Open coniferous forests of West	I/F
Townsend's Solitaire, *Myadestes townsendi*	Boreal Zones of Western North America	I/F
Rufous-sided Towhee, *Pipilo erythrophthalmus montanus*	Southern Rocky Mts. & Southwest	I/S
P. e. curtatus	Mountains of Great Basin Desert	I/S
Western Meadowlark, *Sturnella neglecta confluenta*	Pacific Northwest (west of Cascades)	I/S
(Giant) Red-wing, *Agelaius phoeniceus artolegus*	Far north of North America	I/S
(San Diego) Red-Wing, *Agelaius phoeniceus neutralis*	Coastal southern California	I/S
Evening Grosbeak, *Cocothraustes v. vespertinus*[1]	Coniferous forests of northern and northeastern North America	S
Purple Finch, *Carpodacus purpureus californicus*	Coniferous forests of Pacific Slope	S
Cassin's Finch, *Carpodacus c. cassinii*	Coniferous forests of southern Rocky Mts.	S
Cassin's Finch, *Carpodacus c. vinifer*	Coniferous forests of Northwest	S
Pine Siskin, *Carduelis p. pinus*	Coniferous forests of northern part of North America	S
American Goldfinch, *Carduelis t. tristis*	Eastern half of North America	S
American Goldfinch, *Carduelis t. pallidus*	Interior West	S
Lawrence Goldfinch, *Carduelis lawrencei*	California	S

[1]Taken 15 miles (24 km) north of reservation.

The Black-and-white Warbler, *Mniotilta varia,* and the Ovenbird, *Seiurus aurocapillus,* were both seen in late October, when stray eastern migrants are more likely to be encountered (Austin, 1971). The Wilson's Warbler was the Pacific race; the Myrtle Warbler and apparently the female Common Yellowthroat were northwest Canadian races. At least two male yellowthroats were present also (one of an interior western race collected), but males winter not infrequently in the southern part of the state. All the remainder are eastern species. Five of the species were found at no other time during my fieldwork.

Table 11. Casuals on Lower Salt, Winter 1977

Symbols: Documentation, A = specimen, B = sight record only; Status, 1 = only reservation record, 2 = several records, 3 = only winter record.

Species	Date	Documentation	Status
Black-and-White Warbler, *Mniotilta varia*	28 October	B	1
Prothonotary Warbler, *Protonotaria citrea*	27 December	A	1
Tennessee Warbler, *Vermivora peregrina*	28 December	A	2
Myrtle Warbler, *Dendroica c. "hooveri"*	28 December	A	1
Chestnut-sided Warbler, *Dendroica pensylvanica*	28 December	A	1
Ovenbird, *Seiurus aurocapillus*	28 December	B	1
Wilson's Warbler, *Wilsonia p. chryseola*	28 December	A	3
American Redstart, *Setophaga ruticilla*	29 December	B	2
Common Yellowthroat, ♀ *Geothlypis t. yukonicola* (?)	31 December	A	3
Common Yellowthroat, ♂ *Geothlypis t. occidentalis*	31 December	A	3

The Wilson's Warbler and the Common Yellowthroat (both females) are the only winter records for the reservation; both are rare in winter in southern Arizona (Phillips et al., 1964). The numbers of newly wintering warblers (Black-throated Gray, Townsend's, Hermit) were especially high that year (see Changes in Wintering Status, p. 91). No corvids or unusual cardueline finches were found, as would be expected in a true flight year.

CHAPTER 6

Interrelationships of Birds and Reservation Habitats

The history of the avifaunal changes of the Gila River Indian Reservation has been largely dependent on habitat changes. Birds utilize the vegetational community for foraging, shelter, and nesting sites. Industrial man has had, and continues to have, an enormous impact on natural communities. In xeric regions particularly, the weakening and destruction of habitats may be irreversible. Damage to the Gila River watersheds led to an eventual loss of the lower two-thirds of the Gila River as a permanent stream with its riparian timber. This study has attempted to document, from various sources, the consequences such radical changes have had on bird distributions and numbers along one section of the Gila Basin.

Habitat Utilization by Birds

The eleven major types of habitat presently occurring on the reservation are described in Chapter 3. The eleven lists below enumerate the bird species *regularly* utilizing each habitat. Rare, casual, accidental, erratic, and irregular species are omitted. All the species on each list should readily be found every year in the given habitat.

A bird species might utilize a habitat at three distinctively different times: during breeding, in wintering, or during migration. Emphasis in contemporary studies of habitat utilization is often placed on the reproductive aspect. This unfortunately neglects two other highly important functions of habitats in the lives of birds. Suitable habitats for resting, watering, and feeding are essential requirements in the migration of continental birds. And migration is an important aspect in the life histories of many temperate species. Lacking suitable habitats during fall migration (July through October), some migratory birds must presumably pass over unsuitable areas without stopping, or failing that, perish. Where proper habitats are found, certain species may linger for days (possibly even weeks) before moving on.

The critical value of suitable habitats for migrants may not at first be obvious until viewed from an historical perspective. Present migratory patterns presumably arose during the Pleistocene Era. Birds funneling across the Southwest from the Pacific Coast, the Great Basin, and the Rocky Mountains (as known from the study of subspecies) have been confronted, at least in the last 10,000 years, with extensive hot lowland deserts. The Gila River, spanning in excess of 300 miles (480 km) across this desert, formerly with continuous riparian timber, must have been an important oasis stopover during migrations.

Habitat utilization for wintering is likewise an important aspect of avian biology. (In Chapter 5 I summarized evidence demonstrating significant changes in wintering status and departure times within this century.) For purposes here, wintering will mean using the habitat beyond the usual period of fall migration, at least through the end of December.

In the lists below, birds breeding in a given habitat are marked with an asterisk (*). A bird only suspected of breeding there is marked (?). A bird present in summer but not breeding is marked (S). Use of the particular habitat for resting and feeding during migration is indicated by (M) and wintering by (W). A species may use a number of different habitats during migration but be more restricted in choice of wintering habitats (e.g., Green-tailed Towhee and Lincoln's Sparrow). The data are only qualitative. Great numbers of certain warblers and sparrows pass through the reservation during migrations, with relatively few individuals of these species remaining to winter. In a few instances a fourth category of habitat utilization, roosting, must be considered. In winter probably tens of thousands of icterids (*Agelaius, Euphagus, Xanthocephalus, Molothrus*) arrive on the reservation at dusk to roost in saltcedars and in cattail marshes and leave again at early dawn for the feedlots and fields of the Phoenix area.

Some species do not fit conventional categories. For instance, Costa's Hummingbird arrives in December or January, soon begins courtship and breeding, and departs by April. Technically, it leaves for its "wintering" area from May to November. In each case details are given in the individual species accounts.

During breeding a species may utilize several different habitats in different ways. White-winged Doves nest primarily in saltcedar thickets and mesquite bosques along the Gila and Santa Cruz rivers. However, they feed regularly on saguaros growing in the bajadas and mountains and travel long distances to grain fields and feed lots for food. Thus, they are marked on the list as utilizing certain habitats for breeding where they seldom place actual nests.

The first list contains birds that are ubiquitous over the reservation from the floodplains to the ridges of the mountains. Six of these breed throughout areas of Lower Sonoran vegetation. The ubiquitous species utilize the saltcedar channel vegetation and runoff ponds to a limited extent and the remaining eight habitats with greater regularity. None of the species on this list (with the exception of Turkey Vultures and certain seed-eating species) utilize mechanized farming areas with any regularity.

Ubiquitous (Except Mechanized Farming)

Turkey Vulture, *Cathartes aura* (?, S, W)
Cooper's Hawk, *Accipiter cooperii* (W)
Sharp-shinned Hawk, *Accipiter striatus* (W)
Mourning Dove, *Zenaida macroura* (* W)
Ladder-backed Woodpecker, *Dendrocopos scalaris* (* W)
Western Wood Pewee, *Contopus sordidulus* (M)
Verdin, *Auriparus flaviceps* (* W)
House Wren, *Troglodytes domesticus* (M, W)
Cactus Wren, *Campylorhynchus brunneicapillus* (* W)
Black-tailed Gnatcatcher, *Polioptila melanura* (* W)
Ruby-crowned Kinglet, *Regulus calendula* (M, W)
MacGillivray's Warbler, *Oporornis tolmiei* (M)
Wilson's Warbler, *Wilsonia pusilla* (M)
Western Tanager, *Piranga ludoviciana* (M)
Black-headed Grosbeak, *Pheucticus ludovicianus* (M)
Green-tailed Towhee, *Pipilo chlorurus* (M)
Chipping Sparrow, *Spizella passerina* (M, W)
Brewer's Sparrow, *Spizella breweri* (M, W)
Desert Sparrow, *Amphispiza bilineata* (* W)
House Finch, *Carpodacus mexicanus* (* W)
Lesser Goldfinch, *Carduelis psaltria* (?, S, W)

Floodplains (*Atriplex-Lycium* Flats)

Gambel's Quail, *Callipepla gambelii* (* W)
Greater Roadrunner, *Geococcyx californianus* (* W)
Great Horned Owl, *Bubo virginianus* (W)
Burrowing Owl, *Athene cunicularia* (*)
Lesser Nighthawk, *Chordeiles acutipennis* (*)
Northern Mockingbird, *Mimus polyglottos* (* W)
Bendire's Thrasher, *Toxostoma bendirei* (* W)
Le Conte's Thrasher, *Toxostoma lecontei* (* W)
Loggerhead Shrike, *Lanius ludovicianus* (* W)
Western Meadowlark, *Sturnella neglecta* (W)
Sage Sparrow, *Amphispiza belli* (W)

Floodplains: Mesquite Bosques (Second Growth)

Gambel's Quail, *Callipepla gambelii* (* W)
White-winged Dove, *Zenaida asiatica* (*)
Common Ground Dove, *Columbina passerina* (* W)
Greater Roadrunner, *Geococcyx californianus* (* W)
Barn Owl, *Tyto alba* (?, W)
Common Screech-Owl, *Otus asio* (* W)
Great Horned Owl, *Bubo virginianus* (W)
Northern Flicker, *Colaptes auratus* (W)
Yellow-bellied Sapsucker, *Sphyrapicus varius* (W)
Ash-throated Flycatcher, *Myiarchus cinerascens* (* W)
Northern Mockingbird, *Mimus polyglottos* (* W)
Crissal Thrasher, *Toxostoma crissale* (* W)
Phainopepla, *Phainopepla nitens* (* W)
Loggerhead Shrike, *Lanius ludovicianus* (* W)
Bell's Vireo, *Vireo bellii* (* locally)
Orange-crowned Warbler, *Vermivora celata* (M)
Lucy's Warbler, *Vermivora luciae* (* locally, M)
Yellow-rumped Warbler, *Dendroica coronata* (M)
Black-throated Gray Warbler, *Dendroica nigrescens* (M)
Townsend's Warbler, *Dendroica townsendi* (M)
MacGillivray's Warbler, *Oporornis tolmiei* (M)
Wilson's Warbler, *Wilsonia pusilla* (M)
Northern Oriole, *Icterus galbula* (* M)
Brown-headed Cowbird, *Molothrus ater* (*)
Northern Cardinal, *Cardinalis cardinalis* (* W)
Abert's Towhee, *Pipilo aberti* (* W)

Chandler Boundary Cottonwoods

Red-tailed Hawk, *Buteo jamaicensis* (W)
American Kestrel, *Falco sparverius* (W)
White-winged Dove, *Zenaida asiatica* (*)
Common Screech-Owl, *Otus asio* (?, W)
Lesser Nighthawk, *Chordeiles acutipennis* (S)
Northern Flicker, *Colaptes auratus* (W)
Gila Woodpecker, *Melanerpes uropygialis* (*formerly, W)
Yellow-bellied Sapsucker, *Sphyrapicus varius* (W)
Western Kingbird, *Tyrannus verticalis* (* formerly)

Chandler Boundary Cottonwoods (Continued)

Common Starling, *Sturnus vulgaris* (* W)
Yellow-rumped Warbler, *Dendroica coronata* (M)
Northern Oriole, *Icterus galbula* (* formerly)
Western Meadowlark, *Sturnella neglecta* (* W)
Brewer's Blackbird, *Euphagus cyanocephalus* (W)
Brown-headed Cowbird, *Molothrus ater* (*)
Bronzed Cowbird, *Molothrus aeneus* (*)
Dark-eyed Junco, *Junco hyemalis* (W)
House Sparrow, *Passer domesticus* (* W)

Pima Rancherias (Exclusive of Second Growth Mesquite)

American Kestrel, *Falco sparverius* (* W)
Gambel's Quail, *Callipepla gambelii* (* W)
Killdeer, *Charadrius vociferus* (* W)
White-winged Dove, *Zenaida asiatica* (* W)
Common Ground Dove, *Columbina passerina* (*W)
Inca Dove, *Scardafella inca* (* W)
Greater Roadrunner, *Geococcyx californianus* (* W)
Barn Owl, *Tyto alba* (* W)
Great Horned Owl, *Bubo virginianus* (W)
Burrowing Owl, *Athene cunicularia* (*)
Gila Woodpecker, *Melanerpes uropygialis* (* W)
Yellow-bellied Sapsucker, *Sphyrapicus varius* (W)
Western Kingbird, *Tyrannus verticalis* (*)
Say's Phoebe, *Sayornis saya* (* W)
Northern Mockingbird, *Mimus polyglottos* (* W)
Bendire's Thrasher, *Toxostoma bendirei* (* W)
Phainopepla, *Phainopepla nitens* (* W)
Loggerhead Shrike, *Lanius ludovicianus* (* W)
Common Starling, *Sturnus vulgaris* (* W)
Orange-crowned Warbler, *Vermivora celata* (M)
Yellow-rumped Warbler, *Dendroica coronata* (M, W)
Townsend's Warbler, *Dendroica townsendi* (M)
Northern Oriole, *Icterus galbula* (* M)
Western Meadowlark, *Sturnella neglecta* (* W)
Red-winged Blackbird, *Agelaius phoeniceus* (* W)
Brewer's Blackbird, *Euphagus cyanocephalus* (W)
Brown-headed Cowbird, *Molothrus ater* (*)
Bronzed Cowbird, *Molothrus aeneus* (*)

Northern Cardinal, *Cardinalis cardinalis* (* W)
Pyrrhuloxia, *Cardinalis sinuatus* (* W)
Blue Grosbeak, *Passerina caerulea* (?, M)
Abert's Towhee, *Pipilo aberti* (* W)
Lark Bunting, *Calamospiza melanocorys* (W)
Savannah Sparrow, *Ammodramus sandwichensis* (W)
Dark-eyed Junco, *Junco hyemalis* (W)
House Sparrow, *Passer domesticus* (* W)

Mechanized Farming

Horned Lark, *Eremophila alpestris* (* W)
Water Pipit, *Anthus spinoletta* (W)
Western Meadowlark, *Sturnella neglecta* (W)
Yellow-headed Blackbird, *Xanthocephalus xanthocephalus* (W)
Red-winged Blackbird, *Agelaius phoeniceus* (W)
Brewer's Blackbird, *Euphagus cyanocephalus* (W)
Savannah Sparrow, *Ammodramus sandwichensis* (M, W)
House Finch, *Carpodacus mexicanus* (W)

Salt River (Through 1979)

Green Heron, *Ardeola virescens* (* W)
Black-crowned Night Heron, *Nycticorax nycticorax* (?, W)
Least Bittern, *Ixobrychus exilis* (* formerly)
Cinnamon Teal, *Anas cyanoptera* (?, M, W)
Blue-winged Teal, *Anas discors* (M)
Ruddy Duck, *Oxyura jamaicensis* (* formerly)
Gambel's Quail, *Callipepla gambelii* (* W)
Clapper Rail, *Rallus longirostris* (?)
Common Gallinule, *Gallinula chloropus* (* W)
American Coot, *Fulica americana* (* W)
Killdeer, *Charadrius vociferus* (* W)
Common Snipe, *Capella gallinago* (W)
Spotted Sandpiper, *Actitis macularia* (M, W)
Least Sandpiper, *Erolia minutilla* (M, W)
Western Sandpiper, *Ereunetes mauri* (M)
American Avocet, *Recurvirostra americana* (M)
Black-necked Stilt, *Himantopus mexicanus* (* formerly, M)
White-winged Dove, *Zenaida asiatica* (* W)
Common Ground Dove, *Columbina passerina* (*)

Salt River (Through 1979) (Continued)

Yellow-billed Cuckoo, *Coccyzus americanus* (*)
Greater Roadrunner, *Geococcyx californianus* (* W)
Great Horned Owl, *Bubo virginianus* (W)
Lesser Nighthawk, *Chordeiles acutipennis* (*)
Black-chinned Hummingbird, *Archilochus alexandri* (?, S)
Gila Woodpecker, *Melanerpes uropygialis* (W)
Yellow-bellied Sapsucker, *Sphyrapicus varius* (W)
Ash-throated Flycatcher, *Myiarchus cinerascens* (*)
Black Phoebe, *Sayornis nigricans* (S, W)
Long-billed Marsh Wren, *Cistothorus palustris* (W)
Crissal Thrasher, *Toxostoma crissale* (*, W)
Bell's Vireo, *Vireo bellii* (*)
Orange-crowned Warbler, *Vermivora celata* (M, W)
Lucy's Warbler, *Vermivora luciae* (*)
Yellow-rumped Warbler, *Dendroica coronata* (M, W)
Black-throated Gray Warbler, *Dendroica nigrescens* (M, W)
Townsend's Warbler, *Dendroica townsendi* (M, W)
Common Yellowthroat, *Geothlypis trichas* (*)
Yellow-breasted Chat, *Icteria virens* (*)
Summer Tanager, *Piranga rubra* (*)
Abert's Towhee, *Pipilo aberti* (*, W)
Lincoln's Sparrow, *Passerella lincolnii* (W)
Song Sparrow, *Passerella melodia* (* W)
Northern Oriole, *Icterus galbula* (*)
Yellow-headed Blackbird, *Xanthocephalus xanthocephalus* (M, W)
Red-winged Blackbird, *Agelaius phoeniceus* (* W)
Brown-headed Cowbird, *Molothrus ater* (* W)

Barehand Lane (Former Marsh)

Green-winged Teal, *Anas crecca* (M)
Cinnamon Teal, *Anas cyanoptera* (M)
Blue-winged Teal, *Anas discors* (M)
Gambel's Quail, *Callipepla gambelii* (* W)
Virginia Rail, *Rallus limicola* (?, W)
Sora, *Porzana carolina* (?, M, W)
Common Gallinule, *Gallinula chloropus* (* W)
Common Snipe, *Capella gallinago* (W)
White-winged Dove, *Zenaida asiatica* (*)
Common Ground Dove, *Columbina passerina* (* W)

Greater Roadrunner, *Geococcyx californianus* (* W)
Great Horned Owl, *Bubo virginianus* (W)
Ash-throated Flycatcher, *Myiarchus cinerascens* (*)
Long-billed Marsh Wren, *Cistothorus palustris* (W)
Northern Mockingbird, *Mimus polyglottos* (* W)
Crissal Thrasher, *Toxostoma crissale* (* W)
Hermit Thrush, *Catharus guttatus* (W, M)
Phainopepla, *Phainopepla nitens* (* W)
Common Starling, *Sturnus vulgaris* (roost)
Bell's Vireo, *Vireo bellii* (*)
Orange-crowned Warbler, *Vermivora celata* (M, W)
Lucy's Warbler, *Vermivora luciae* (*)
Yellow-rumped Warbler, *Dendroica coronata* (M, W)
Townsend's Warbler, *Dendroica townsendi* (M)
Common Yellowthroat, *Geothlypis trichas* (*)
Yellow-breasted Chat, *Icteria virens* (*)
Northern Cardinal, *Cardinalis cardinalis* (* W)
Blue Grosbeak, *Passerina caerulea* (M)
Abert's Towhee, *Pipilo aberti* (* W)
Lincoln's Sparrow, *Passerella lincolnii* (W)
Song Sparrow, *Passerella melodia* (* W)
Northern Oriole, *Icterus galbula* (*)
Yellow-headed Blackbird, *Xanthocephalus xanthocephalus* (M, W, roost)
Red-winged Blackbird, *Agelaius phoeniceus* (* W, roost)
Brown-headed Cowbird, *Molothrus ater* (*)
Brewer's Blackbird, *Euphagus cyanocephalus* (roost)

Gila River Channel

White-winged Dove, *Zenaida asiatica* (*)
Lucy's Warbler, *Vermivora luciae* (?, recent local)
Yellow-headed Blackbird, *Xanthocephalus xanthocephalus* (W, roost)
Red-winged Blackbird, *Agelaius phoeniceus* (W, roost)
Brewer's Blackbird, *Euphagus cyanocephalus* (W, roost)

Bajadas and Arroyos

Red-tailed Hawk, *Buteo jamaicensis* (* W)
American Kestrel, *Falco sparverius* (* W)
Gambel's Quail, *Callipepla gambelii* (* W)
White-winged Dove, *Zenaida asiatica* (*)

Bajadas and Arroyos (Continued)

Common Screech-Owl, *Otus asio* (* W)
Great Horned Owl, *Bubo virginianus* (* W)
Lesser Nighthawk, *Chordeiles acutipennis* (?)
Costa's Hummingbird, *Archilochus costae* (* W)
Gilded Flicker, *Colaptes auratus* (* W)
Gila Woodpecker, *Melanerpes uropygialis* (* W)
Ash-throated Flycatcher, *Myiarchus cinerascens* (* W)
Say's Phoebe, *Sayornis saya* (* W)
Western Flycatcher, *Empidonax difficilis* (M)
Cañon Wren, *Catherpes mexicanus* (* W)
Rock Wren, *Salpinctes obsoletus* (* W)
Curve-billed Thrasher, *Toxostoma curvirostre* (* W)
Loggerhead Shrike, *Lanius ludovicianus* (* W)
Common Starling, *Sturnus vulgaris* (* W)
Orange-crowned Warbler, *Vermivora celata* (M)
Western Meadowlark, *Sturnella neglecta* (W)
Dark-eyed Junco, *Junco hyemalis* (W)

Mountains

Turkey Vulture, *Cathartes aura* (?)
Red-tailed Hawk, *Buteo jamaicensis* (* W)
Prairie Falcon, *Falco mexicanus* (*)
White-winged Dove, *Zenaida asiatica* (?)
Nuttall's Poor-will, *Phalaenoptilus nuttallii* (*)
White-throated Swift, *Aeronautes saxatalis* (*)
Gilded Flicker, *Colaptes auratus* (* W)
Say's Phoebe, *Sayornis saya* (W)
Cañon Wren, *Catherpes mexicanus* (* W)
Rock Wren, *Salpinctes obsoletus* (* W)
Curve-billed Thrasher, *Toxostoma curvirostre* (* W)
Phainopepla, *Phainopepla nitens* (*)
Canyon (Brown) Towhee, *Pipilo fuscus* (* W)
Rufous-crowned Sparrow, *Aimophila ruficeps* (* W)

Runoff Ponds

Cinnamon Teal, *Anas cyanoptera* (W, M)
Green-winged Teal, *Anas crecca* (M)
Shoveler, *Anas clypeata* (M, W)
Ruddy Duck, *Oxyura jamaicensis* (M)

Killdeer, *Charadrius vociferus* (* W)
Spotted Sandpiper, *Actitis macularia* (M)
Least Sandpiper, *Erolia minutilla* (M)
Western Sandpiper, *Ereunetes mauri* (M)
American Avocet, *Recurvirostra americana* (M)
Black-necked Stilt, *Himantopus mexicanus* (M)
Rough-winged Swallow, *Stelgidopteryx ruficollis* (?, M)
Northern Mockingbird, *Mimus polyglottos* (M)
Water Pipit, *Anthus spinoletta* (W)
Red-winged Blackbird, *Agelaius phoeniceus* (M, W)

Vulnerable Habitats

Among developed countries of the world there appears to be a beginning awareness of man's role of stewardship of the natural environment. It is futile to discuss the protection of birds without taking into account what is ultimately the most important issue: the protection of habitats. Four major habitats on the Gila River Indian Reservation may be singled out because of their vulnerability and critical importance to the avifauna.

Aquatic Habitats

The Gila, Santa Cruz, and Salt rivers were three living streams that just a generation ago converged on the Gila River Reservation. As described by early explorers, their confluences formed mesic communities attractive to birds as well as other forms of life. In spite of their eventual loss, local riparian and marsh habitats recently have been re-established by wastewater derived from sewage effluent and irrigation tailings. Two such areas (one with cottonwood-willow overstory, the other with understory only) were studied for a number of years during my fieldwork. These two areas have demonstrated that with habitat amelioration most locally extirpated species are still capable of recolonization as breeding species: Pied-billed Grebe, Least Bittern, Green Heron, Cinnamon Teal, Ruddy Duck, Yellow-billed Cuckoo, Virginia Rail, Bell's Vireo, Lucy's Warbler, Yellow-breasted Chat, Common Yellowthroat, Summer Tanager, and perhaps Black-crowned Night Heron and Sora. Reoccupation of such habitats by winter residents is even more prompt. It is fairly certain that all these species had been extirpated from the reservation, none being found during the earlier years of my fieldwork.

The vulnerability of such habitats is demonstrated by the recent eradication of the Barehand Lane marsh in order to reclaim several acres of farmland that could be irrigated by runoff water. It is likewise conceivable that demands for water could lead to confining the Phoenix sewage effluent to concrete pipes for industrial or agricultural use downstream from the reservation. This would eliminate several miles of marsh and riparian habitat along the Salt River from 91st Avenue to the Salt-Gila confluence and drastically reduce the present avian diversity.

Upper Mountain Habitats

The isolated upper slopes of the Sierra Estrella, particularly the north-facing slopes, have important and apparently relictual flora and fauna (see Chapter 3). These areas in the past have been invulnerable to exploitation and destruction by human agency. The vegetation of some whole upper basins is still naturally protected from predation by Eurasian herbivores. Not only the birds, but the entire biota of these desert islands should be protected from disturbance.

The Saltbush-Bursage Community

To the casual observer, the great stretches of *Atriplex-Ambrosia* community on the upper terraces of the Gila River are a wasteland that might best be converted into agricultural lands, irrigated by fossil water that is still available beneath tribal lands. Indeed, the bird-species diversity of this habitat is relatively low. However, this open, barren-looking habitat is the home of the Le Conte's Thrasher, a bird that is in fact (though not "officially") a threatened species. Phillips et al. (1964:124) observed, "Intolerant of man and his activities, it has retreated from the newly farmed areas of central Arizona." This rarely observed, extreme desert-adapted thrasher seems not to object to proximity with the Pima Indians. But how much further exploitation of its habitat the species might tolerate is unknown. Non-Indian agencies are presently pushing the "development" of the saltbush-bursage community and threatening the local survival of Le Conte's Thrasher.

Traditional Pima-Maricopa Rancherias

As shown in Chapter 3 and earlier in this chapter, as well as in the species accounts, the traditional Pima and Maricopa farming methods greatly increased both bird-species diversity and the numbers of individuals. Traditional farming communities (rancherias) had fruit trees, truck gardens, Athel tamarisks (in the twentieth century) for shade, and grain or forage crops. Before the introduction of barbed wire, fields were confined by broad brush fences. Subsequently, thorny shrubs were permitted to grow forming fence rows (Figs. 3.6, 3.7). Irrigation ditches were mostly unlined earth, often with a cottonwood overstory and some thin understory (Figs. 3.5, 3.8). Such traditional farming methods increased avian diversity because of augmented availability of food (insects and seeds), shelter, and nest sites (fence rows and ditch bank microhabitats), and water. Where such farms can be found today, usually there are high breeding densities of Gambel's Quail, Western Kingbirds, Verdins, Northern Mockingbirds, Bendire's Thrashers, Phainopeplas, Northern Orioles, Northern Cardinals, and Pyrrhuloxias. In the searing heat of July, August, and September, migrants from montane regions find shade and reduced temperature on rancherias. By contrast, large-scale mechanized farms (usually on lands leased to non-Indians) have almost no breeding birds and few wintering species. With acculturation fewer Indians are turning to traditional farming as a means of livelihood.

Fig. 6.1. Habitat changes on the Gila.

Not all habitat changes on the Gila River occur slowly. On 24 May 1982 I stopped at a marsh between Tacna and Roll in Yuma County, Arizona. Abundant Red-winged and Yellow-headed Blackbirds as well as Common Yellowthroats, Song Sparrows, Common Gallinules, Virginia Rails and Clapper Rails were nesting in the emergent vegetation (above). In less than an hour miles of marsh were destroyed in a "flood control" project (below). Only the Clapper Rail territory was left undisturbed. While one governmental agency severely regulates the taking of two or three specimens of each species to be preserved in research collections for posterity, another agency destroys thousands of acres of critical desert wetland habitat at the height of the breeding season. (Photographed by A. M. Rea.)

The future of the avifauna of the Gila River Indian Reservation is dependent on the protection of existing habitats from further destruction. A serious question must be faced by the political leaders of the tribe: should they capitulate to the expansionist demands of the Phoenix metropolitan area to exploit reservation habitats in exchange for immediate short-term benefits—often at the expense of nonrenewable natural resources such as fossil water and ancient (perhaps relictual) ecological communities?

PART II

Species Accounts

Contents

Plan of Species Accounts 115
Order Podicipediformes 125
Family Podicipedidae: Grebes 125
Order Pelecaniformes 126
Family Pelecanidae: Pelicans 126
Family Phalacrocoracidae: Cormorants 127
Order Ciconiiformes 127
Family Vulturidae (Cathartidae): New World Vultures 127
Family Plataleidae (Threskiornithidae): Spoonbills and Ibises 130
Order Accipitriformes (Falconiformes) 131
Family Accipitridae: Hawks, Eagles, Kites 131
Family Pandionidae: Ospreys 134
Family Falconidae: Falcons and Caracaras 135
Order Galliformes 139
Family Phasianidae: Quail, Turkey, and Grouse 139
Order Ardeiformes 141
Family Ardeidae: Herons, Egrets, and Bitterns 141
Order Gruiformes 147
Family Gruidae: Cranes 147
Family Rallidae: Rails, Gallinules, and Coots 148
Order Charadriiformes 150
Family Charadriidae: Plovers 150
Family Scolopacidae: Sandpipers and others 151
Family Recurvirostridae: Avocets and Stilts 153
Family Laridae: Gulls and Terns 154
Order Anseriformes 155
Family Anatidae: Ducks, Geese, and Swans 155
Order Columbiformes 159
Family Columbidae: Doves and Pigeons 159
Order Psittaciformes 162
Family Psittacidae: Parrots and Macaws 162
Order Cuculiformes 163
Family Cuculidae: Cuckoos, Roadrunners, and Anis 163
Order Strigiformes 165
Family Tytonidae: Barn Owls 165
Family Strigidae: Typical Owls 166
Order Caprimulgiformes 172
Family Caprimulgidae: Goatsuckers 172
Order Apodiformes 174
Family Apodidae: Swifts 174
Family Trochilidae: Hummingbirds 174
Order Coraciiformes 177
Family Alcedinidae: Kingfishers 177
Order Piciformes 178
Family Picidae: Woodpeckers 178
Order Passeriformes 182
Family Tyrannidae: Tyrant Flycatchers 182
Family Alaudidae: Larks 191
Family Hirundinidae: Swallows and Martins 194
Family Corvidae: Jays, Crows, and Magpies 197
Family Laniidae: Shrikes 201
Family Paridae: True Tits 202
Family Aegithalidae: Long-tailed Tits 202
Family Remizidae: Verdins and Penduline Tits 203
Family Sittidae: Nuthatches 204
Family Certhiidae: Treecreepers 204
Family Troglodytidae: Wrens 205

Order Passeriformes (Continued)
Family Mimidae:
Thrashers and Mockingbirds 208
Family Turdidae: Thrushes,
Bluebirds, and Solitaires 212
Family Sylviidae:
Old World Warblers, Kinglets,
and Gnatcatchers 216
Family Motacillidae: Pipits 217
Family Bombycillidae:
Waxwings and
Silky Flycatchers 217
Family Sturnidae: Starlings 219
Family Vireonidae: Vireos 219
Family Parulidae:
Wood Warblers 221
Family Thraupidae: Tanagers 228
Family Emberizidae:
Emberizine Finches 229
Family Icteridae: Blackbirds,
Orioles, and Meadowlarks 241
Family Cardeulidae:
Cardueline Finches 250
Family Passeridae (Ploceidae):
Weaver Finches 252

Plan of the Species Accounts

> Because the human race is incurably addicted to seeing things—usually super-rare birds and flying saucers in the twentieth century, and various kinds of ghosts in preceding centuries—it is necessary to lay down rules for inclusion of species.
>
> Edgar B. Kinkaid, Jr.
> (*in* Oberholser et al., 1974)

A species is admitted to the reservation list only if it is substantiated by specimen evidence (study skin, skeleton, or diagnostic feather). Hypothetical species based only on sight records are indicated by brackets around the species heading. A careful distinction must be made between bird watching as a pastime and ornithology as an *empirical science*. Science must be based on verifiable evidence preserved for reexamination.

Field data from a number of men, regularly collecting in the Arizona lowlands and regularly handling specimens, have been used in detailing the species accounts: Their names, with other relevant data, are listed in Table 12.

Common names in English are generally those of the American Ornithologists' Union Check-list (1957; 1973). In a few instances I have considered some provincial or ill-chosen and so have substituted names more appropriate. For example there are almost thirty species of robins in the Americas, a dozen breeding in North America and the West Indies. Northern Robin is a better name than "American" Robin for the northernmost species, *Turdus migratorius*.

Each species account is subdivided, where appropriate, into six categories: archaeological, ethnographic, historic, modern, taxonomy, and change.

Table 12. Field Ornithologists Working On or Near the Reservation

Name	Time Period	Location of Specimens*
Breninger, G. F.	1900–01	MCZ, MNA & others
Dickerman, R. W.	1953	CU
Gilman, M. F.	1907–15	SBC, MVZ; others?
Hargrave, L. L.	1933, 1940	LLH (at MNA)
Housholder, V. H.	1919–69	KANU, ARP
Johnson, R. R.	1952–76	JSW (at MNA)
Monson, G.	brief trips	GM (at ARIZ)
Phillips, A. R.	brief trips	ARP (at DEL)
Rea, A. M.	1963–81	AMR, ARP, MVZ, UWO, SD
Simpson, J. M.	1949–76	JSW (at MNA)
Werner, J. R.	1950–61	JSW (at MNA)
Yaeger, L. D.	1933–50	MNA, ARP, LLH

*For abbreviations see Table 13, p. 119.

Archaeological

Included here are avian osteological remains excavated from two Hohokam sites (300 B.C. to A.D. 1450) on the reservation: Snaketown (1964–1965 excavation) and the South Santan Salvage, both identified by Charmion R. McKusick. I have verified certain critical identifications. I have also included my identifications from the Las Colinas Site (Classic Hohokam) in western Phoenix and Pueblo Grande in eastern Phoenix, though these are beyond the political boundary of the reservation (see Rea, 1981b). Unworked bird bones from the first Snaketown excavation (Gladwin et al., 1937) cannot now be relocated (Emil W. Haury, personal communication), and several of the identifications are suspect. I have therefore excluded the unverifiable 1937 data.

Ethnographic

This section is headed by the vernacular (Piman) name of the species, its plural (if any), and, where possible, a gloss of the native name. Some names cannot be translated into English because they are primary words, are onomatopoetic, or are compounded of archaic words whose meaning no one remembers. To aid readers already familiar with systems used locally for writing Piman, I have generally used the orthography of Alvarez and Hale (1970), except where Riverine Pima distinguishes additional phonemes. A brief discussion of my phonemic analysis, together with several comparative orthographies, is given in Appendix A. The ethnotaxonomy is summarized in Appendix B.

All the data following in this section are the ethnoscience relating to that particular species as recounted by native consultants. The oral history presumably extends at least to 1873, when the first western village was reestablished (Russell, 1908:54). Brief references to two other genres of Piman oral literature (song and prose legends) are mentioned, but no claim is made for completeness. In most cases these genres are presumed to be more ancient. There is a richness of symbolism only hinted at here. Explication of the deeper cultural significance of animals will require a separate work.

I worked on the ethno-ornithology of the Pima over a fifteen-year period. Major consultants were Joseph Giff (1907–1982), Ruth Giff (born 1910), Sylvester Matthias (born 1911), all of Komatke, and David Brown (1917–1981) and Rosita Brown (1923–1977) of Santa Cruz village. Of all the Pima with whom I am acquainted, these five had the most thorough knowledge of field biology; all are native Piman speakers.

To add perspective to this section, I have occasionally included some comparative ethnotaxa from other mutually intelligible lowland Pima groups. Between 1977 and 1979 I worked with Pedro Estrella and María Córdova, the last two native speakers at Onavas, a lowland Pima Bajo settlement on the lower Río Yaqui, Sonora. Fairly complete plant and animal ethnotaxonomies were obtained. I also did some work with the Papago. Major consultants were Frank Jim, formerly of Anegam, Arizona, and Delores Lewis of Gu Oidak (Big Fields), Arizona, as well as Luciano Noriego of the ancient oasis village of Quitovac, Sonora. Though Northern Tepehuan is not a mutually intelligible language, the ethnotaxa are often cognate or even sometimes identical with Upper Pima. Pennington's (1969) work on the Northern Tepehuan frequently gives enough descriptive information with a native name that the biological species can be inferred. Some are included here, as are a few from Tepecano, the southernmost Southern Tepehuan dialect in Jalisco (Mason, 1917). Curiously, the Riverine Pima and the Tepecano, geographically at the very extremes of the Piman languages (900 airline miles [1520 km] apart), appear to be the nearest in ethnotaxonomies. A comprehensive comparative ethnobiology of all the Tepiman groups (Northern Tepehuan, Southern Tepehuan, Mountain Pima, and Lowland Pima with its various dialectical groups) remains to be assembled.

I have made no attempt to investigate the ethnobiology of the Yuman-speaking Maricopa Indians living on the Gila River.

Historic

The data in this section are from two primary sources. George F. Breninger (1901) spent four days (18–19 and 25–26 September 1901) on the reservation. His observations resulted in the earliest paper published on the local avifauna. Though Breninger was a collector and added considerably to the knowledge of Arizona ornithology in the late nineteenth century (see Phillips et al., 1964), he apparently collected no specimens on the reservation. His paper discussed "86 species seen on a tract probably not more than five miles [8 km] long by two miles [3 km] wide," and includes an important early description of the habitat (see Chapter 2). Two of his records are undoubtedly misidentifications and several common species apparently were missed on his list. Otherwise his account is reliable. It is unfortunate that he did not segregate the dates for migrants in his list.

The biologist and government-employed educator M. French Gilman lived on the reservation from 1907 to 1915, teaching at Blackwater, Sacaton, and Santan. Though primarily a botanist, he published eleven papers (1909 through 1915) on the reservation bird life. His accounts of doves, owls thrashers, and woodpeckers remain the classic contributions to the natural history of Lower Sonoran Desert birds. Unfortunately, Gilman's comprehensive manuscript on the reservation avifauna, written in 1915 after he left the reservation (cited here as MS), was never published, but parts were utilized by Phillips (1946) and Phillips et al. (1964) in preparing the state bird book. I have incorporated the greater part of Gilman's unpublished manuscript into the historic sections.

Modern

This section begins with a description of the present status of the species on the reservation. "Present" means approximately 1960–1980. The subjective phrases used to describe the status are those given by Marshall (*in* Phillips et al., 1964):

abundant: in numbers
common: always to be seen, but not in large numbers
fairly common: very small numbers or not always seen
uncommon: seldom seen but not a surprise
rare: always a surprise, but not out of normal range
casual: out of normal range
accidental: far from normal range and not to be expected again.

To these I have added two qualifying terms:

irregular: not to be found every season
erratic: found only a few seasons each decade

Next follow details of habitat, outside dates, specimen records, nesting, and so forth. Where I judged a particular specimen to be of critical importance to the local or state avifauna such that an investigator might wish to verify the specimen evidence, I have given the original collection number and present museum location, followed by the age, sex, and locality if important. Collection abbreviations are listed in Table 13. Specimen dates given in brackets indicate that the bird was found already dead for some time, perhaps mummified. For the sake of uniformity, I have converted all dates into the military or continental system of day, month, year. Migrations are complex phenomena; I have attempted to avoid oversimplifications, though these are inevitable where data are few. Where the information permits, I have suggested

Table 13. Abbreviations for Collections

AMNH	American Museum of Natural History, New York, NY
AMR	Amadeo M. Rea Collection, San Diego, CA
ARIZ	University of Arizona, Tucson, AZ
ARP	Allan R. Phillips Collection, Delaware Museum
CAS	California Academy of Sciences, San Francisco, CA
CGR	Casa Grande Ruins National Monument, Coolidge, AZ
CM	Carnegie Museum of Natural History, Pittsburgh, PA
CU	Cornell University, Ithaca, NY
DEL	Delaware Museum of Natural History, Greenville, DE
DM	Denver Museum of Natural History, Denver, CO
GCN	Grand Canyon National Park, Grand Canyon, AZ
GM	Gale Monson Collection, University of Arizona
JSW	Johnson–Simpson–Werner Collection, Museum of Northern Arizona
KANU	University of Kansas, Lawrence, KS
LAM	Los Angeles County Museum, Los Angeles, CA
LBSU	Long Beach State University, Long Beach, CA
LLH	Lyndon L. Hargrave Collection, Museum of Northern Arizona
MCZ	Museum of Comparative Zoology, Harvard, Cambridge, MA
MFG	M. French Gilman Collection (part of MVZ and SBC; rest?)
MNA	Museum of Northern Arizona, Flagstaff, AZ
MVZ	Museum of Vertebrate Zoology, Berkeley, CA
NAU	Northern Arizona University, Flagstaff, AZ
NMS	New Mexico State University, Las Cruces, NM
PC	Phoenix College Collection, Museum of Northern Arizona
ROM	Royal Ontario Museum, Toronto, Ontario
RWD	Robert W. Dickerman Collection (at KANU, AMNH, CU)
SBC	San Bernardino County Museum, San Bernardino, CA
SBM	Santa Barbara Museum of Natural History, Santa Barbara, CA
SD	San Diego Natural History Museum, San Diego, CA
SWAC	Southwest Archaeological Center, Globe, AZ (now Western Archeological Center, Tucson)
UCLA	University of California at Los Angeles, Los Angeles, CA
UNLV	University of Nevada at Las Vegas, Las Vegas, NV
UNM	University of New Mexico, Albuquerque, NM
US	United States National Museum of Natural History, Washington, DC
UT	University of Utah, Salt Lake City, UT
UTEP	University of Texas at El Paso, El Paso, TX
UW	University of Washington, Thomas Burke Museum, Seattle, WA
UWO	University of Western Ontario, Toronto, Ontario
WF	Western Foundation of Vertebrate Zoology, Los Angeles, CA
WSU	Washington State University, Pullman, WA

differences in migratory patterns of the ages, sexes, or subspecies within the species. Unless otherwise noted, all observations are my own.

Several criteria have been used to determine whether a bird is a breeding species on the reservation. In most cases this was documented by discovery of nests with eggs or young. At other times this was based on collecting juvenals in a very immature state with flight feathers still ensheathed. Some juvenals (Horned Larks, for example) may wander far from their breeding grounds once wing and tail feathers are fully grown. The third criterion was females collected in breeding condition (ovary with yolks or ruptured follicles; active brood patch). I have not considered males with enlarged testes sufficient evidence for breeding, as males may be in evident readiness while females are not ovulating (see Blue Grosbeak account). Similarly, mere presence on the reservation during the breeding season does not prove breeding (ibis, herons, egrets). However, I assume that certain sedentary members of the resident avifauna must reproduce (Brown Towhee, Rufous-crowned Sparrow), though I have found no active nests. Other questionable cases are discussed in the species accounts.

Evidence for breeding in the historic past is based primarily on Gilman's records and on ethnographic accounts. In several cases former breeding under aboriginal riparian conditions is inferred by analogy with other areas on the Gila and its tributaries. Inferred breeding is clearly distinguished from documented breeding in each species account.

Three age categories are used throughout. *Juvenal* indicates a recently fledged young bird that has not yet undergone its fall (first prebasic) molt. *Immature* indicates first-year birds after first prebasic molt, distinguishable in some taxa for the first twelve to fifteen months of life. *Adult* means a bird that has achieved at least the first definitive plumage. Age criteria are based on plumage, molt, skull pneumatization, and ovary condition.

Brackets around the sex indicate that the specimen was sexed by size or plumage rather than dissection of gonads.

Where reservation data are in accord with Arizona lowlands in general as described in Phillips et al. (1964), that work is not cited.

Taxonomy

Since the publication of the last American Ornithologists' Union Check-list (1957), three general works have discussed various taxonomic levels of birds occurring in Arizona: Phillips et al. (1964), Mayr and Short (1970), and Oberholser (1974). Additional papers have dealt with the taxonomy of particular species occurring on the reservation.

For the most part I have followed the sequence of species as given in the A.O.U. Check-list (1957) and have confined my taxonomic discussions to the generic, specific, and subspecific (racial) levels. Problems of nomenclature are included. Where questions of priority are involved, I have given the date of publication following the author of the taxon; these are not literature citations.

Though there is an "official" inertia with regard to making changes in traditional higher taxonomic categories of birds, there is no justification for following check-list arrangement of families into orders where anatomical and other studies have shown a group to be polyphyletic. Several essential, well-documented changes are incorporated into this work. I have separated the Ardeidae from the Ciconiiformes, placing them alone in the order Ardeiformes. The Vulturidae (Cathartidae *auctorum*) are removed from any association with the polyphyletic Falconiformes (Jollie, 1976, 1977; de Boer, 1976) and placed instead in the order Ciconiiformes, together with the Plataleidae (Threskiornithidae *auctorum*), Ciconiidae, and extinct Teratornithidae, following the order Pelecaniformes (see Rea, in press).

The family Phalaropodidae 1831 is probably best merged in the Scolopacidae 1825 (see Jehl, 1968:39–40), but I have not compared the osteology of the sandpiper complex. The main anatomical characters of the phalaropes are femoral, undoubtedly related to swimming (see Campbell, 1979:120, who maintains familial status for the three species).

The Anseriformes are placed after the Charadriiformes, following the suggestion of Olson and Feduccia (1980). The essentially Old World family Laniidae I consider most closely related to the unspecialized genera of Corvidae; therefore, I have juxtapositioned these families. The correct allocation of the apparently intermediate Old World genus *Platylophus* has not been satisfactorily determined. The Corvidae are kept early in the oscines because they are osteologically primitive. The Paridae are not osteologically allied to the Corvidae, as sometimes suspected.

The cranial as well as the postcranial osteology of Ptilogonys, Phainopepla, and Bombycilla are quite similar, constituting a definable family (see also Arvey, 1951); I have therefore included the Ptilogonatidae 1858 in the Bombycillidae 1832. I have not seen skeletons of the probably related *Hypocolius* and *Phainoptila* or the possibly related *Dulus*. *Myadestes* has nothing whatsoever to do with this family.

The nine-primaried New World passerines present taxonomic problems that are not satisfactorily resolved in currently used American sequences (e.g., Wetmore, 1960; A.O.U. Check-list, 1957). Anatomically, the Parulidae, Coerebidae, Thraupidae, Cardinalinae, Emberizidae, and Icteridae appear to grade into each other when Neotropical forms are considered, so that at best these are probably only subfamilies. Paynter and Storer (1970) suggested some possible grouping of genera. In an anatomical study of the nine-primaried complex, Raikow (1978:31) concluded, "As a whole, [the Cardinalinae] cannot be distinctly separated from either the emberizine finches or the tanagers on the basis of limb myology." The coerebid-thraupid part of the complex contains so many diverse genera that are so poorly known anatomically that the most practical course in our present state of ignorance may be to leave these as families and to combine the somewhat better-known cardinal finches and sparrows in the Emberizidae. The Icteridae is usually considered a definable assemblage; yet, it encompasses species ranging from primitive to highly derived. Because of the constraints of a linear sequence, I put the Icteridae at the end of this complex. The one group that I consider legitimately deserving of separate familial status, comparable to other oscine families, is the Carduelidae. Their osteology, proportions (small eyes and short legs), nest sanitation, seed diet even during breeding, and propensity for vagrancy all suggest distinct familial status. This leaves the Ploceidae, which I have placed after all these so as not to interrupt the sequence of related groups, but not because of any close relationship to the Carduelidae (see Raikow, 1978:33). They might alternatively be placed before the wood warblers. According to Brodkorb (1978), the proper name for the Ploceidae is Passeridae (Illiger). On the basis of limb myology (Raikow, 1978:26, 41) and osteology (personal observation), the vireos and their allies are not closely related to either the Laniidae or the Parulidae. Certainly, the relationships of oscines deserve further serious study, which is greatly hampered now by lack of comprehensive osteological and myological collections.

A genus should include all closely related species believed to be of monophyletic origin. However, as pointed out by Mayr and Short (1970:105), "By the 1920s a generic name was available for almost every good [avian] species (except sibling species)." But during the past half-century the function of the avian genus to show close evolutionary relationship has had growing acceptance as more and more untenable genera have been trimmed away. A recent regressive trend is found in Oberholser (1974), where virtually every species capable of being defined by external "morphology" is considered a genus. I am unable to accept any of Oberholser's newly proposed or revived genera as being of value in demonstrating evolutionary relationships.

Conversely, I am reluctant to overturn long-standing genera until careful comparisons, particularly of osteology, pterylography, myology, behavior, and fossil history, have been made. "Skin" taxonomy has led to blind alleys in ornithology, and relying on this type of evidence today to solve taxonomic problems above the subspecific level will only lead to more errors. Therefore, I have retained certain questionable genera (e.g., *Megaceryle, Stelgidopteryx, Catherpes, Amphispiza, Zonotrichia*) where I have been unable to examine osteologically all the pertinent related species.

At the species level I have followed most recent authors in applying the biological species concept to populations. Populations that regularly and freely interbreed (even though the contact zone may be limited by habitat) are considered conspecific. A few instances where the actual field situation remains to be determined are pointed out.

I consider the towhees of the interior, *Pipilo fuscus,* specifically distinct from those of the Pacific slope, *P. crissalis,* and the interior gnatcatchers, *Polioptila melanura,* almost certainly specifically distinct from coastal *P. californica* (see species accounts.)

My philosophy of subspecies (races) is that, if they are to be formally recognized, they should be useful. I maintain that a primary function of continental avian subspecies is to teach about migration. The essential criterion is: Will a member of a named population (subspecies) be identifiable as such when it moves (as migrant or vagrant) to a different geographical region and be distinct from individuals of other populations of the same species? Races distinguishable merely on "average" differences of size or color on their breeding grounds, but not identifiable on the basis of *individual specimens* taken on migration or in winter, are useless and should not be taxonomically recognized. Some species show good racial variation only in one sex (as in some *Pipilo, Agelaius,* and *"Hesperocichla"*). In these cases it is the duty of the field ornithologist to collect taxonomically useful specimens. The same applies to species whose plumage becomes so worn and faded by spring that the subspecies are no longer distinguishable. In some taxa from very xeric habitats, specimens are identifiable for only several months following the prebasic molt. Other species undergo postmortem color changes ("foxing") over a period of years, so that migrant specimens can be correctly identified to subspecies only by comparison with specimens of comparable museum age. As old museum collections become increasingly foxed, it becomes impossible to identify newly taken migrants without reference to fresh, unfoxed specimens from the breeding ranges of the various subspecies, one of the facts of which far too many biopoliticians in charge of issuing scientific collecting permits are utterly ignorant!

Another function of subspecies is to show the local (endemic) evolution of a species, even though it may be sedentary. These races tell us about centers of differentiation as well as the origin of occasional vagrants. Examples of this type of subspecies are found in the species *Otus asio, Auriparus flaviceps,* and *Pipilo aberti.*

It has not been possible for me personally to examine the subspecific taxonomy of all of the more than 125 polytypic species found on the reservation. Where comparative materials were not available, I have relied on specialists working with the taxon (Phillips, Parkes, Dickerman, Hubbard, Banks, and others). Where I have not been responsible for the ultimate subspecific identification, the particular ornithologist is acknowledged in the appropriate taxonomic account. And in a few places where local material was not satisfactory, I have omitted the subspecific discussion entirely (e.g., *Ardea herodias*).

For subspecies which occur on the reservation only as transients or winter visitants I have given a brief statement of their breeding range, inclusive specimen dates, and, where exceptional for the area or state, the individual collection numbers as an aid to future workers.

In this work I use the term *hybrid* in its strict sense for crosses between full species. Such interspecific crosses are rare among birds in nature, resulting from a breakdown in isolating mechanisms. If the offspring survive, often they are sterile and show physical abnormalities. To use the term hybrid in the loose sense of the geneticist to indicate crosses between any two individuals of unlike genotypes vitiates the taxonomic meaning of the term because nearly all sexually reproducing animals could be considered "hybrids." True hybrids are encountered mostly among such birds as hummingbirds, pheasants, and ducks where the secondary sexual characteristics of the males are highly elaborated but the females are relatively little differentiated interspecifically. The frequency of hybridization under natural conditions can argue for relatedness at the generic level. I use the term *secondary intergrade* for specimens resulting from the interbreeding of previously differentiated and geographically isolated gene pools within a species (see Raitt, 1967). Secondary intergradation must be inferred from discordant phenotypic variation without abrupt environmental change between clines and from the paleontologic and historic distributional records. The larks and grackles discussed here are examples (pp. 191 and 249). I use the term *intermediate* for specimens originating from areas of *primary intergradation* where two adjacent subspecific populations are in contact and the distinguishing characters are in transition. Subspecific intermediates are common in nature. Subspecies are, after all, only usefully named geographic aggregates; they should not be thought of as little species. The middle Gila drainage is an area of primary intergradation for a number of polytypic birds that are differentiated within the Sonoran Desert (for example, see the breeding distribution map of Northern Orioles, p. 243). The terms hybrid, secondary intergrade, and primary intergrade or intermediate are not a priori categories. In all cases it is only by study of the behavior and vocalization of the birds in the field and of the morphology and coloration of geographic series in the hand that the appropriate status can be inferred.

All specimen measurements are in millimeters (mm). The wing measurement is the chord of the folded wing. I measure both wings and give the measurement of the larger. The plus sign within parentheses, (+), indicates that a wing or tail shows some wear; the same sign without parentheses, +, indicates significant wear. Enclosure of the entire measurement within brackets, [ ]+, indicates fraying to a degree that the measurement is no longer useful in comparisons except to show a minimum length. The following notations are used in presenting statistical data: ♂♂ = males, ♀♀ = females, $\bar{x}$ = mean, R = range of measurements, n = sample size (number of individuals).

Each symbol on the maps represents *specimens* I have examined, or, in a few instances, have had measured for me. A small numeral next to a symbol indicates that several specimens of the same form were taken at that locality. I personally aged and sexed most of the verdin, oriole, and grackle specimens that were mapped. A question mark may indicate there is no satisfactory material yet from a critical area or the specimen could not be definitively identified.

The subject of taxonomy and the value of specimens would not be complete without commenting on that much neglected but critical part of every specimen, its label. In addition to the usual date, locality, sex, and collector on the face of the label, generous supporting data should be recorded on the obverse. Weight, soft part colors, habitat notes, and behavior should be recorded at the time of collecting. The preparator should record notes on fat, molt, and skull pneumatization. Careful attention should be given to gonads. In most birds immatures in their first winter can be distinguished by the filmy rather than granular surface of the ovary. Breeding females are indicated by enlarged oviducts, yolks, or ruptured follicles. These data belong *with the specimen,* not in the collector's field notes, which are seldom available to the working taxonomist.

Change

This section summarizes any change in the species' status, distribution, or numbers as indicated by a comparison of the other sections (archaeological, ethnographic, historic, or modern). Where possible, correlations between known changes in particular habitats and the bird's present status are suggested.

The Species

Podicipedidae

Eared Grebe, *Podiceps nigricollis* C. L. Brehm

Modern: Rare. One seen alive (in breeding plumage) at the Riggs Road pond and another found there dried (AMR 4499), both on 27 April 1973.

Change: As this grebe requires open water of some considerable extent, it is expected only accidentally, as at the temporary but extensive pond formed at Riggs Road. Gilman did not record the species.

Western Grebe, *Aechmophorus occidentalis* (Lawrence)

Modern: Rare. One was found dead [28 November 1974] (AMR 4872) on the irrigation runoff pond southeast of Bapchule. I saw one individual briefly above the Salt-Gila confluence 27 November 1977. I have no further records for this species, which requires more extensive open water than found presently on the reservation.

Taxonomy: Dickerman (1963) has demonstrated that this species, long thought to be monotypic (A.O.U. Check-list, 1957; Palmer, 1962; Mayr and Short,

1970), has a well-marked subspecies, *A. o. clarkii,* breeding on the Mexican Plateau. The one reservation specimen is the larger, nominate race. Its measurements are: wing chord 206; bill from nares 61, total culmen 74.3, tarsus 75.

Pied-billed Grebe, *Podilymbus podiceps* (Linnaeus)

Modern: Formerly fairly common locally. All records (R. Johnson, Simpson, and Werner MS) are from the Salt River from 91st Avenue to 115th Avenue, where the species has not been seen in summer since 1971. Two families at the Salt-Gila confluence 8 May 1970 consisted of four adults, seven juvenals, and one immature. Three were observed 11 March 1971, two giving territorial calls. Two chicks with an adult were at this location 14 March 1971. An immature bird was seen farther upstream 30 July 1969. Sightings of nonbreeding birds include August, September, November, and December, all from 1970 to 1972. I observed one or two small grebes (this species?) at the open marsh at Barehand Lane.

Change: This grebe requires more extensive open water and emergent vegetation than found on the reservation since the pond above the Salt-Gila confluence was washed out by the floods of 1973. Phillips et al. (1964) note: "The few individuals seen during the summer in south-central Arizona may be belated migrants or may belong to a sparse nonbreeding population." Thus the breeding on the reservation for three summers appears to be a response to exceptional conditions.

Pelecanidae

White Pelican, *Pelecanus erythrorhynchus* Gmelin

Modern: Rare. A single specimen (LLH 3071) was found on the Salt River east of 91st Avenue 1 June 1970 by S. R. Demaree. There were about fifty White Pelicans present in a group. The dead pelican, an adult female, had been banded 3 August 1967 as a flightless juvenal at Tule Lake, California. On 7 February 1982 I watched an apparently fully adult White Pelican flying up and down the Salt above the confluence.

[Brown Pelican, *Pelecanus occidentalis* Linnaeus]

Ethnographic: *chuawgiakam vákoiñ (chuawgia,* 'netted' [from *giáwha,* 'burden net'], *-kam,* 'has' [attribution or possession marker], *vákoiñ,* 'heron' [hence 'heron with a net' in reference to the gular pouch])

The name and all related information come from one consultant, Matthias. Pelicans were described to him by his grandmother. The method of diving for fish was related. The species was never common but apparently of regular occurrence; hence, it merited a Pima name of 'netted heron.' Only dark ones, like the Great Blue Heron, were described, never white ones. The birds were relatively tame, and a Pima could walk up to one and throw it a fish, which it would eat.

Modern: Rare. I observed one Brown Pelican flying over the Sierra Estrella near Komatke on 16 July 1972, just before a powerful dust storm without rain. Jon N. Young (personal communication) saw one near Gila Buttes (Bapchule) in June or July 1957 or 1958.

Change: The status given by Phillips et al. (1964) is: "A few enter the state in summer and fall, rarely staying until winter or even spring; most records are for the Colorado Valley, but stragglers reach most of the state. . . ." There appears to be no change in actual status. Stragglers may have been more concentrated about the Pima villages when the Gila was running, and the fish-loving Pima were willing and able to spare a few fish for the odd visitors, as they did later to the Europeans (Kino, 1919; Bartlett, 1854 [2]:242).

Phalacrocoracidae

[Double-crested Cormorant, *Phalacrocorax auritus* (Lesson)]

Historic: Gilman (MS) records one caught in an irrigation ditch, Sacaton Experimental Station, 9 October (year not given). I have not located the specimen. I have no further observations.

Vulturidae (= Cathartidae)

Turkey Vulture, *Cathartes aura* (Linnaeus)

Ethnographic: *ñúi;* [pl.] *ñúñui*

Ñúi is one of the two moieties of the Pima and Papago and a highly important figure in legends and songs. The birds are reported to nest in caves. Their feathers, used to fletch arrows, were obtained from beneath a roost (usually a dead tree). Otherwise the species is rigorously tabooed (Rea, 1981). The ñúi will vomit on people. The California Condor, *Gymnogyps californianus* (Shaw), is unknown to the Pima.

This ethnotaxon is Pan-Piman with slight variants occurring among other Piman speakers: *ñúvi* or *núwi* (Papago), *núi,* pl. *núnui* (Pima Bajo), *nuí* (Northern Tepehuan; Pennington, 1969), *Nu:í* (Tepecano; Mason, 1917).

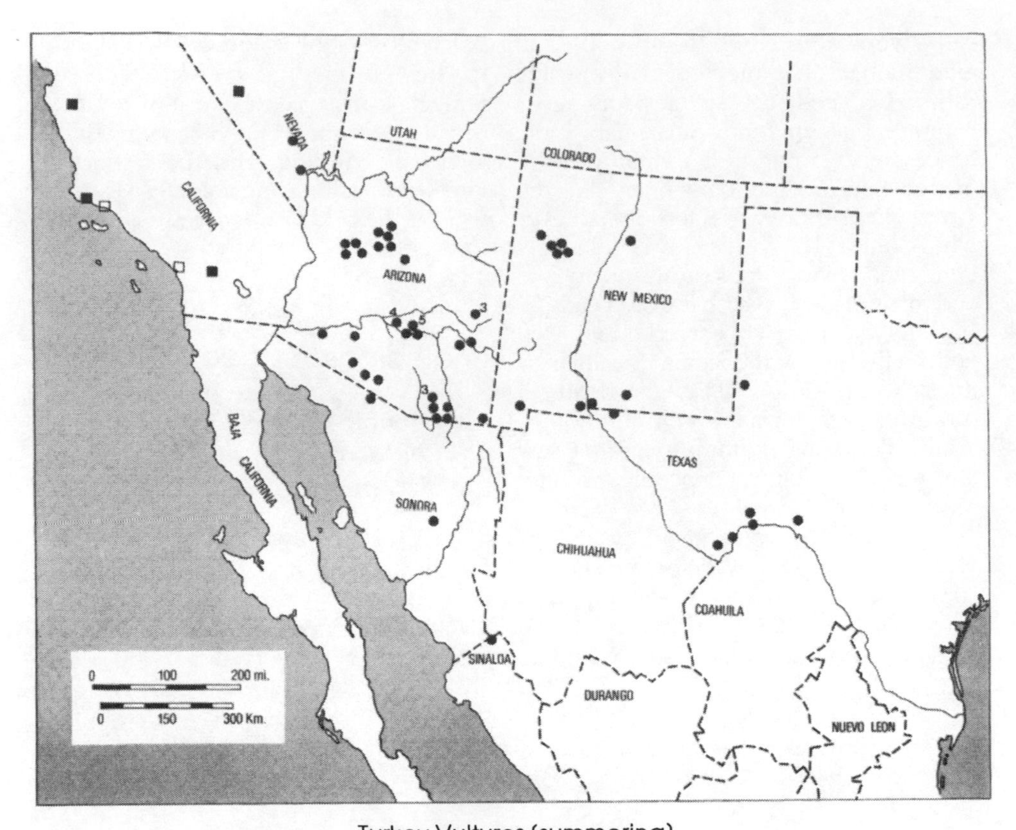

Turkey Vultures (summering)

● = Cathartes aura aura ■ = C.a. meridionalis □ = intermediate

Historic: Breninger (1901) recorded this species as "Always in sight from daylight till dark," in September. Gilman (MS) observed it from "Fall to spring only, no nests. Seen in small numbers." Specific dates given are 11 October 1914, 2 and 9 January 1915, and 12 February 1915. Records kept at Casa Grande Ruins National Monument (files, 1935–1944) indicate large fall migratory flocks passing in early October and the first spring birds returning from mid-February to early March.

Modern: Common in summer, fairly common in winter. Through 1972 the summer population (June to August) on the reservation consisted of twenty-six to thirty (possibly thirty-one) Turkey Vultures. The resident winter population (November to April) was comprised of a maximum of usually seven (twice eight) birds.

Frost flights of sixteen (5 December 1970, Simpson) and thirteen (31 December 1975, Rea) have appeared in the face of a storm, evidently pushed down from the north, but the birds were gone the following day. In the afternoon of 28 December 1977 there were thirteen or fourteen Turkey Vultures at Barehand Lane. Five were perched on burnt saltcedars and mesquites. All were gone between 5:30 and 6:00 P.M. It rained that night and the next morning, but the storm was warm. On 4 January 1980 I found a primary, a pellet, and whitewash in the Enos tamarisk grove, Komatke, indicating vultures had roosted there the previous night. At dusk one adult caracara and at least fifty-four vultures came to roost. Two may have been immatures. The birds were wary, evidently itinerants stopping over. Two weeks later, when Albert Pablo rechecked the grove, all were gone. On 7 February 1982 I counted sixty-four vultures coming to roost between Gila Crossing and Komatke. There is evidence of such apparent "frost flights" in the past. On 27 December 1954, forty-seven vultures were counted (some duplication?) on the Phoenix Christmas Count, twenty-nine of these together northwest of Phoenix. The following day Simpson and Werner found only one. On 16 November 1957, Johnson and Werner counted about a hundred vultures in a grove at Tal Wi Wi Ranch, north of the west end of the reservation. Werner and Simpson saw a similar number the following evening, but a week later there were only six individuals, evidently the winter residents. Otherwise, from 1973 to 1981 usually seven (but once eleven in late August) vultures have been counted on the reservation, both summer and winter (32 observations, average 3.5 birds).

Taxonomy: Wetmore (1964) revised the entire species. The two subspecies in western North America are the small Mexican *C. a. aura,* coming north approximately to the International Boundary, with the remainder of the West occupied by the larger *C. a. meridionalis* Swann (*C. a. teter* Friedmann 1933 best being considered a synonym of the nominate race, perhaps somewhat intermediate toward *meridionalis*). Nineteen

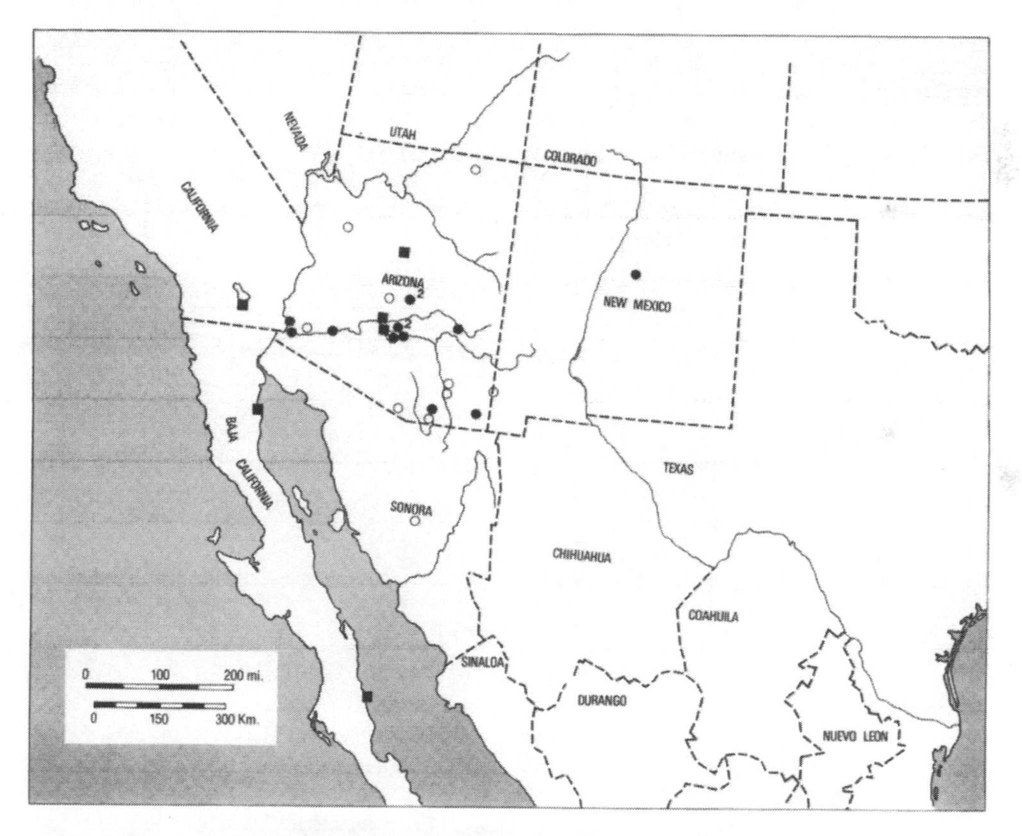

Turkey Vultures (wintering and northern migrants)

■ = Cathartes aura meridionalis (wintering) ● = C.a. meridionalis (transient)
○ = C.a. meridionalis x aura (transient)

Turkey Vulture specimens (most of them found dead) were obtained on the reservation. Small *aura,* the breeding race of Arizona, New Mexico, southern Nevada, and southern California (Rea, unpublished notes), are represented by specimens from 24 April to 21 October (and another found dead apparently between 31 October and 28 November). Their measurements are: wing chord, 471–500; tail, 242–265; humerus, 136.5–145; ulna, 166–175. Migrant and wintering *meridionalis* are from 26 August to [13 February]. Their measurements are: wing chord, 505–525; tail, 271; humerus, 153–155; ulna 179–183. (Because of protracted molt, flight feather wear, and changes in primary curvature in specimens, wing bones are more reliable than feather measurements in racial determinations of these large soaring birds.) Winter birds do not use the Santan roost area formerly occupied by the summer residents. The presence of two different populations on the reservation conforms with my data for the rest of Arizona. All small summer *aura* apparently migrate.

Change: The summer population has declined. See discussion, p. 83 in Chapter 5, regarding decline in this species.

[Black Vulture, *Coragyps atratus* (Bechstein)]

Ethnographic: None of my regular consultants knew of the Black Vulture. However, at an outdoor feast on 25 November 1976, Ambrose Juan (brother of Ruth Giff) pointed to three Black Vultures flying low over Komatke, calling them *s'chuk ñúi* ('black'; 'vulture'). He said they were common about Santa Cruz village where he farmed. Apparently we had witnessed the coining of a new taxonomic category. Black Vultures have been on the Papago Reservation for over half a century. Delores Lewis of Gu Oidak called them *á'ali ñúñuwi* ([sing.] *áli ñúwi,* 'little vulture'). At Onavas, the lowland Pima Bajo village, where presumably the Black Vulture has had a longer history, it is called *stuk núi* ('black vulture'). Pedro Estrella further distinguished the Turkey Vulture as *keli núwi* ('old man vulture').

Modern: Irregular. Not recorded by Gilman (MS) on the Gila River Reservation, though Black Vultures were first discovered in Arizona by him below Sells, on the Papago Indian Reservation in June 1920. Apparently the first observations in the Pima country are those of Housholder starting in 1944 (twelve at a Sacaton roost; also about twenty-three counted south of Chandler 21 May 1947). My sightings of Black Vultures on the reservation, beginning 2 June 1964, span from 25 November (Komatke) through 30 July and 31 August (both Barehand Lane) with none seen from September through mid-November. The maximum number was 21 (average = 5.3 individuals, 11 observations). There is no evidence yet that they nest in the area.

Taxonomy: There is no specimen from the reservation, but Holocene Black Vultures I have measured from Sonora and Arizona (AMR, LLH, ARIZ, MNA) are all the small *C. a. brasiliensis* (Bonaparte), according to the latest revision of the species (Wetmore, 1962). Some larger specimens (wing chord, 420–428) measured by Phillips (personal communication) from Tucson and northern Sonora appear to be the nominate race, suggesting perhaps an invasion from the east as well as from the south. This needs to be verified by more skeletal measurements from Arizona and Sonora.

Plataleidae (=Threskiornithidae)

White-faced Ibis, *Plegadis chihi* (Vieillot)

Ethnographic: Possibly jujupul.

Historic: Gilman (MS) saw the species in small numbers in September and October.

Modern: Uncommon. An immature (AMR 3075, nearly complete skeleton) was found long dead [26 November 1970] along the Salt above 91st Avenue. Sight records, usually of flocks passing high over the Pima villages, extend from 23 April to 20 September. The summer records for the lower Salt (21 June to 1 August, maximum flock of fifteen individuals) are for bare rocky flats, where the species does not breed.

[Roseate Spoonbill, *Platalea ajaja* (Linnaeus)]

Historic: Phillips (1946:20) noted, "Harold Moore of Phoenix writes me, under date of 27 February 1939, an accurate description of one killed 'about November 1916' at the mouth of the Salt River, west of Phoenix, but the bird was not preserved." This location is on the reservation.

Modern: Casual. A few have been seen in winter on the Gila below the confluence of the Salt. The Pima have no oral tradition of this species.

Taxonomy: I strongly doubt that the genus *Ajaia* Reichenbach (1853) can be maintained separate from Old World *Platalea* Linnaeus (1758), in spite of the striking coloration of the New World species. In 1974 at the San Diego Zoo I filmed a nesting pair consisting of a *P. ajaja* and a *P. leucorodia* which produced an apparently normal offspring that survived.

Accipitridae

Sharp-shinned Hawk, *Accipiter striatus* Vieillot)

Historic: Breninger (1901:45) recorded "Several seen dashing after sparrows" on either 18–19 or 25–26 September 1901. Gilman (MS) noted that the Sharp-shinned Hawk "arrives in small numbers about the middle of October and remains through the winter."

Modern: Fairly common in winter. Four specimens (AMR) are from 21 October to 14 March, with a sigh, record of a male 2 October. The only adult I saw on the reservation was in January 1980. *Accipiter* hawks vary greatly in numbers from winter to winter. Some years they seem to be in every village and other years virtually or completely absent. Field separation of female Sharp-shinned Hawks from male Cooper's is hazardous, so that relative numbers of the two species are difficult to estimate.

Cooper's Hawk, *Accipiter cooperii* (Bonaparte)

Archaeological: An ulna of the Cooper's Hawk was recovered from Snaketown.

Historic: Breninger (1901:45) noted "one seen patiently watching a chance to pounce on a Coot or Gallinule" in mid or late September. Gilman (MS) recorded this species as "numerous in fall and spring and not scarce during the winter. None seen in midsummer and no nesting records."

Modern: Fairly common in winter. Specimen records span from 19 Setember to 7 January. Their numbers vary from year to year. They seem so infrequent in mid and late winter that I suspect that individuals hunt an area for a while then move on. My only observations of an adult are 22 October 1968 (Komatke) and 21 February 1976 (Salt River).

Red-tailed Hawk, *Buteo jamaicensis* (Gmelin)

Archaeological: This species was recovered from South Santan Salvage and Pueblo Grande.

Ethnographic: *háupal*

This hawk nests in the arms of saguaros. Tail feathers of the young were the most prized for fletching arrows. Young were lifted down from the nest with a *kwí:pat* ('saguaro rib'), the tail feathers plucked, and the birds replaced. Until the early 1900s young were raised in captivity, held in cages constructed of saguaro ribs. Boys who hunted food (rodents, lizards, small fish) for the birds expected some feathers in return. Feathers were pulled out at the time the birds normally molted. The following year new nestlings were sought in early May and the old birds released, as it was considered a very bad omen for the owners to have a hawk die in captivity. Feathers were used also for decorating the hair. (For a confirmation of the oral record, see Russell, 1908:86, 115–117.) The same lexeme occurs in Papago and Pima Bajo.

Historic: Breninger did not mention Red-tails, but Gilman (MS) recorded them as "Numerous resident, nesting in cottonwood trees along the river, in tall ironwood up toward the foothills, in saguaro...and on the cliffs in hills. Nesting begins in March, and April sees most of it done, May nesting being the exception."

Modern: Common. An adult male (AMR 833) was obtained from a Pima 30 September 1965. He had shot it in the Sierra Estrella across the river from Komatke. A pair has had a nest, evidently in use for many years, on a cliff in the Sierra Estrella above Santa Cruz village. I observed a guileless immature in the nearby bajada 3 August 1967. An immature specimen (AMR 2569) was shot by a Pima at Maricopa Colony 11 December 1968.

Taxonomy: The expected breeding race is *B. j. calurus* Cassin (A.O.U. Check-list, 1957). The adult specimen has been identified by Phillips as *B. j. fuertesi* Sutton and Van Tyne, a paler and more immaculate race breeding normally across southern Texas and northern Chihuahua but apparently intergrading in extreme southeastern Arizona (Phillips et al., 1964).

During most of 17 March 1973 I watched an exceedingly pale broad-winged hawk about Vahki and Casa Blanca. The bird appeared to be *B. j. harlani* (Audubon). A Harlan's specimen, taken 10 January 1962 near the reservation, is cited by Phillips et al. (1964). One or more melanistic *Buteos* usually winter in the cottonwood rows along the reservation boundary below Chandler.

Swainson's Hawk, *Buteo swainsoni* Bonaparte

Archaeological: Recovered from both Snaketown (between 300 B.C. and

A.D. 100) and from South Santan Salvage.

Historic: Gilman (MS) saw this species occasionally on migration and a single bird 10 January 1910. Monson and Phillips (personal communication) saw one at Santan 17 June 1939, and Phillips observed another near Upper Santan 26 July 1947.

Modern: Rare. I have seen the Swainson's Hawk but once on the reservation, a completely melanistic bird (AMR 4525) taken 22 September 1973 at Casa Blanca.

[Zone-tailed Hawk, *Buteo albonotatus* Kaup]

Modern: Rare transient. I have but one sight record for this species. What I thought was a Turkey Vulture flying low over Komatke 6 April 1968 wheeled to reveal the diagnostic tail pattern and black, feathered head. This date is in accord with the spring migration period for Arizona.

[Ferruginous Hawk, *Buteo regalis* (Gray)]

Modern: Uncommon and irregular. There are no specimens, archaeological or recent, from the reservation. I have seen one about every winter, with dates ranging from 22 November (possibly 30 September) to 4 March. Careful observation was possible because most of the reservation birds did not flush readily.

Harris' (Bay-winged) Hawk, *Parabuteo unicinctus* (Temminick)

Ethonographic: *vá:kaf*

On a trip to The Buttes, Sylvester Matthias identified this striking and rather tame hawk. He had not seen it since about the mid 1920s. "It used to nest in the cottonwoods when the river was still running and the people were still farming. Feathers were not used for anything—not even by the *mámakai* [medicine men]." Delores Lewis, a Papago from Big Fields, also identified this hawk in the field, calling it *wá:kaw.*

Historic: Gilman (MS) found the Harris' Hawk resident in small numbers, nesting annually in a partly decayed tall cottonwood about 5 miles (8 km) up the river from Sacaton. He noted several along the river in December and February and collected an adult 22 December 1909 at Blackwater. I reexamined the specimen at San Bernardino County Museum.

Modern: Rare. Werner (Johnson et al., MS) saw a Harris' Hawk flying over the Salt at 107th Avenue 25 June 1955, but I know of no other recent report for the reservation. This boldly marked bird could hardly be overlooked. I have seen as many as ten individuals between Florence and The Buttes, east of the reservation.

Change: The cause of extirpation is unknown, but apparently the species is local elsewhere in Arizona, the northwest periphery of its range.

Golden Eagle, *Aquila chrysaetos* (Linnaeus)

Ethnographic: *ba:k* (sometimes qualified *skúguch ba:k; skúguch,* 'handsome,' 'pretty'; *ba:k,* 'eagle')

As with the Red-tailed Hawk, the eagle was kept in the villages, but in a stouter cage (made of mesquite limbs) with a little house. Eagles used the same nests each year. Ropes were used to descend to the nest. All the young were taken. Nests were known from the north side of Montezuma Peak, in the Sierra Estrella, and at Four Peaks. The Maricopas were the primary ones to capture Sierra Estrella eagles. Ruth Giff's grandfather had a captive eagle. The *mákai* (shaman) used two matched eagle primaries for brushing away disease in curing sessions. Arrows intended for shooting Apaches (*áw'awp*) were fletched with eagle feathers.

The Golden Eagle is of considerable cultural significance throughout Pimería Alta (see, for instance, Underhill, 1946: 243–252). The ethnotaxon is broadly distributed. I obtained *bá'ak* from Papago and *bá:ag* from Onavas Pima Bajo (probably more correctly transcribed by Ken Hale as *bá'aag*). The sixteenth-century Onavas vocabulary recorded *vaagui,* which is similar to cognate *báágai* of Northern Tepehuan (Pennington, 1968, 1979), suggesting a temporal sound shift between upland and lowland Pimans.

Historic: Gilman (MS) noted an eagle high over the Gila at Sacaton, 5 August, commenting "[it was] apparently outside of the breeding range."

Modern: Rare. A dried specimen (later discarded) was picked up by Jim I. Mead near Gila Butte 12 June 1973. Bones of another, identified by L. L. Hargrave as a male, were found 17 December 1972 spread along several hundred meters of a wash in a higher slope of the Sierra Estrella; additional elements were found in subsequent months.

Change: There are no active eagle nests on the reservation and I have seen but one live eagle (species not determined because of poor light) 3 January 1976 at Bapchule.

Northern Harrier (Marsh Hawk), *Circus cyaneus* (Linnaeus)

Archaeological: Four specimens recovered from Snaketown.

Ethnographic: *se:p wéhadum* (*se:p,* 'cold'; *wéhadum,* 'to put things down' [hence, 'it drops down the cold'])

When the Pima see this harrier over their fields in the fall, they know that winter is coming. They say that the bird, flying low over the ground with occasional dips, is bringing the cold down from the north and spreading it over the country. Joe Giff believes there is another name for the Marsh Hawk, perhaps *vákut.*

Historic: Breninger (1901:45), on either 18/19 or 25/26 September noted, "One seen; an early migrant." Gilman (MS) saw it "occasionally during the fall, winter, and early spring months." He gave early dates of one on 30 September and three on 10 October, the last being for Sacaton in 1914.

Modern: Fairly common winter resident. Specimen evidence rests on some bones and feathers (AMR 4806) picked up above the Salt-Gila confluence. Sight records extend from 2 October to 26 March (four or five circling, apparently on migration) and 27 March.

Pandionidae

Osprey, *Pandion haliaetus* (Linnaeus)

Ethnographic: *vákoiñ ba:k (vákoiñ,* 'heron' [hence, 'fishing']; *ba:k,* 'eagle')

Sylvester Matthias was the only consultant who knew this species, which he termed "just a low grade eagle" that lives on fish. The name does not apply to the Bald Eagle, *Haliaeetus leucocephalus* (Linnaeus).

Luciano Noriego, at the Quitovac oasis in Sonora, called the Osprey *gú'u ví:sak,* which he glossed *'gavilán grande.'* On the Río Yaqui at Onavas, Sonora, Pedro Estrella called it *shúdgit bá:ag,* 'water eagle.'

Historic: Gilman (MS) recorded "One seen perched on top of flagpole at Sacaton in April." The year is not given.

Modern: Rare transient. At the Salt-Gila confluence Simpson observed the species 25 April 1971, S. R. Demaree 5 April 1976, and Demaree recovered an immature male (AMR 4873) found here injured September or October 1975. Phillips et al. (1964) give September and April as the usual months for migration of Ospreys.

Taxonomy: The reservation specimen (wing chord 460) is the expected darker *P.h. carolinensis* (Gmelin).

Change: That the Pima, who themselves relied so heavily on fish for animal protein, had a formal name for the Osprey, indicates some familiarity with the species, which presumably nested along the river in aboriginal conditions, as it did not far to the north (Granite Reef Dam) until 1951 (Johnson et al., MS). Conditions are unsuitable for it today, even as a migrant, and must have been poor around Sacaton in Gilman's day.

Falconidae

Crested Caracara, *Caracara cheriway* (Jacquin)

Ethnographic: *oam ñúi (oam,* 'yellow'; *ñúi,* 'vulture')

Matthias was the only consultant who knew *oam ñúi* ('yellow buzzard') as more than a mythical character, and he readily picked immature caracara skins from among others as representing this species. It was last seen in the early 1920s (perhaps 1922 or 1923). Before that it was common, especially after a rain, when it would sit in a dead cottonwood to dry its wings. Some Pima wondered if the pale area on the back [of the immature] was an indication of old age. These birds were considered stupid. When a deer was being butchered, *oam ñúi* would walk right up and eat the blood, but the regular *ñúi* [Turkey Vulture] was afraid. When *ñúñui* are eating at a dead horse and *oam ñúi* comes, there is a fight. There were usually only a couple of *oam ñúi,* but maybe five would come to a dead horse, with about a dozen bald-headed *ñúñui. Oam ñúi* makes a noise with its wings when it comes down, but not when it goes up. Regular *ñúñui* nest in caves, but Matthias never saw the caracara nesting, nor did he ever see the black adult plumage. (The Pima name, of course, is most appropriate for the first-year plumage.) Delores Lewis of Gu Oidak (Big Fields), who regularly encounters the caracara in Papago country, calls it *kú:sijim*. I suspect this is actually the more universal Piman name. Luciano Noriego of Quitovac, Sonora, gave the cognate *kúbichum*.

Historic: Gilman (1914b:261) recorded the caracara "seen at Sacaton three times in early spring between 1908 and 1914." His MS notes "One seen 17 March and an Indian showed me the head, wings, and tail of one he had shot in early spring." Phillips (1946) records one trapped alive at Casa Grande sometime prior to 1919. Mearns collected a specimen (AMNH, 19 November 1886) on the Salt River near Phoenix. Housholder saw one in Centennial Wash 8 miles (13 km) north of Gillespie Dam 16 September 1925. Yaeger caught another in a steel trap and released it in March 1935 near (or on?) the reservation.

Modern: At the Enos grove of old tamarisks on 4 January 1980, while I was watching over fifty Turkey Vultures assemble at dusk, an adult caracara entered the roost. After about five minutes of interspecific jostling for perch sites, the caracara finally settled low in the crown, though most of the vultures eventually left for other groves. This is my only observation on the Gila.

Change: Southern Arizona is the northern periphery of the species' range. Thus, the periodic ebb and flow over time is expected. But caracara numbers must have seriously declined in the past half-century or more (Phillips et al., 1964). Heermann (1859), following the Gila from Yuma to the "Pima Villages," reported seeing one or more every day in mid-winter of 1854. Phillips (1968:135) believes "this tame and conspicuous bird was probably largely shot out, aided by poison put out by government killers." It is significant that the ethnographic account mentions only younger individuals, which would be expected beyond the periphery of the breeding range. It still breeds locally (1976) on the Papago Reservation to the south.

Prairie Falcon, *Falco mexicanus* Schlegel

Ethnographic: *vi:suki*

This falcon nested on cliffs in the Pima territory. Its manner of hunting and swiftly diving on its prey are well known. One native consultant believed that the *ví:sukĭ* killed its prey with an impact of its clenched fist. Another Pima taxon, *chúduk ví:suk,* 'blue falcon,' might refer to the Peregrine Falcon, *Falco peregrinus* Tunstall. All the above behavioral characteristics would apply to this species as well. The Peregrine might well have occurred and even nested on the middle Gila under aboriginal conditions and is still to be expected as a migrant or vagrant.

Though the lexeme may be applied to different raptors, it appears to be Pan-Piman: *visagui* (seventeenth-century Névome or Pima Bajo), *ví:sa:g* (modern Pima Bajo), *wísag* (Papago), *ví:sag* (Quitovac Papago), and possibly cognate *pitca:g* (Tepecano; Mason (1917:329).

Historic: Gilman (MS) recorded the species, "Resident and nesting in the cliffs in the hills. A pair nested at least two seasons in the hills just north of the river but the nest was high on a cliff and inaccessible."

Modern: Rare in lowlands. Johnson collected a female (ROM 114102) 13 December 1970 at the Salt-Gila confluence. Simpson observed one (possibly two) there 19 September 1971. I have seen the bird on the reservation only three times (twice in mid-winter, reservation boundary below Chandler; once at Barehand Lane, 31 August 1981). Dr. E. Linwood Smith (personal communication) found six nests in the Sierra Estrella between 15 March and 8 April 1976. He could see into three nests, and at three others, less accessible, he was actively harassed by the pair.

Taxonomic: Recently Oberholser (1974:977) has recognized two subspecies of Prairie Falcon, a nominate "eastern and southern race" and a "northwestern" one, *F. m. polyagrus* Cassin. He gave no diagnostic characters, and even the application of the two names is tenuous. My examination of ten southwestern winter specimens (ARIZ) indicates that, when segregated according to age, there are still darker and lighter birds in each group. Among twenty fall and winter specimens (SD) from farther west, those from southern California and Baja California show a tendency to be somewhat paler than northern skins, particularly on the crown. However, K. C. Parkes (personal communication), with the more extensive Carnegie Museum series, could not make such a separation. Unless it can be shown conclusively that the birds from the Northwest differ constantly, I recommend considering the species monotypic, *contra* Oberholser.

[Merlin (Pigeon Hawk), *Falco columbarius* Linnaeus]

Modern: On 27 November 1977 I watched an adult male Merlin for an extended period sitting half way up a large, mostly dead cottonwood at the intersection of Elliot Road and the reservation boundary, east of Co-op Colony. This is my only record for the species.

American Kestrel (Sparrow Hawk), *Falco sparverius* Linnaeus

Archaeological: Three specimens represented from Snaketown.

Ethnographic: *sísik*

Nests in saguaro or cottonwood holes. Found in *s'háwshunek* ('place of many saguaros'). Its voice and foraging methods are known. This bird supposedly will not live in captivity and should not be handled lest it die. In the Pima mind the kestrel is considered related to owls rather than hawks (because a hole-nester?). The feathers were used for hair dressing but apparently not for arrows.

A Pima song, published by Joseph Giff (1980:137), captures the kestrel in Hopkins' "Windhover" style:

Toha hohokimal
toha hohokimal
da:m ka:cim wui cem al hi:

sisik u'uhig
hewel ku:g ab hahiwa e-nagia
wakwan u'uhig
wi'al ke:kkhim
k eda am hihinnag.

White butterfly
white butterfly
sky towards tries to go
sparrow hawk [bird]
wind's end-on hangs
heron bird
east standing [dawn]
just then honking.

The ethnotaxon is widespread. Papago Delores Lewis identified this kestrel in the field as *sí:sik*. Surely the "hawk" *shishikamuli* or *chichikamuli* recorded by Pennington (1969:115) "distinguished by a reddish tail and back" could only be this falcon. In the seventeenth century Névome (lowland Pima Bajo) vocabulary, *sisica* is given as one form of *gavilán* (Pennington, 1979).

Historic: Breninger (1901:45) found the kestrel already "Common along fences and about fields," in September. Gilman (MS) considered it "Very common, nesting in Gilded Flicker nest holes in cottonwood, willows and saguaro. Four eggs...the usual number...."

Modern: Common in winter, uncommon in summer. Wintering birds appear in September in places where the species does not breed, becoming common by the end of the month. Breeding birds are few, usually in extensive saguaro stands. Perhaps a thorough search might reveal half a dozen breeding pairs on the reservation today, though I have found no more than two pairs in the same year, and some years not even that many.

Taxonomy: Two races on mainland North America can be identified on the basis of size (and perhaps color in females): small *F. s. peninsularis* Mearns of Baja California and western Sonora, and the large nominate race over the remainder of the continent (see Bond, 1943; Friedmann, 1950). Phillips (1946) originally believed that *peninsularis* bred along the Gila drainage but later considerably modified his position (Phillips et al., 1964:27) to "either occurs at or influences the [breeding] population around Yuma."

Though the bulk of the wintering birds from on or near the reservation are *sparverius*, winter specimens include both subspecies. A male (AMR 630) taken at Komatke 21 November 1964 is *peninsularis* (wing 178, tail 112.8). An adult female (SD 41241), found shot at the Laveen crossroads 1 September 1980, is *peninsularis* (wing 180, tail 113 [+], 89.7 grams). A male from the Ajo Mountains farther southwest in the desert is likewise *peninsularis* (SD 18570, 12 March 1939, wing 174.4, tail 113.4). There are few measurable specimens of breeding kestrels from anywhere along the Gila Valley. An ovulating female (AMR 4744) found 28 April 1971 near Phoenix (E. Radke) is intermediate (wing 182+). An unworn adult male (CGR 56) taken 3 May 1938 two miles (3 km) north of Casa Grande Ruins is *peninsularis* (wing 176, tail 114.5, including white terminal bar). Gilman collected four July kestrels at Sacaton (orig. MFG 272, 274–276, now MVZ). One of these is a grown juvenal male, unworn, *peninsularis* or intermediate (wings 175.5 and 179, tail 120.9). The adult male is very small *peninsularis* (wing 165, tail 110.4) even considering some wear. The one measurable adult female is *sparverius* or perhaps intermediate (wing 186.5+).

Gilman's birds were the *only* breeding specimens Bond (1943) had available from the entire Gila River. He plotted these on the map (Fig. 49, p. 178) with a symbol indicating in the size range of both *sparverius* and *peninsularis*. He concluded (Bond, 1943:181),

> In an area comprising the Colorado and Mojave deserts, the whole watershed of the Gila River and its

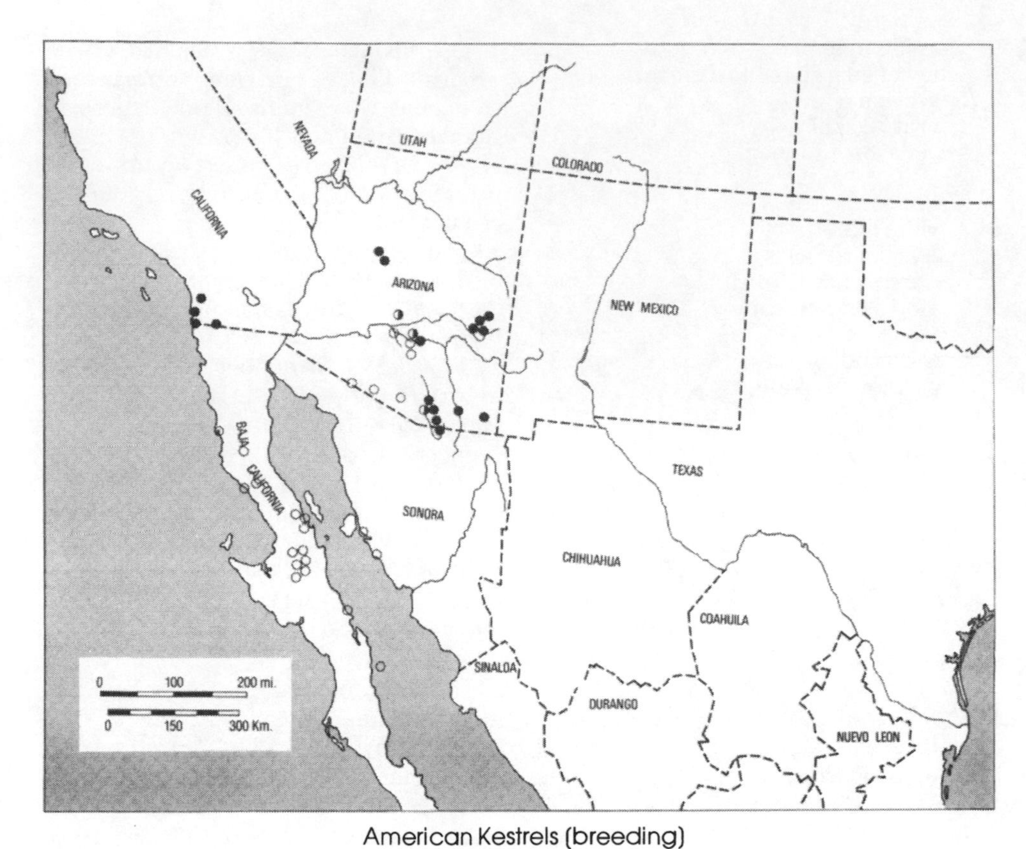

American Kestrels (breeding)

● = Falco sparverius sparverius ○ = F.s. peninsularis ◑ = intermediate

> tributaries (at least as far east as the New Mexico line), and south at least to the Mexican boundary, the breeding sparrow hawk population is extremely variable.... The situation seems more nearly to represent the sort of "hybridizing," or secondary interegradation, to be expected from the interbreeding of fully differentiated parental races....

However, I find that all five of the breeding specimens (ARIZ) from the southeastern section of Arizona are good *sparverius* showing no approach to *peninsularis.*

Change: With the destruction of the cottonwood gallery forest, the population size of breeding kestrels has decreased drastically. The taxonomic situation needs clarification based on unworn specimens taken early in the breeding season. With the northward advance of Lower Sonoran Desert vegetation (including saguaros) since late Pleistocene (Van Devender, 1973, 1977), perhaps *peninsularis* of the south has reached and interbred with larger *sparverius* of higher elevations, the riparian woodlands of major rivers being avenues of secondary contact. This would parallel the situation found with Gilded and Red-shafted Flickers in Arizona (see Short, 1965), but with contact zones lower in the desert. Miller (1932) and Hargrave (1939) have commented on the size of kestrel bones from prehistoric sites.

Phasianidae

Gambel's Quail, *Callipepla gambelii* (Gambel)

Archaeological: One of the most abundantly represented species from Hohokam sites (Snaketown, South Santan Salvage, Las Colinas).

Ethnographic: *kákachu;* [pl.] *ká:kachu*

(The name is probably onomatopoetic.) Both the eggs and the birds themselves were commonly eaten. The heads were first plucked off and discarded to prevent blindness to the handler. Quail were secured in traps called *útpaw* constructed of arrowweed bound with the inner bark of willow. Consultant Sylvester Matthias made a Pima quail trap for the Arizona State Museum, Tucson.

Sylvester Matthias described another quail, *sádam kákachu,* 'sticky quail,' which is all gray, without distinctive markings on the sides or the black topknot. They are larger than "regular quail" and live in higher places. On the Pimas' October pilgrimage to Magdalena, Sonora, they hunted these birds in the Baboquivari Mountains and farther south in Mexico. This is the Scaled Quail, *Callipepla squamata* (Vigors).

The quail lexeme is found throughout the lowland Piman languages. The Papago say *kákaichu* (Saxton and Saxton, 1969, 1973) or *kákachu.* At Onavas, the Pima Bajo village, Ken Hale recorded *kakatu:v* and I *kákatu:*. During the 1660s it was transcribed here by Spanish Jesuits as *cacatuba* (*=kakatuvă*). When two or more biological species occur, this folk generic is modified by specifics. But apparently the quail name in Northern and Southern Tepehuan is not even cognate.

Historic: Breninger (1901:45) found the bird "Numerous and unsuspicious of man." Gilman (MS) recorded it, "Resident in great numbers.... The Indians do not harm them very much and they become quite tame...." He found the species nesting late, the earliest eggs being 1 April, with young rarely before mid-May.

Modern: The species is abundant today throughout the reservation and is seldom hunted by the Pima, somewhat more frequently by the Maricopa.

Taxonomic: On the basis of both external morphology (Raitt and Ohmart, 1966) and osteology (personal observation), the proposed genus *Lophortyx* (1838) is not separable from *Callipepla* (1832). The genera have been merged by Phillips et al. (1964) and Mayr and Short (1970). Perhaps even the expanded genus *Callipepla* should be merged into *Colinus* (1820), which is osteologically closely related (personal observation).

The reservation quail are the nominate race. Southern Arizona specimens average slightly darker on the flanks than topotypical birds from southern Nevada, but less than half of the specimens can be distinguished on this single character.

[Common Turkey, *Meleagris gallopavo* Linnaeus]

Archaeological: Fewkes' (1912) casual mention of antelope, turkey, rabbit, and bear bones from Casa Grande Ruins does not inspire confidence. The specimen reported from Snaketown (Gladwin et al., 1937) apparently has been lost (Haury, personal communication), so that the identification cannot be verified. Bones of *Meleagris* and *Grus* are frequently confused (Hargrave and Emslie, 1979). No turkey bones were recovered from the second Snaketown excavation. I identified one small turkey coracoid from the Las Colinas site (Rea, 1981b). This coracoid is smaller than in modern domestic breeds and smaller, even, than in wild females of the local race, *M.g. merriami.* It is assigned to the Small Indian Domestic, a very early puebloan breed that appears very sparingly in upper Santa Cruz sites around A.D. 1300 (McKusick, 1980). Unlike the more northerly puebloan cultures of the Southwest, the Hohokam must not have regularly kept domestic turkey, as evidenced by the almost total absence of turkey bones from their sites.

Ethnographic: *tóva;* [pl.] *tótova*

The Pima never domesticated the bird, but it figures in legends and the morpheme is still familiar. The feathers were used for fletching hunting arrows. The word *tóva* is the same in Papago, Pima Bajo, and at least as far south as the Northern Tepehuan, another Piman-speaking group. The Maricopa went up the Salt to hunt turkey.

Historic: Emory (1848:78) passed the mouth of the San Pedro River, 42 miles (67 km) east of the present reservation, and noted on the Gila, "Flights of geese, and myriads of blue [Gambel's] quail, and flocks of turkies, from which we got one. The river bed, at the junction of the San Pedro, was seamed with tracks of deer and turkey." I do not doubt that turkeys, at least in winter, were once within easy reach of hunting Pimas (see also G. Davis, 1982). But the species was extirpated from the lowlands before Gilman's time.

Modern: For the most part, the turkey is now absent from the Sonoran zones and survives only in mountainous pine and fir refugia. However, Robert Schmalzel recovered a road-killed grown young male (SD, June 1980) from a very dense mesquite bosque near the mouth of the San Pedro, and Alice Carpenter reported a family near Oracle about the same year. This is evidence that the southwestern race, *M.g. merriami* Nelson, has the potential for recolonizing mesquite bosques where predation is not excessive.

Change: No change is implied in the immediate area, but turkeys were eliminated from adjacent areas historically (Phillips, 1968).

Ardeidae

Green Heron, *Ardeola virescens* (Linnaeus)

Ethnographic: Though this species is known to the Pima and no doubt nested near their farms, I have not been successful in eliciting a name for it.

Historic: Gilman reported the Green Heron as a summer visitant only, nesting annually in his yard at Sacaton.

Modern: Fairly common but local on the Salt; absent elsewhere. Beginning in July 1969 (postbreeding?) the Green Heron was seen among young willows and cottonwoods of the Salt River from 91st to 115th avenues. By the following year the grove at the Salt-Gila confluence was apparently sufficiently mature to attract a breeding pair; Simpson and Johnson (personal communication) observed nuptial display here 8 May 1970 and the birds have been present each summer since, though no actual nests have been discovered. I have seen only adults on the reservation. Three individuals heard in the confluence grove 3 June 1978 were undoubtedly a family group. Wintering birds (December and January) have been noted from 1970 through December 1979 and again in 1981–1982.

Taxonomy: I have compared the postcranial osteology of the nominal genera *Ardea, Casmerodias, Leucophoyx, Florida, Hydranassa, Butorides, Ardeola,* and *Bubulcus,* as well as *Heterocnus, Nycticorax, Nyctanassa,* and *Cochlearius.* Of the true herons, the Green Heron is both sufficiently distinct from *Ardea* in overall pelvis shape, in sternum shape, and in the characters of the distal end of the femur (popliteal area) to be considered generically separate. However, slight differences found in *Casmerodius* (1842), *Leucophoyx* (1894), *Egretta* (1817), *Florida* (1859), *Hydranassa* (1858), and *Bubulcus* (1855) are no more than specific characters, and I would include all these in the oldest genus *Ardea* (1758), as do Phillips et al. (1964). I recognize *Dichromanassa* on behavioral as well as osteological grounds (head of humerus, dorsal configuration of pelvis). *Butorides* (1849) on the other hand fits right into *Ardeola* (1822) on both cranial and postcranial osteological characters (e.g., lachrymals, internal and external furcular spine, lateral margin iliac crest, peculiarly bowed tarsometatarsal shaft). The very slight differences are of specific value only. These all have strongly streaked immatures with dark backs. Although the two night heron species which are broadly sympatric in the New World, *Nycticorax nycticorax* and *Nyctanassa violacea,* indeed have numerous cranial as well as postcranial characters suggesting their generic distinctiveness (see Adams, 1955; Campbell, 1979; Payne and Risley, 1976), I think the exclusively Old World species of *Nycticorax,* particularly *N. leuconotus* and *N. magnificus,* must be examined critically before the generic status of *violocea* can be resolved. *Cochlearius,* a specialized night heron, closely resembles other Nycticoracinae except for the peculiar complex of cranial adaptations. Of the genera examined, *Heterocnus* is the most divergent in postcranial osteology. While distinct from the bitterns, the Tigrisomatinae share enough skeletal characteristics with the Botaurinae to justify calling the group tiger bitterns rather than tiger herons.

Ardeola virescens and *A. striata* may be conspecific (Payne, 1974; and others), but Wetmore (1965:88) advised caution in this merger.

An adult female Green Heron (AMR 4830) was found rather recently shot at the Salt-Gila confluence 1 January 1976. It is a large example of the race *A. v. anthonyi* (Mearns). The bird was exceedingly fat, but the stomach was empty.

Change: The Green Heron was apparently the last breeding heron to be extirpated from the reservation with the loss of riparian habitat. With the local redevelopment of willow-cottonwood thickets along the Salt, the species returned as a breeding and wintering bird.

Great Blue Heron, *Ardea herodias* Linnaeus

Archaeological: One specimen recovered from the recent Snaketown excavation.

Ethnographic: *kó:makĭ vákoiñ* (*kó:makĭ*, 'gray'; *vákoiñ,* 'heron'; [pl.] *vápkoiñ*)

Though *kó:makĭ* is usually glossed 'gray,' Joseph Giff explained that it is really a more indefinite color better described as 'blurred,' 'dirty,' or 'dusty.' Sometimes a flock would nest in the same tree. Joe Giff notes that sticks of considerable size were carried to the nest. Sylvester Matthias recalls at least a dozen nests in one dead cottonwood across the river from Komatke. These herons used the same colony tree from year to year. The people did not bother them or their young. The nest tree was eventually washed away in a flood.

Vákoiñ, in its various forms, is probably an ancient Pan-Piman ethnotaxon. From the two lowland Pima Bajo speakers I obtained *vá:kwun* or *vá:koan.* Northern Tepehuan is *vakóñi* (Pennington, 1969) and Tepecano (a Piman language in northern Jalisco) is *vákon* (Mason, 1917). But with Papago I found apparently only noncognate *káwkut* or *kókḍ.* The Pima folk generic contains the 'water' root, *va-*, as do many other animals, plants, and objects associated with aquatic situations.

Historic: Breninger noted several in September 1901. Gilman (MS) found this heron only occasionally from October to early spring. Phillips et al. (1964) reported that Ruth and Harry Crockett found an active colony of about sixty nests in 1930 just a few miles downriver from the Salt-Gila confluence. This was about the time the rookeries at the lower Pima villages were still active.

Modern: Fairly common on the Salt River, irregular elsewhere. A single specimen (AMR 2978, partial skeleton, immature) found 29 March 1970 on the Salt River above 91st Avenue. Singles or small groups, usually from two to eight, are seen on the lower Salt from May through February, with the bulk of the birds occurring from late July to mid-September. May and June birds are nonbreeding singles.

Change: The species has been extirpated from the reservation as a breeding species with the loss of permanent water for fish and the cottonwood gallery forest for nesting.

Common Egret, *Ardea alba* Linnaeus

Ethnographic: *tóa vákoiñ* (*tóa,* 'white'; *vákoiñ,* 'heron')

The Piman ethnotaxon may apply to this species or the next, but more likely to both, without distinction. Sylvester

Matthias believed two sizes of *tó a vakoiñ* were present and one probably bred near Komatke.

Historic: Gilman (MS) observed three on the Gila in April (year not recorded).

Modern: Uncommon on the Salt. Simpson (Johnson et al., MS) saw four at the Salt-Gila confluence 28 July 1973 and two with a flock of seventeen Great Blue Herons below 91st Avenue 16 November 1974. I have seen a single individual near 91st Avenue 17 July 1972 and 19 June 1976. But a flock of twelve was here from at least late December 1981 to early February 1982. Salome R. Demaree collected a specimen here 29 September 1968 (skeleton).

Taxonomy: I can find no osteological justification other than size to separate the larger day herons from the smaller ones presently retained in the genus *Egretta* by Payne and Risley (1976), Payne (1979), and others (see generic comments above about Green Heron). That the smaller ones generally have specialized breeding plumes (aigrettes) I think is of no great taxonomic significance. Payne and Risley (1976:106) considered the aigrettes in *Ardea alba* to be independently evolved from those in *"Egretta"* but noted (p. 77) that the only osteological difference between the two "groups" was in the posterolateral projection of the palatines. I regard these all as constituting a single genus, with those having specialized breeding plumes probably derived within the genus.

Snowy Egret, *Ardea thula* Molina

Ethnographic: *tóa vákoiñ* (see preceding species). Joseph Giff said *tóa vákoiñ* (probably this species) used to nest in dead cottonwoods just across the Gila from Komatke, where the people used to farm: "lots of cottonwoods and just rough-made nests."

Modern: Rare. Simpson took an immature specimen (JSW 846) on the Salt below Maricopa Colony 27 July 1969. Additional sight records (Simpson and Rea) are: two on 20 May 1970, one on 17 July 1972, two on 28 March 1976, four on 19 June 1976, one on 25 June 1977, and two on 6 June 1978.

Taxonomy: Oberholser (1974:971) has restricted the nominate race *A. t. thula* Molina, based on a bird from Chile, to the smallest population, occurring over most of South America. On the first revisor principle the somewhat larger birds of the eastern United States through Central America to northern South America become *Ardea t. candidissima* Gmelin, based on a bird from northern Colombia. K. C. Parkes (personal communication) thinks the use of *candidissima* can be upheld for the northeastern bird, even if the type were only a migrant, as Oberholser notes.

The traditional separation of Snowy Egrets into two North American subspecies has been considered not entirely satisfactory (see Bailey, 1928; Palmer, 1962; Phillips et al., 1964). Not to be outdone, Oberholser (1974) erected an additional race, *A. t. arileuca,* for the supposedly intermediate population of the Great Basin. These birds, for the most part, fall within the range of overlap between the largest race, *A. t. brewsteri* (Thayer and Bangs) of Baja California and the smaller *A. t. candidissima* of the eastern United States but are nearer the latter. Individual specimens cannot be identified off their breeding grounds, and the slightly larger birds, in one or several dimensions, are best labeled "*candidissima* approaching *brewsteri.*" Perhaps the overlap in published measurements between *brewsteri* and eastern *candidissima* is partially an artifact caused by missexed specimens (e.g. Willett #2363 in Bailey, 1928) and by the inclusion of wandering, nonbreeding specimens of eastern or northern birds within samples of *brewsteri.* Though not mentioned in the original description or subsequent racial discussion, one of the best characters identifying true *brewsteri* is its more massive bill

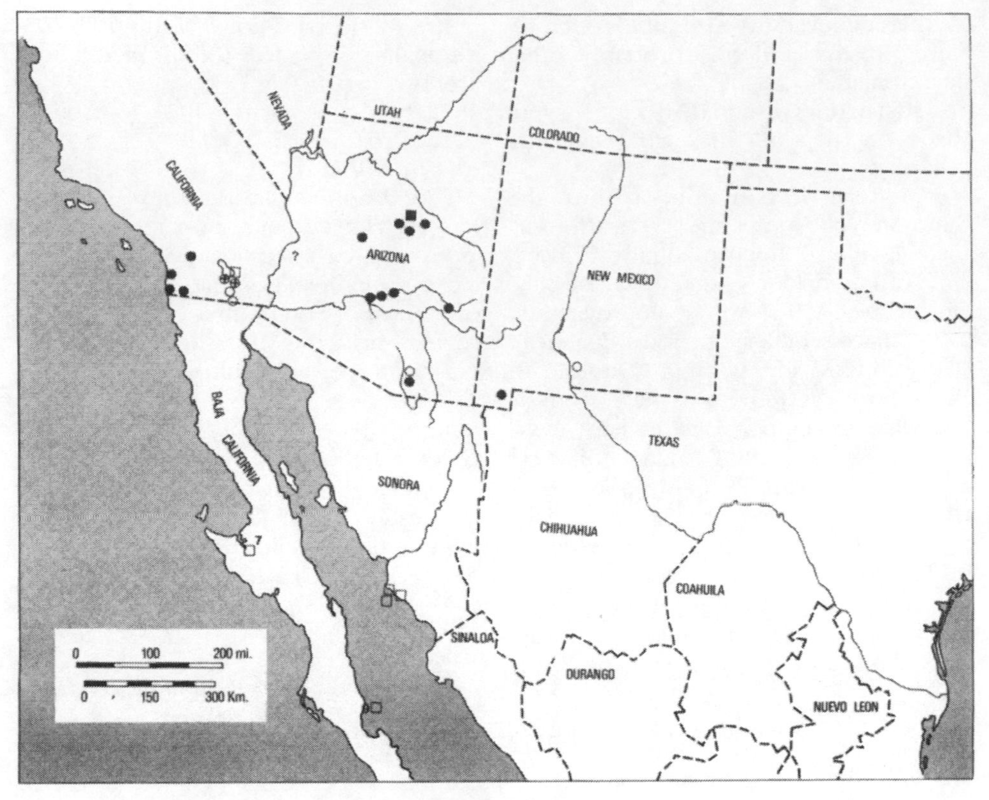

Snowy Egrets

□ = Ardea thula brewsteri (breeding) ■ = A.t. brewsteri (nonbreeding)
○ = A.t. candidissima (breeding season) ● = A.t. candidissimi (nonbreeding)
⊕ = A.t. candidissima x brewsteri (breeding)

(breadth at base and overall depth). Wing length is least useful. I find the following range of overlap between true *candidissima* of eastern United States and *brewsteri* of Baja California:

	wing chord	tarsus
males	261—270	103.6—106
females	255—279	93.5—102
	culmen length	bill depth
males	86—89	9.95—10.3
females	83—89	(none)

Phillips (in Phillips et al., 1964) referred all Arizona birds to *brewsteri* as did Grinnell and Miller (1944) for southern California, and the A.O.U. (1957) for New Mexico. The only southwestern specimen that appears properly referrable to *brewsteri* (see Table 14) is the late April male taken at a small pond near Flagstaff, Arizona. I identify all the remainder as *candidissima* or *"arileuca."* Even the summer adult female from Las Cruces, Doña Ana County, New Mexico, presumably a representative of the local breeding population, is quite small except in bill depth. Breeding specimens (SD) collected by Mary Platter-Rieger from the Salton Sea area of California are mostly small, but one is *brewsteri.* It would be of interest to determine the racial status of the egrets breeding on the lower Colorado River.

Table 14. Nonbreeding Southwestern *Ardea thula* Specimens

Collection	Sex	Date	County, State	W. Chord[a]	Tarsus	Culmen	Bill depth (medially)
LLH 666	♂	16 April 1945	Pima Co., AZ	260	102.7	83.5	9.5
JSW 846	[♀]	27 July 1969	Maricopa Co., AZ	247	89.4	72.6	9.0
JSW 845	♀	13 Oct. 1956	Maricopa Co., AZ	240	89.0	80.5	9.0
AMR 751	♂	9 May 1965	Gila Co., AZ	265	98.4	83.6	9.7
ARIZ 98	♂	23 Sept. 1928	Yavapai Co., AZ	251	94.5	73.5	9.3
ARIZ 4497	♂	1 May 1896	Pima Co., AZ	257	89.0	82.2	9.1
NAU 2	[?]	1 April 1962	Coconino Co., AZ	263	102.5	80.1	10.6
NAU 1014	♂	24 April 1972	Coconino Co., AZ	265	110.1	89.2	11.1
NAU 1473	[?]	7 May 1975	Coconino Co., AZ	238	95.9	81.0	10.7
NAU 1391	♀	14 May 1974	Coconino Co., AZ	250	102.4	82.9	9.4
CM 125434	♂	13 May 1940	Pima Co., AZ	236+	92.1	80.1	(12.2)
SD 126	♀	24 April 1878	Riverside Co., CA	246.7	90.5	85.0	9.3
SD 38443	♀	1 April 1973	San Diego Co., CA	257.2	96.2	81.7	9.1
SD 35074	♂	10 Jan. 1964	San Diego Co., CA	245.2	88.5	81.2	10.7
CM 21619	♂	23 Sept. 1886	Hidalgo Co., NM	246	96.5	74.7	(11.4)
CM 124250	♀	6 July 1939[b]	Doña Ana Co., NM	243+	84.6	72.4	10.9

[a]For comparison with Palmer's (1962) flattened wing measurements, add 5 mm to wing chord.

[b]Postbreeding local?

[Cattle Egret, *Ardea ibis* (Linnaeus)]

Modern: Rare and erratic winter transient. My only sight records for the area are two groups consisting of nine and twenty-five birds in an alfalfa field and along the road edge just north of the Salt River 14 December 1974. The birds were very tame.

Change: This Old World species invaded the New World perhaps in the latter part of the nineteenth century and was authentically documented for the Caribbean part of South America by the 1930s and in Florida by the 1950s (G. Crosby, 1972). It appeared in Arizona as a winter visitant by 1966 (Monson, 1968) but is still irregular. Browder (1973) discussed long-distance movements of the species.

Taxonomy: Although some authors (e.g., Morony et al., 1975) place *Bubulcus* Bonaparte 1855 in the genus *Ardeola* Boie 1822, the Cattle Egret has no osteological resemblance whatsoever to *Ardeola* (see also Wetmore, 1965:95). Rather, its osteological characters (e.g., pelvis, particularly the iliac crest, internal furcular spine, lachrymal shape, and triosseal canal of coracoid) fit quite well with other species of *Ardea, sensu lato* (see Green Heron account).

Black-crowned Night Heron, *Nycticorax nycticorax* (Linnaeus)

Ethnographic: *tash cúnam vákoiñ* (*tash,* 'sun'; *cúnam* [?]; *vákoiñ,* 'heron')

Matthias remembered that a number of night herons roosted at a different place from the Great Blue Herons, using instead leafy cottonwoods. His statement: "Usually come in gray, but different kinds," apparently refers to the well-marked distinction between first-year and adult herons. The Pima considered this to be a very mean bird. A Pima child, climbing in a tree, fell into a nest and was attacked by night herons.

Historic: Breninger (1901) found the species "Numerous among flooded willows," in September. Gilman (MS) recorded them several times from fall to 27 May (no year given).

Modern: Fairly common locally. All recent records are from along the Salt River beginning 19 July 1969 (Simpson) to the 1980s. I found a partial skeleton of an immature (AMR 4637) above the Salt-Gila confluence. Sightings (maximum of thirteen individuals) are in different months, including May of three years. Both Johnson and Simpson (personal communication) believe the species probably nested in the riparian groves developing west of 91st Avenue until about 1980. Simpson and I observed adult and first-year herons here 19 June 1976 but have not located nests. I found ten immatures and an adult in very thick willow-cottonwoods here 1 September 1980, before the groves were bulldozed for "flood control."

Change: The species has been extirpated from the Pima villages, where it bred, according to ethnographic data, before the destruction of the cottonwood gallery forest. Phillips et al. (1964) recorded its status as breeding "formerly on Salt River and Verde Rivers, but present status there uncertain."

[American Bittern, *Botaurus lentiginosus* (Rackett)]

Historic: Gilman (MS) noted the bittern several times, giving March and October as the months.

Modern: Rare. Lewis D. Yaeger collected a male (LLH 317) at a desert tank 1 mile (1.6 km) north of Laveen 15 May 1940. This is a short distance off the reservation. I have one sight record at Barehand Lane 7 January 1967. The species is "chiefly known as a rare transient in Arizona. . . ." (Phillips et al., 1964:7).

Least Bittern, *Ixobrychus exilis* (Gmelin)

Modern: Formerly fairly common locally. Vic Housholder and Harold Yost saw six Least Bitterns on the Salt River at the 107th Avenue marsh 20 June 1943. Housholder (*fide* Simpson) said the marsh was then still of good size with abundant cattails. James Werner collected a juvenal (PC 375, now MNA) at this marsh 25 June 1955. This record, mentioned by Phillips et al. (1964), appears to be the only specimen from the reservation. At the former Salt-Gila confluence pond Johnson and Simpson recorded Least Bitterns from 25 April to 27 July (1969–1971). Along the effluent channel west of 91st Avenue, Simpson and I flushed one into a cattail thicket 25 June 1977.

Change: I have no historic or ethnographic account of Least Bitterns. The species has not been recorded at the confluence pond since the extensive cattail growth was washed out in the winter flood of 1972–1973 or west of 91st Avenue since the floods of early 1978.

Gruidae

Sandhill Crane, *Grus canadensis* (Linnaeus)

Archaeological: Two occurrences from Snaketown and one from South Santan Salvage.

Ethnographic: *haiñ júlshap (haiñ,* 'something broken'; *júlshap* [?])

(The word *júlshap* is an archaic expression the meaning of which no one remembers.) This taxon might refer to the Wood Stork, *Mycteria americana,* but is more likely the crane. According to Sylvester Matthias' account, these dark grayish birds stand tall and always manage to elude one approaching for a better look. About half a dozen would come into a field someone was irrigating. Whether they sought fish transported in the water or flooded-out gophers was not known. At other times these birds were seen on the sandbars along the river. (Interestingly, the Pima generic *vákoiñ* does not form part of this ethnotaxon as it does for herons, egrets, pelicans, and ospreys.)

Historic: Gilman (MS) recorded: "Large flocks seen the last of February and the first week in March nearly every season." He specifically noted a large flock at Sacaton 5 March 1916.

Modern: None observed.

Taxonomy: The two measurable archaeological bones represent the smaller and more widespread race, *G. c. canadensis* (*fide* McKusick).

Change: Phillips et al. (1964:30) note, "Virtually unknown as a migrant [in Arizona] in recent years." Drewien et al. (1975:300) observe, "Major problems associated with Sandhill Cranes relate to lack of overall management goals, lack of reliable population, recruitment and harvest data for the two hunted races [which reach Arizona], and continued deterioration of breeding and wintering habitats due to man's activities."

Rallidae

[Clapper Rail, *Rallus longirostris* Boddaert]

Modern: Rare and local. Richard L. Todd (personal communication), of the Arizona Game and Fish Department, found a "pair" which answered tape-recorded calls at the marshes of 107th Avenue and Salt River 3 June to 23 September 1970. The habitat has since been destroyed.

Virginia Rail, *Rallus limicola* Vieillot

Historic: In September Breninger (1901:45) recorded the Virginia Rail: "Often heard, and by watching a spot from where the calls came, I discovered a bird, posed perfectly motionless." The species was not found by Gilman. It survived somewhat longer about the northwestern end of the reservation, with specimens taken near Laveen 22 December 1919 (US) and 17 October 1920 (ARIZ 3836, lost).

Modern: Formerly uncommon locally. At a Salt River slough west of 91st Avenue Robert W. Dickerman collected an immature male (RWD 898, CU) in heavy molt on 22 August 1953. Werner found the species still here in May and November 1956. The last Virginia Rails heard on the Salt were in 1970 (Johnson et al., MS). I observed several at the open east end of Barehand Lane 2 November 1969. West of here on 20 May 1974 a small black rail (presumably an immature Virginia but possibly a Black Rail, *Laterallus jamaicensis* [Gmelin]) flew up in response to my tape recording of Virginia Rail. This was in a solid stand of *Typha* measuring 220 × 480 feet (67 × 148 m). But the following year the marsh vegetation at Barehand Lane was destroyed. By October 1979 once again the vegetation was rank and flooded and at least one Virginia Rail (transient or wintering?) was present. In April 1981 no rails responded to tape recordings, although there was extensive growth of *Rumex* with some *Typha*.

Change: The species has occurred on the reservation wherever there has been sufficient expanse of cattails and other emergent vegetation. All the habitats where Dickerman, Werner, Johnson, Simpson and I have recorded the species apparently breeding have since been destroyed. It is no longer found on the lower Salt even as a winter bird.

Sora, *Porzana carolina* (Linnaeus)

Historic: Gilman (MS) saw the species three times: late April and 2 and 3 May, the latter at Sacaton.

Modern: Uncommon. I netted an immature male Sora (AMR 1707) 12 September 1967 in the open marsh at Barehand Lane. It was in heavy molt and might have been a local bird. I collected another male at the same spot (AMR 2851) 27 November 1969. My only other observations there are 1 August, 29 September, and 29 October (each a single bird) in 1972. Simpson (personal communication) saw the species 14 March 1971 on the Salt.

Change: Circumstantial evidence (Phillips et al., 1964) indicates that the Sora might breed in the Salt River Valley, central Arizona. But there is no conclusive evidence yet for its breeding on the reservation. November records indicate limited wintering of Soras since Gilman's time.

[Yellow Rail, *Coturnicops noveboracensis* (Gmelin)]

Historic: Gilman (1910:46) recorded the species for 28 March 1901. "[A friend] had caught it out on the desert about eight miles [13 km] west of Sacaton and several miles from water. As I was flat on my back suffering with erysipelas I was unable to do anything with the bird, and the man releast [sic] it on the bank of a small stream."

Change: This is the only state record for the species, which must be considered accidental.

Common Gallinule, *Gallinula chloropus* (Linnaeus)

Historic: Breninger (1901:45) noted, "A number seen along willows; the only place I have ever found gallinules." Gilman did not find the species in his day.

Modern: Abundant along the Salt year round, formerly fairly common at Barehand Lane, rare elsewhere. The species was not yet nesting at Barehand Lane marsh the summers of 1966 and 1967. A pair was here the summer of 1968 and a nest was found in the cattails, but no eggs or young were ever seen in it. However, we observed one fully grown young (usually at dawn and at dusk) starting 1 September 1968. A wing of the species (AMR 2445), victim of some predator, was found at the pond 22 October 1968. In July 1972 two families of about eight birds each were on the pond and a nest with two chicks was found 1 August 1972. By 1975 there was no suitable habitat remaining, but by April 1981 at least one gallinule was again at Barehand marsh. The species became exceedingly abundant in the sewer effluent and low vegetation from 91st Avenue to the Salt-Gila confluence. On 25 June 1955 J. Werner saw only one gallinule and one coot on the Salt. A specimen (JSW, lost?) was found dead 26 November 1956 at 107th Avenue. By 30 July 1969 Johnson and Simpson found fifteen nests and collected one chick between 91st Avenue and 107th Avenue. For the next decade the species was virtually colonial at the Salt-Gila confluence and above, with the family groups staying together far into winter. At Sacaton I saw four gallinules among sixteen coots on the deep sewer pond 23 February 1976 (Fig. 3.12).

Change: The species was extirpated from the dryer upper section of the reservation by Gilman's day (1907–1915). An irrigation water storage pond in the Salt River bed near Tempe had some gallinules in the early 1930s, but the habitat was gone by the end of the decade (L. D. Yaeger, personal communication; Phillips et al., 1964). The spectacular comeback along the reservation part of the lower Salt probably occurred in the 1960s. About the same time the species was first reported breeding in southern Utah (Wauer and Russell, 1967) and southern Nevada at Tule Springs Park (Austin, 1968). This bird is a strong colonizer when local conditions are suitable.

American Coot, *Fulica americana* (Gmelin)

Ethnographic: *váchpik (vach, 'to dive* in the water' [*va* = aquatic stem])

This ethnospecies, which was known to all my native consultants, is probably distinct from the gallinule. The coot, as the Pima taxon states, escapes by diving into the water like a grebe, whereas the noisy gallinule flutters clumsily into the nearest clump of emergent vegetation. The fishy savor of the *váchpik* is described as *s'dáwhiuk.* The species was not eaten by Riverine Pima.

Historic: Breninger (1901:45) recorded the coot "Seen along with the last [gallinules]" in September. Gilman (MS) noted it "Seen at various times during March but not common. Seemingly stops a few days during migration to rest or feed." The only specific date from his notes is 7 March 1911, when three were seen at Sacaton.

Modern: Uncommon for most of the reservation, perhaps fairly common on the Salt, but common in the deeper channel effluent below 91st Avenue. I observed five coots on the Riggs Road pond 3 March 1973 and six there on 21 June 1973. I have occasional records for other more open catchment areas as at the Casa Grande pond (AMR 4689 [28 November 1974]). Coots bred (regularly?) along the Salt. Johnson took a young juvenal (MNA 3186, female, all rectrices ensheathed) 30 July 1969, and I found a dead male (AMR 4561) with broodpatch and enlarged testes (18 × 10 mm) 12 May 1974.

Change: The species was gradually extirpated with the riparian habitat destruction. A very limited return as a breeding resident is shown (at least some years) for portions of the lower Salt River.

Charadriiformes

Charadriidae

[Snowy Plover, *Charadrius alexandrinus* (Linnaeus)]

Modern: Rare transient. One sight record, 27 April 1969, identified by Phillips and seen by Hargrave, Johnson, and me at the Casa Grande pond. There is a September specimen (JSW) from near Peoria (north of the reservation) taken by Johnson. Otherwise the species is unknown for this part of Arizona.

Killdeer, *Charadrius vociferus* Linnaeus

Ethnographic: *chívichuich;* [also] *síwichich*

The second variant is used by people from the Bapchule area. The Pima believe this species was formerly much more common.

Historic: Breninger (1901:45) recorded "A few seen; evidently migrants, for they appeared worn and tired." Gilman (MS) found the species "Resident. Great numbers spend the winter.... Young seen 15 April to the last of June."

Modern: Common, at least in winter. A preovulating female (AMR 2879) was found shot on the Salt, 29 March 1970. Today on the reservation there are relatively few locations suitable for Killdeer breeding, such as the gravel flats of the Salt River bed.

Change: As indicated by the ethnographic and historic accounts, there has been a decline in Killdeer numbers with the loss of the Gila and lower Santa Cruz as living streams.

[Mountain Plover, *Charadrius montanus* Townsend]

Historic: Gilman (MS) found the species "...but once when five were noted in a field 11 December 1914." There are no subsequent observations and apparently no specimen.

Taxonomy: The merger of *Eupoda* (1845) into *Charadrius* (1758) follows Bock (1958), Jehl (1968), and the A.O.U. Check-list Committee (1973).

Scolopacidae

Common Snipe, *Capella gallinago* (Linnaeus)

Historic: "Seen at various times during fall, winter, and early spring. Winter dates 12 December; 8 and 20 January" (Gilman, MS).

Modern: Locally common in winter. I took a female (AMR 456) at a tiny *charco* in Komatke 12 March 1964. In winter snipes occur regularly in low aquatic vegetation along the Salt channel and at least formerly in the open, meadowlike marsh at the west end of Barehand Lane, starting by 17 October.

Long-billed Curlew, *Numenius americanus* Bechstein

Modern: Rare transient. I observed thirty-five or more 2 April 1966 in a flooded lower-terrace alfalfa field just north of the Salt River at 91st Avenue. Two unsexed immature specimens (AMR, ARIZ, skeletonized) were found rotting near the reservation boundary below Chandler 7 September 1973. Their bills measure 92.2 and 115.8 and wing chords 261 and 253, respectively. If races be recognized, both specimens would be *N. a. parvus* Bishop, according to the measurements given in the original description. Phillips et al. (1964) questioned the utility of this race.

Sandpiper (all species)

Ethnographic: *shú:dakĭmat;* [pl.] *shú:dakĭmamat* (*shú:dakĭ*, 'water'; *-mat,* 'offspring of,' 'child'; *-mamat,* 'children')

The Piman name for sandpipers is literally 'water children.' No individual species of 'water children' are distinguished in the Pima lexicon. This is Sylvester Matthias' Sandpiper Song, which he and Richard Purcell, O.F.M., transcribed and freely translated:

Shu:dagĭ i ma:math
shu:dagĭ i ma:math
i vo'i hai-chun agit
kaij chua-pa i o'op
hiosig motk s-ge'e li.

Water Children
Water Children
Came and told me something
They told me that a prostitute
With a burden of flowers
Winded her way away from me.

(The Sandpiper Song must be sung four times.)

Spotted Sandpiper, *Actitis macularia* (Linnaeus)

Ethnographic: All sandpipers are *shú:dakĭmamat,* 'water children.'

Historic: Gilman (MS) recorded the Spotted Sandpiper "Seen rarely and one secured 28 September."

Modern: Fairly common transient; possibly uncommon in winter along Salt. The two specimens I collected (AMR 604, 20 September 1964, Komatke; AMR 2211, 12 May 1968, Vahki) are both taken during dates of expected migration for Arizona (Phillips et al., 1964). The entire ranges of recent dates for the reservation are 11 to 30 May in spring and 28 July to 20 September for fall. On the Salt below 91st Avenue, Simpson and I saw an individual 21 February 1976 which may have been wintering. On 11 May 1978 we observed a flight of Spotted Sandpipers on the lower

Salt. There were individuals and small groups of up to five and six birds all along the river.

Taxonomy: Burleigh (1960a) described a race, *A. m. rava.* Though Oberholser (1974) considered it valid, Parkes (personal communication) believes it is not recognizable. I have not examined the matter.

Solitary Sandpiper, *Tringa solitaria* Wilson

Historic: Gilman (MS) recorded, "One secured 19 September 1914, Sacaton, and three others seen the rest of month."

Modern: Rare transient. A specimen was collected 3 May 1938 two miles (3 km) northwest of Casa Grande Ruins. Robert W. Dickerman saw a Solitary Sandpiper 22 August 1953 on the Salt at 91st Avenue. I have not seen the species.

Taxonomy: The one available specimen (CGR, unsexed, wing chord 134.8) is the larger *T. s. cinnamomea* (Brewster), breeding in western North America.

Greater Yellowlegs, *Tringa melanoleuca* (Gmelin)

Historic: In September Breninger (1901:45) recorded the species simply as "Seen along water course of river," with no indication of numbers. Gilman (MS) recorded yellowlegs as "One seen 26 September 1914 and three on 5 October 1914, one of which was shot." I have not relocated this specimen.

Modern: Uncommon transient. I collected a female (AMR 2019), one of two, 7 March 1968 at a small pond of irrigation runoff water below Sacate. My only sight records away from the Salt are one at the junction of Casa Blanca Road and Interstate 10 on 15 October 1972 and three at Riggs Road pond a week later. Occurs with more regularity during migrations on the lower Salt.

Pectoral Sandpiper, *Erolia melanotos* (Vieillot)

Modern: Rare. One specimen (JSW 978, male) collected 22 September 1970 by Simpson at the 91st Avenue marsh. It was taken from a flock of six. I have no further records.

Taxonomy: Although the plumages are similar and the external structural differences seem trivial, I concur with Campbell (1979:113–115) that *Erolia, Ereunetes,* and *Calidris* are best maintained as separate genera on osteological grounds. *Ereunetes* may be more closely related to *Calidris* than size would suggest.

Least Sandpiper, *Erolia minutilla* (Vieillot)

Historic: Breninger (1901:45) saw "A small bunch flying about a mud-flat," in September. Gilman (MS) recorded Least Sandpipers "Seen during migration and several [ten or twelve individuals] seen 10 January 1901 and 14 December 1914."

Modern: Abundant outside breeding season. Specimens for the reservation are from 30 July 1969 (JSW) to 6 January 1973 (AMR). The only habitats now suitable for the species are along the lower Salt River and at several runoff ponds.

Semipalmated Sandpiper, *Ereunetes pusillus* (Linnaeus)

Modern: Casual. I collected one specimen (DEL 18528, male) on the Casa Grande pond southeast of Bapchule 27 April 1969. Johnson, Hargrave, Phillips, and I had stopped to look over the migrant water birds. This individual, with its distinctive call note and general aloofness from the numerous Western Sandpipers, caught our attention. This is the second state specimen (see Phillips et al., 1964).

Western Sandpiper, *Ereunetes mauri* Cabanis

Modern: Common during migration. The Western Sandpiper appears to be much more numerous in fall than in spring, though the span of fall records on the reservation is only 22 August to 22 September.

Long-billed Dowitcher, *Limnodromus scolopaceus* (Say)

Modern: Uncommon fall migrant, wintering some years. One observation each for the five months from July to November, four dowitchers being the maximum seen on any one date. A male was taken 20 August 1975 (AMR 4787). Spring records restricted to 27 April 1969, when ten were seen at Casa Grande pond (Johnson, Hargrave, Phillips, Rea). My only winter observations of dowitchers (presumably this species) were at the Salt-Gila confluence: four on 26 December 1976 and sixty to seventy on 27 December 1981.

Wilson's Phalarope, *Phalaropus tricolor* Vieillot

Historic: Gilman (MS) noted "Two seen on a pool after a shower 15 April 1901." Housholder took a specimen (KANU) on the Salt above the Salt-Gila confluence 1 October 1921.

Modern: Uncommon on migration. Seen in spring on 27 April (1969 and 1973) at runoff ponds. A female (AMR 2612) taken in the former year and several late August skeletons (AMR, ARIZ) preserved. Fall migration is from 11 August to 22 September. There are seldom ponds of sufficient extent on the reservation to attract migrating phalaropes.

Taxonomy: The genus *Steganopus* Vieillot, 1818 (and perhaps *Lobipes* Cuvier, 1817) are probably best included in the oldest genus *Phalaropus* Brisson, 1760.

Northern Phalarope, *Lobipes lobatus* (Linnaeus)

Historic: Gilman (MS) secured a pair 19 September 1914 on a pond formed by irrigation drainage. There are no subsequent records for the reservation.

Recurvirostridae

American Avocet, *Recurvirostra americana* (Gmelin)

Archaeological: A partial humerus was recovered from Snaketown.

Historic: In September Breninger (1901:45) noted "A flock seen on a sand-bar in river." The avocet was unrecorded by Gilman.

Modern: Fairly common, spring and fall migrations. Spring records span from stilt, and Gilman (MS) recorded "Nine-late August. Specimens (AMR) are skeletons of birds found dead. At least in some years the species nests or attempts to nest on the gravel beds of the Salt above 91st Avenue. Salome R. Demaree (personal communication) found ten adults and one nest with four eggs there 20 May 1970. Also on the Salt, but somewhat east of the reservation, she found a female (SWAC) 20 June 1968 with yolks and an egg in the oviduct. The following year at 35th Avenue and the Salt River she found a nest and three eggs of this species among those of stilts, but the nest was drowned out. (This locality is near the reservation.)

Black-necked Stilt, *Himantopus mexicanus* (Muller)

Ethnographic: *chúvul ú'uhik* (*chúvul,* 'tall'; *ú'uhik,* 'bird')

Pima consultants are familiar with this species. This native taxon does not include the avocet, which is apparently unnamed in the Piman lexicon.

Historic: Breninger did not see the stilt, and Gilman (MS) recorded "Nineteen September 1914 one was seen at the same time and on the same pond where the Northern Phalaropes were seen. . . ."

Modern: Fairly common during migration. Migrations (birds seen at nonbreeding areas) extend from 26 April to 30 May and from late August through September. A flock of about thirty stilts above the Salt-Gila confluence on 7 February 1982 may have been wintering. When conditions were favorable, stilts nested or attempted to nest on the gravel beds of the Salt River above 91st Avenue (formerly) and below (till 1978). Demaree (personal communication) observed six adults and one juvenal at the sewer ponds 4 August 1968. Simpson (personal communication) and Demaree saw about twenty stilts below the plant between 18 and 30 July 1969. The following year Demaree observed forty adults and nests on 5 May, but Simpson found only two stilts remaining on the gravel breeding area on 20 May. By 22 June I found between two hundred and four hundred adults there, paired and territorial. They flew low over my head crying until I left the area completely, but I observed no broken-wing distraction behavior. I found no eggs or young. I have no observations for 1971. When I visited the colony area 17 July 1972 about a hundred adults were present, vociferous, giving broken-wing display. Only adults were observed. Johnson and Simpson (personal communication) saw an adult here with two partially grown juvenals 11 August 1973. The sewer settling areas were dry in subsequent seasons, and no water birds of any sort were seen except Killdeer.

Laridae

[Gull, *Larus* sp?]

Historic: Gilman (MS) noted "A large gull seen flying down the Gila River one evening, 5 October, but it was not possible to even guess at the specific rank." I have never seen a gull on the reservation.

[Tern, *Sterna (forsteri?)*]

Modern: Over Riggs Road pond 27 April 1973 I saw a tern with a large, deeply forked tail, black cap, and orange bill. Phillips (personal communication) believes Forster's is the most likely species. I have no further observations of terns for the reservation.

[Black Tern, *Chlidonias niger* (Linnaeus)]

Historic: Gilman (MS) wrote, "On 19 August, when the Gila River was in flood, eight terns were seen flying down stream. They were the size of the Black Tern and looked like that bird in winter plumage. However, this might not be possible at this season of the year and as no specimen could be taken, the identification is doubtful." He reported another at about the same place 21 August. Both of these records are from Sacaton in 1914. Gilman's identification is highly probable for this date in central Arizona. There are no recent records.

Anseriformes

Anatidae

Ethnographic: *vápkukĭ; also vápkuch*

This Pima taxon includes all ducks, swans, and geese in general, without distinction. "We saw the differences, but they are all just *vápkuch.*" There is uncertainty in the consultants' minds that any ducks or geese were hunted before the acquisition of shotguns and the influence of Anglos.

[Whistling Swan, *Olor columbianus* (Ord)]

Historic: In mid-November 1846 Griffin (1943) noted ducks and geese of several species and Whistling Swans below the Salt-Gila confluence. Farther down the Gila River, at Painted Rocks, Emory's (1848:92) entry for 19 November 1846 reads: "The pools in the old bed of the river were full of ducks, and all night the swan, brant, and geese were passing."

Gilman (MS) obtained a specimen (preserved?) 21 November 1910 near Santan, opposite Sacaton. His account (Gilman, 1911a:35) reads:

> November 21, while driving across the desert, I found a Whistling Swan with a crippled wing. He could fly and half run and it took quite a chase to run him down. It was a long ways from water so I gave him a drink from my canteen which he seemed glad to get, and putting him in the wagon took him home. Here I placed him in a big irrigating ditch with grassy banks...but he grew weaker and died on the fourth day.

Change: No doubt formerly more common along the Gila, swans, "white brant" (Snow Geese), geese, and ducks were mentioned in the accounts of many explorers crossing the state. Today the species is "uncommon and rather irregular late fall transient" away from the lower Colorado Valley (Phillips et al., 1964:9). I have not found the species on the reservation; there are no places suitable for it today.

Canada Goose, *Branta canadensis* (Linnaeus)

Archaeological: One specimen (humerus) recorded from Snaketown.

Modern: Rare. Johnson et al. (MS) saw twelve Canada Geese 4 February 1973 above the Salt-Gila confluence. The Salt was then in flood.

White-fronted Goose, *Anser albifrons* (Scopoli)

Archaeological: Four specimens from the 1964–1965 excavations at

Snaketown represented by a humerus, an ulna, two radii, and a carpometacarpus. I have identified a worked furcula (pendant) from the 1930s excavation.

Historic: Gilman (MS) recorded several shot by an O'Brien 4 January 1909, but I have not been able to locate a specimen, if one was saved.

Change: I am aware of no modern record for the reservation. Today the species is rare away from the Colorado Valley (Phillips et al., 1964). Johnson and others (MS) saw one on the Salt west of Phoenix (but east of the reservation) from 9 to 11 October 1970.

Snow Goose, *Anser caerulescens* Linnaeus

Archaeological: Four records from Snaketown consisting of three carpometacarpi and a humerus.

Ethnographic: *biá:rak*

Sylvester Matthias described *biá:rak* as a flocking white water bird with black wings that flies in formation. It once occurred on the Gila and lower Salt. The name refers to the sound these birds make when flying over. This appears to be the only anseriform distinguished to a Linnaean species in the Pima lexicon.

Modern: Neither Gilman nor I saw this species on the reservation. Phillips et al. (1964:11) considered it "Formerly common winter visitant to Colorado and Gila Valleys, now reduced in numbers."

Taxonomy: I doubt that *Chen* can be maintained as a genus separate from *Anser,* which has priority; see also Delacour and Mayr (1945). The Snow Goose was originally described in the latter genus by Pallas. The taxonomic status of the Snow and Blue Goose has received varying treatment. Although they are usually considered two species (A.O.U. Check-list, 1957), Cooch (1961) and Cooke and Cooch (1968) have demonstrated that the two forms are best considered dichromatic polymorphs of a single subspecies, *A. c. caerulescens.* This has been accepted by the A.O.U. Check-list Committee (1973). The Snaketown bones, listed as *hyperborea* (McKusick *in* Haury, 1976), belong to this smaller, interior and western form.

Mallard, *Anas platyrhynchos* Linnaeus

Archaeological: Two specimens from Snaketown represented by a humerus, ulna, and carpometacarpus.

Modern: Rare. An Indian gave me an adult male Mallard which he shot 12 April 1968 at Barehand Lane. The skin and skeleton (AMR 2091) were preserved. I saw one male and two females above the Salt-Gila confluence 27 November 1977. I have no further observations from the reservation.

Taxonomy: The single specimen is *A. p. platyrhynchos.* It is considerably smaller, both externally and osteologically, than the domestic breed. Oberholser (1974) considered the mainland North American population, *A. p. neoborea* Oberholser, separable on the basis of size from the Old World nominate. However, I find his measurements of "small" Old World Mallards completely bracketed by measurements of New World specimens given by Brodkorb (1968).

Pintail, *Anas acuta* Linnaeus

Archaeological: One record from Snaketown, an ulna.

Historic: Gilman (MS) saw a flock of Pintails 18 August 1914 at Sacaton.

Modern: Uncommon transient and winter. I observed one Pintail at Barehand Lane marsh 23 December 1972. All other sightings are by Simpson (personal communication) near the Salt-Gila confluence. They span from 28 July to 4 February, with flocks of up to fifty birds. Werner (Johnson et al., MS) saw a solitary Pintail here 26 May 1956.

Green-winged Teal, *Anas crecca* Linnaeus

Archaeological: One coracoid of a Green-wing was recovered from Snaketown.

Modern: Fairly common transient and winter. Specimens consist of an adult (AMR 4301, nearly complete skeleton) found rotting in a canal near Barehand Lane 16 October 1972 and three apparent botulism victims at the Casa Grande pond (AMR, ARIZ) found [30 August 1975]. I saw five at the Sacaton sewer pond 23 February 1976. Sight records for the Salt-Gila confluence (five to ca. twenty-four teal) are 14 March 1971, 16 September 1972, and 6 January 1973.

Taxonomy: Various authors (Delacour and Mayr, 1945; Gabrielson and Lincoln, 1959; Johnsgard, 1965; Mayr and Short, 1970; and others) have considered the American Green-winged Teal only subspecifically distinct from Old World *crecca.* This has now been adopted by the A.O.U. Check-list Committee (1973). Reservation specimens are *A. c. carolinensis* Gmelin.

Blue-winged Teal, *Anas discors* Linnaeus

Historic: Gilman (MS) apparently collected a male Blue Wing on 1 May 1910 and observed another 10 May 1914 at Sacaton.

Modern: Uncommon transient. An immature female (AMR 2324) was obtained at Barehand Lane, 1 September 1968. The identification was confirmed by K. C. Parkes. Other records may belong either to *discors* or to *A. cyanoptera,* but the two species are indistinguishable afield in late summer and fall. Even in the hand, identification is difficult, the distinction best being made on the basis of the very different bullae (enlargement of the lower trachea) of males.

Cinnamon Teal, *Anas cyanoptera* Vieillot

Modern: Fairly common. An adult female (AMR 2296) was obtained at the junction of Maricopa and Casa Blanca roads, 15 August 1968. Dr. Parkes confirmed the identification. Male specimens are from 22 February 1968 (Vahki), 19 January 1975 (bulla obtained from hunter on Salt), [17 July 1972] (dried skeleton found long dead on Salt), and 25 August 1981 (raptor-killed first-year bird found at the confluence). Frequent sight records of males in breeding plumage range from 21 February 1976 (eleven birds) to 17 July. Cinnamon Teal bred occasionally along the lower Salt (R. L. Todd, personal communication). The birds I observed in the lush vegetation west of 91st Avenue in April, May, and June 1978 were probably breeding. I found a punctured egg and saw a female and three males at a weedy pond in Co-op Colony, 20 June 1976, but saw no young. The species may have attempted to breed in the thick Johnson grass of the canals there. Additional sight records (as about a hundred on the Salt 17 July 1972) cannot be placed to species, being either *cyanoptera* or *discors.*

Shoveler, *Anas clypeata* Linnaeus

Historic: Gilman (MS) observed this species 11 October 1914 at Sacaton. This is his only mention of the species.

Modern: Fairly common transient. This is the most frequently encountered duck on the reservation today. An adult female (AMR 2090) was obtained from an Indian 17 April 1968. Spring records are from 17 April to 30 May and fall from 27 July to 29 September. I saw two males and three females at the Sacaton sewage pond 23 February 1976. As with the teal, shovelers are tolerant of small, shallow ponds and have been seen away from the Salt River and Barehand Lane.

Taxonomy: The merging of *Spatula* (1822) into *Anas* (1758), proposed by various authors, has been officially sanctioned by the A.O.U. Check-list Committee (1973).

American Wigeon, *Anas americana* Gmelin

Modern: Uncommon transient. I observed two males and one female at the Riggs Road pond, 3 March 1973. On 10 March 1975 I found a freshly killed and partially eaten male (AMR 4704, skeleton) on the Salt above the Salt-Gila confluence.

Taxonomy: The merging of *Mareca* (1824) into *Anas* (1758), proposed by various authors, has been sanctioned by the A.O.U. Check-list Committee (1973).

[Canvasback, *Aythya valisineria* (Wilson)]

Modern: Rare transient. Simpson observed a male and female Canvasback 22 November 1970 on the Salt River at 91st Avenue.

Ring-necked Duck, *Aythya collaris* (Donovan)

Historic: Gilman (MS) obtained a male at Sacaton, 19 February 1910. I reexamined the specimen at San Bernardino County Museum.

Lesser Scaup, *Aythya affinis* (Eyton)

Archaeological: One specimen, a carpometacarpus, was identified by McKusick from Snaketown.

Modern: Rare transient. I observed a male *Aythya,* apparently this species, 3 March 1973 on the Riggs Road pond.

[Bufflehead, *Bucephala albeola* (Linnaeus)]

Modern: Rare transient. I observed this species once on a tiny *charco* (cattle pond) in Komatke village, 21 November 1964. Johnson et al. (MS) have no records for the Salt from 83rd Avenue to the confluence.

[Common Merganser, *Mergus merganser* Linnaeus]

Historic: Gilman (MS) recorded "One shot along the Gila River 16 November" at Sacaton, year not given. I have not relocated the specimen.

Ruddy Duck, *Oxyura jamaicensis* (Gmelin)

Archaeological: One tarsometatarsus recovered from Snaketown.

Historic: Gilman (MS) reported "A pair seen 19 February 1910."

Modern: Uncommon, formerly breeding. I have rarely seen this species on the reservation: about ten to fifteen on the Riggs Road pond 3 March 1973, and two there 21 June 1973; a single male at Sacaton 24 November 1972, and a male and female there 23 February 1976. Conditions were once suitable above the Salt-Gila confluence for breeding Ruddy Ducks. Two males and a female were seen on the Salt 30 July 1969, but their breeding status was undetermined. Two downy young, dying apparently of botulism, were collected at the confluence pond 30 May 1971 (JSW, now MNA 4056; other lost?) and two partially grown young were watched by Johnson and Simpson (personal communication) 5 August 1972. Richard L. Todd (personal communication) saw a female here with at least five ducklings less than one-third grown 9 August 1971 and three nearly grown young 25 August 1971. I collected an adult male road-killed specimen (AMR, skeleton) on the Salt at 51st Avenue in June 1977.

Change: As this species requires extensive areas of cattail marsh for breeding, it is doubtful that there is any longer sufficient habitat for its nesting on the reservation.

Columbidae

White-winged Dove, *Zenaida asiatica* (Linnaeus)

Ethnographic: *'áwkakoi;* [pl.] *'áw'awkakoi*

The name is onomatopoetic and is similar in Riverine Pima, Papago, and lowland Pima Baja. This large dove was and to some extent continues to be a food item with the Pima. It was either trapped or shot with arrows. Frank Russell (1908:86) records that "Sonora doves were and are yet confined in log-cabin cages built up out of arrow rods [*Pluchea*]." Curiously, Gilman (1911b:52) said the Pima do not hunt this or the next species. According to Sylvester Matthias, the call of these doves is rendered *gogoks shap kaich,* 'the dogs said it.' The Papago Delores Lewis gave the call as *gigik kik ku:k.*

Historic: Breninger (1901) did not mention seeing this species during his mid to late September visits to the reservation. Gilman (1911b:53) found White-wings arriving on the reservation about 20 April, with egg dates from 10 May to 2 August, with peaks in May and June in 1908 and 1909, but extending well into July in 1910:

> Along in August the big flocks begin to grow less, the birds probably scattering out and seeking feeding grounds more distant from the breeding grounds. Toward the first of September they begin to thin out in earnest and by the 15th of the month very few are seen. Individuals may linger a little longer, as in 1909 I saw one as late as October 12, and in 1910 the last was on September 25. A few lingered on a sorghum field up till September 10 of this year but were not seen any later.

Modern: Abundant summer resident. White-wings nest abundantly in the saltcedar thickets that now choke the Gila and lower Santa Cruz rivers from above Santa Cruz village to the Salt-Gila confluence. My earliest dates are a single bird at Komatke 1 February 1964 and two starting to call there 4 March 1966. My latest date, well away from sorghum and wheat fields where they remain somewhat later, was 3 September 1966, when I saw three doves in the bajada of the Sierra Estrella.

Change: Phillips et al. (1964:42) presented evidence that "The White-wing has spread in comparatively recent times northward to Safford and the Gila River generally, to the Verde Valley, and up the Colorado River from Yuma to Needles." Phillips (1968:137) believed that by 1885 "it has occupied most of its present Arizona range." Though the Mourning

Dove is relatively common in archaeological sites (Hargrave, files), there is but one record for the White-wing (Haury, 1950). The bone was recovered from the surface deposit of Ventana Cave, southern Arizona. Charmion R. McKusick (personal communication) kindly reexamined the specimen and found it correctly identified but quite recent in appearance. There is no direct evidence as to when the species arrived on the reservation. An onomatopoetic name no doubt enters the folk taxonomy rapidly when the species is conspicuous and vocal. The idea that White-wings have hastened their fall departure date in response to modern hunting seasons (opening in September) is unsupported by the historic data (Breninger and Gilman, above). However, they abandon nesting when shooting begins (Neff, 1940a:123).

Taxonomy: The subspecies is *Z. a. mearnsi* (Ridgway), with type locality 5 miles (8 km) north of Nogales, Arizona. Saunders (1968) revised both North and South American populations of this dove.

Mourning Dove, *Zenaida macroura* (Linnaeus)

Ethnographic: *háwhi;* [pl.] *háwhawhi*

This species is also sometimes referred to as *kómarukdam háwhi,* 'gliding mourning dove' from its courtship flight. The species was and still is eaten by the Pima. Boundary Commissioner Bartlett (1854:237), who spent two weeks in July 1852 at the Pima villages, observed: "It is quite common for them [Pima and Maricopa boys] to shoot doves with their arrows, and to bring in half a dozen of these birds after a ramble among the cotton-woods." However, Gilman (1911b:52) said, "The Indians never hunt them [Mourning Doves] and they are quite tame." Nor does Russell (1908) mention any dove species in his food chapter. The name is the same in Papago and Pima Bajo and is onomatopoetic.

Historic: Gilman (1911b:52) recorded:

> During November and December they are fewest in number, but in January and February many more arrive, and in breeding months they are everywhere. Nesting begins the first part of April, my first find being dated 12 April, the nest containing eggs partly incubated. Many nests were noted during April, May, June, and July, with no attempt made to keep a correct census of them. August 13 was the latest date, and on that day two nests with eggs were seen.

Gilman's distribution of nests was: April, 12 percent; May, 16 percent; June, 20 percent; July, 44 percent; August, 8 percent. He observed, "It seems strange that so large a proportion of the nesting is so late, considering the locality is so warm in late spring and summer."

Modern: Common winter, abundant summer resident. My observations of this species seem to correspond with those reported by Gilman, though I have not observed any noticeable increase in numbers as early as January. I have found eggs by late March.

Taxonomy: Aldrich and Duvall (1958) reviewed the subspecies. Specimens I have examined are the paler race *Z. m. marginella* (Woodhouse) which occurs in the West.

Common Ground Dove, *Columbina passerina* (Linnaeus)

Ethnographic: *chéhupĭ;* [also] *vúhigam (vúhi-,* [onomatopoetic?]; *-gam* [attributive])

That this dove has two commonly known names, one of which (*chéhupĭ*) appears not to be onomatopoetic, is perhaps suggestive of a long cultural association. Rosita Brown described it as "the small dove that doesn't go anywhere." Matthias identified the call of *chúhupĭ* at the Salt-Gila confluence in June. Delores Lewis, a Papago from Gu Oidak, called this dove *huchigam* (cognate with Pima *wuhigam*). Four hundred miles to the south on the lower Río Yaqui, Pedro Estrella and María Córdova, lowland Pima Bajo, gave me *tehop* (cognate with river Pima *chehupĭ*) for both Ground and Inca Dove.

Historic: Breninger (1901) noted only one pair seen during two September weekends in the field. Gilman (1911b:54) believed that "They are absent from this locality during the winter months, usually making their appearance about the middle of March. November 18 is the latest I have seen them but they are rarely seen as late as October 20." But in his subsequent manuscript Gilman (MS) noted "a few seen" in December and February of the winter of 1910–1911 and again in 1914–1915, so surmised that they might occur each winter. His nest dates were from 7 July to 8 October. Birds in this latter nest hatched 16 October but died two days later (Gilman, 1911b).

Modern: Fairly common. I have made no special study of this localized species to discover if it might be partially migratory or completely absent in some winters, as suggested by Phillips et al. (1964). I have mid-winter (November and December) specimens for 1963 and 1964 and observations for December 1976 and 1977.

Inca Dove, *Scardafella inca* (Lesson)

Ethnographic: *gúgu*

The name is onomatopoetic. This dove is well known to the Pima and can be found about every village and cluster of houses.

Historic: Phillips et al. (1964:43) believed the Inca Dove was "probably absent from the state prior to 1870." Scott (1886) found it common in the newly established Florence 9 miles (14 km) east of the reservation, but Breninger (1901) did not find it on the reservation. Gilman (MS) called it "resident and increasing." Gilman (1911b) noted its close proximity to human habitation. He found nests of eggs from 11 April to 25 September.

Modern: Common. This species is almost never found at any distance from Pima homes or other buildings, in spite of the fact that these almost never have lawns.

Taxonomy: Though Johnston (1961) proposed the merger of *Scardafella* (1855) into *Columbina* (1825), I find the postcranial osteology of most major elements of *Scardafella* so distinctive that I recognize it as a valid genus. A comprehensive osteological study of the ground doves throughout the world needs to be undertaken to assess correctly their evolutionary relationships.

Mayr and Short (1970) and S. M. Russell (personal communication) considered *S. inca* conspecific with allopatric *S. squamata.* Since the latter has priority, the North American form would become *S. s. inca.* Goodwin (1967) maintained the two as separate species.

Change: It is difficult to interpret the history of this species for the Pima villages. Breninger's failure to find it and Gilman's cryptic remark "resident and increasing" from 1907 to 1915 suggest that the dove was becoming established early this century. However, if Mearns found it already well established at Phoenix in 1885 (Phillips et al., 1964:44), then it should have arrived also at the Pima villages before Breninger's time. Perhaps Breninger did not work close to houses where he would more likely encounter the dove. Possibly, the species was slower in colonizing the Pima villages which were then suffering from a drought period.

Psittaciformes

Psittacidae

Parrots (all species)

Ethnographic: *s'náwkadam ú'uhik* (*s'*, 'very' [intensifier]; *nawk,* 'talk'; *-dam* [attributive]; *ú'uhek,* 'bird' [life form])

This is the only contemporary name. There is no recollection of the capture or use of parrots. This taxon is glossed 'talking bird.' However, Russell (1908:222) recounted a legend where the parrot appears as *scu-utûk ó'ofîk* which is literally 'green bird.' In 1716 Padre Luís Velarde (1931:129) recorded that "at San Javier del Bac and neighboring *rancherías,* there are many macaws, which the Pimas raise because of the beautiful feathers of red and of other colors, almost like those of the peacock, which they strip from these birds in the spring for their adornment."

The correct upper Piman name for macaw is *ahḍo;* [pl.] *a:'ahḍo.* The Vocabulario en la lengua Névome, a Pima Baja dictionary compiled at Onavas, Sonora, in the mid-seventeenth century, gives *'guacamayo'* (macaw) as *arho* (Pennington, 1979:57). In her Papago studies Underhill (1946:84) correctly translated this lexeme 'macaw.' However, today both Pima and Papago appear to have forgotten the original meaning of the word. All my Pima consultants, when translating the legend of the origin of tobacco, gave only 'peacock' for *a.ḍo* (see also Saxton and Saxton, 1969:83, 167; 1973:267—268, 389). Mathiot (n.d.:361) gives both glosses. This anomalous transferral from a native trade item to an introduced domestic species is understandable. Both birds have very large, brilliantly colored tail feathers (note Velarde's comparison). The trade in macaw feathers, so important in the ceremonialism of southwestern agricultural tribes, was disrupted with the imposition of European power, though according to Velarde, the Pima were still raising macaws early in the eighteenth century. Whereas the Puebloan peoples have continued to obtain macaw feathers for use in dances, Pimans have long ago abandoned actual use of macaws. Being familiar today with the Old World galliform of the Anglo Americans instead of macaws, they have substituted the English gloss 'peacock.'

The legend recounted by Russell (1908:222) regarding green parrots is interesting in its implications. Two relationships are established: between parrots and turquoise (reciprocal trade?) and between the Piman narrator and another class of people (Salado? Hohokam? Puchteca?). In the version of the tobacco origin myth given by Russell (1908:224), the boys were instructed to go east with their newly hatched parrots. "At length they set free the parrots, which flew up into the mountains, where they concealed themselves in the forest." In narrating this story eighty years later, Joseph Giff adds that, from other place names given, the mountains appear to be the Santa Catalinas near Tucson. In his version 'forest' is specifically 'pines.' The location, the habitat, and the green color recorded in the oral literature suggest the Pimans formerly had firsthand knowledge of Thick-billed Parrots.

Archaeological: The Thick-billed Parrot, *Rhynchopsitta pachyrhyncha* (Swainson), was recorded from Snaketown (carpometacarpus and femur) between A.D. 100 and 900. The Scarlet Macaw, *Ara macao,* appeared in Snaketown (various elements) between A.D. 1 and 700. As McKusick (*in* Haury, 1976) pointed out, these are the earliest occurrences for these two psittacids in the Southwest (see also Hargrave, 1970). Several undated specimens are from Pueblo Grande (McKusick, Rea, notes). I identified a Scarlet Macaw humerus from the Las Colinas site (Rea, 1981b).

Change: It is at least conceivable that the Thick-billed Parrot occasionally occurred naturally on the Gila River in prehistoric times. The species continued to irrupt into the pines of southeastern Arizona well into the twentieth century (see Phillips et al., 1964). But I doubt its occurrence naturally at such low elevation as Snaketown.

The Scarlet Macaw, a southern Mexican species, was definitely a trade item in the aboriginal Southwest (Hargrave, 1970; DiPeso, 1976; and others). It is something of a surprise that trade (or at least rearing) of live birds to the north continued at least as late as Velarde's time (1716).

Cuculidae

Yellow-billed Cuckoo, *Coccyzus americanus* (Linnaeus)

Ethnographic: *kádgam*

Matthias mentioned this bird to me late in 1976, saying that it occurred rarely in the former thickets of willows and cottonwoods along the river, although he had never seen it. The behavior, color, size, shape, and distinctive vocalization ("like clapping hands"), as described by Joseph Giff, leave no doubt that *kádgam* is the cuckoo. I know of no reference to this species in any genre of Piman oral literature.

Historic: Gilman (MS) recorded the cuckoo "Seen occasionally in thickets along the river in June and July and one in a dry canyon three miles [5 km] from the river, 13 September 1914."

Modern: Fairly common in summer in very limited cottonwood stands with dense understory and surface water, from late May to at least late August. James Werner (Johnson et al., MS) found the species at 107th Avenue marsh from 20 May to 4 August 1956. My first observation of cuckoos was of two apparently territorial birds flushed eight times 21 June 1970 just northwest of the Sacaton Police Station. The birds were in a thicket along a canal, with laterals and an overstory of mature cottonwoods. This habitat has since been destroyed. (Here also the same day were a Western Flycatcher, a Parula Warbler, and a female grackle carrying food.) At the Salt-Gila confluence a pair resided and apparently nested in the *Populus-Salix-Tamarix* grove the summers from 1972 through 1978. By 23 June 1974 two pairs were at the confluence, but only those in the main grove were carrying food. At least one other pair held a territory in maturing willows about 107th Avenue in 1976 and 1977. I collected an immature from a group of at least three at the confluence 13 September 1976.

Taxonomy: A somewhat weakly differentiated western race, *C. a. occidentalis* Ridgway, is currently recognized (A.O.U. Check-list, 1957). Based on size, it may be useful in separating most birds off their breeding grounds. Van Tyne and Sutton (1937:35) "seriously question the value of maintaining" it. The measurements of the only reservation specimen (AMR 4859, immature female, wing chord 146.4, tail 148.9) are equivocal.

Change: The cuckoo is dependent on tall, broad-leafed riparian timber with well-developed understory (Gaines 1974). Spaced cottonwoods lacking undergrowth, as about Santan and the Chandler-boundary area (see Fig. 3.18), are unsuitable for cuckoo breeding. Limited recent recolonization was occurring with habitat redevelopment along the lower Salt (Fig. 3.15).

Greater Roadrunner, *Geococcyx californianus* (Lesson)

Archaeological: Bones of this species were identified from the Snaketown and Las Colinas sites.

Ethnographic: *táwdai*

The roadrunner is well known to the Pima. It figures predominantly in many songs and legends. The roadrunner's dietary fondness for *s'vekĭ vámat,* 'red racer,' and *koi,* 'rattlesnake,' are related. The Papago name for roadrunner is the same as Pima. The lexeme is *tá:da* with lowland Pima Bajo.

Historic: Gilman (MS) found roadrunners a "Common resident. Nesting early and late and often. Eggs 16 April to 11 July."

Modern: Common. The species is conspicuous in Pima fields, using fence rows for nesting and protection. A specimen (AMR 1898) taken 30 December 1967 was exceedingly fat.

Taxonomy: Oberholser (1974) described the roadrunners inhabiting most of Texas as *G. c. dromicus,* a race smaller than the nominate form from California and Arizona. Though there is overlapping in all measurements (partly caused by missexed specimens?), *dromicus* might prove a useful race.

[Groove-billed Ani, *Crotophaga sulcirostris* Swainson]

Modern: Casual. I watched one individual for some time 13 May 1974 in the cottonwoods on the reservation boundary below Chandler. Though unmolested while feeding in alfalfa fields, the ani was mobbed by starlings when it flew into the cottonwoods. Even the White-winged Doves seemed disturbed by its presence.

Salome R. Demaree (personal communication) relates that Robert Witzeman and Scott Burge found two Groove-billed Anis along the effluent channel just west of 91st Avenue on 21 December 1972. One was photographed three days later. At least one stayed until 8 April 1973.

Change: Though never seen by Gilman and unknown to Pima consultants, anis have been seen by other observers irregularly in the Phoenix area, sometimes just off the reservation along the lower Salt. A specimen (ARIZ 12290, skeleton) was recovered by Chuck Bindner 26 December 1971 north of the reservation (107th Ave. and W. Thomas Rd.). No Arizona breeding known.

Tytonidae

Barn Owl, *Tyto alba* (Scopoli)

Ethnographic: *é:'et váhudam* (*é:'et,* 'blood'; *váhudam,* 'it squeezes out')

This is the usual name, well known to the Pima. In the fourth warpath speech, Russell (1908:388) glosses *U'Ut vaohotam* as 'white blood sucker' in linear translation and simply 'Owl' in the free translation. Joseph Giff said the Barn Owl can also be referred to as *vanam mukatdam,* 'sharpening a knife.' This name is in reference to the bird's vocalizations. According to Sylvester Matthias, Barn Owls formerly nested in holes along the banks of the Gila, as did kingfishers.

Pima generic ethnotaxa are usually relatively short monolexemes that are often unanalyzable (see, for instance, vulture, quail, eagle, turkey, heron, macaw, crow, mockingbird). The more descriptive, polylexemic names I expected would be only locally used, confined to one Piman language or perhaps even to one dialect. It was some surprise, then, when María Córdoba, a lowland Pima Bajo of the lower Río Yaqui, Sonora, identified the *'lechusa'* or Barn Owl as *é:'et váhut,* which she glossed *'sangre sudora.'* During the past three centuries the Pima Bajo have had little or no effective contact with the widely separated Riverine Pima of the north. Linguistic innovations in the interim could not be transferred. The ethnotaxonomy emphasizes the basic unity of early Piman concept and language throughout an extensive geographic area.

Historic: Gilman (MS) reported "28 April, two found in dry well; one in high bank of river and remains of another in mesquite. Evidently rare in this locality."

Modern: Fairly common. I encountered Barn Owls much less frequently than Great Horned Owls. Barn Owls bred at Komatke (St. John's old grammar school building, vacant third floor) at least in 1964, when young were banded. One of these was recovered at Cashion, Arizona, about 15 miles (25 km) northwest. A female (AMR 580) found dead just north of the Salt 9 June 1964 had at least three ruptured follicles in an ovary measuring 16 mm. I have not definitely seen the species from November through January, outside dates being 9 February through 15 October. Perhaps the bulk of the population moves on in mid-winter. Barn Owls are occasionally seen in summer at places where the species does not breed, as one at Barehand Lane in low saltcedars on 22 May 1974.

Strigidae

Great Horned Owl, *Bubo virginianus* (Gmelin)

Ethnographic: *chúkut;* [pl.] *chúchkut*

Though occasionally used in general speech to refer loosely to owls in general (only by younger Piman speakers?), the morpheme applies specifically to the Great Horned Owl. *Húhugam,* 'deceased people,' are reincarnated as *chúchkut.* An owl announcing death is termed *kómalvúpĭdam chúkut,* and the bird is feared.

Pima *chúkut* and Papago *chúkud* appear as *tukut* (?plural *tuktuk)* with lowland Pima Bajo. But *tucurhu* of seventeenth-century Névome, Ken Hale's *tucur* of modern lowland Pima Bajo, Northern Tepehuan *tukurai* (Pennington, 1969), and Tepecano *tuku:r* (Mason, 1917) may be cognate with Upper Pima *kú:kul* (see Common Screech and Ferruginous Pygmy-Owl accounts below). Since only one owl ethnotaxon is given in each of these lists, strict comparisons are not possible.

Historic: Gilman (MS) found Great Horned Owls "Resident and fairly common, nesting in February and March in deserted Red-tail Hawk nests in cottonwoods and saguaros, also in natural cavities in big trees." During the day Gilman (1909b:145) "found [them] mostly in cottonwood trees along the river…[as well as] in bluffs and cliffs on the rocky hills a few miles from the river."

Modern: Fairly common. Great Horned Owls are widespread across the reservation. I have encountered this species often (perhaps the same individuals repeatedly), even in mid-day. A very pale female (AMR 2876), found 29 March 1970 on the Maricopa Road at the Gila River, had at least two ruptured follicles. Other specimens are skeletons found long dead.

Taxonomy: As noted by Phillips (1946), various authors have assigned Arizona specimens of Great Horned Owls to three different subspecies. Phillips (1946) questioned the separation of *B. v. pallescens* Stone, 1897 of the arid Southwest from *B. v. pacificus* Cassin, 1854 of the more mesic parts of California (exclusive of the humid coastal redwood belt). He later (*in* Phillips et al., 1964) synonymized *pallescens* with *pacificus.* I have examined 271 specimens of these two taxa (AMR, ARIZ, ARP, CAS, GCN, JSW, LLH, MNA, MVZ, NMS, SBM, SD, UNLV, UNM, UTEP), and find that *pallescens* is a recognizable subspecies, distinct from *pacificus.* Breeding topotypes of the latter (AMR) were made available by Aldena Stevens. Specimens examined were from California, Baja California, Oregon, Nevada, Arizona, Utah, New Mexico, and western Texas. *Pacificus* has a darker, browner, and richer dorsal and ventral ground color, more numerous barring on the ventrum and feet, and black in the humeral area (concealed by the interscapulars), facial disk edging, and "ear" tufts. As Oberholser (1904) pointed out, *pallescens* is a variable race with three moderately well-defined color phases (pale ochraceous, pale gray, and darker gray), but even the darker individuals are distinguishable from *pacificus* by their paler dorsal ground color, paler legs with few or no bars, and dark umber (not black) areas of the "ears," face, and humeral patch. The two taxa do not differ in size. I find the ranges to be as given by Grinnell and Miller (1944) except that most owls (MVZ) of the xeric interior of the southern and western San Joaquin Valley are *pallescens.* The southern coastal areas are variable (e.g., 5 *pacificus,* 7 intermediates, and 6 *pallescens*

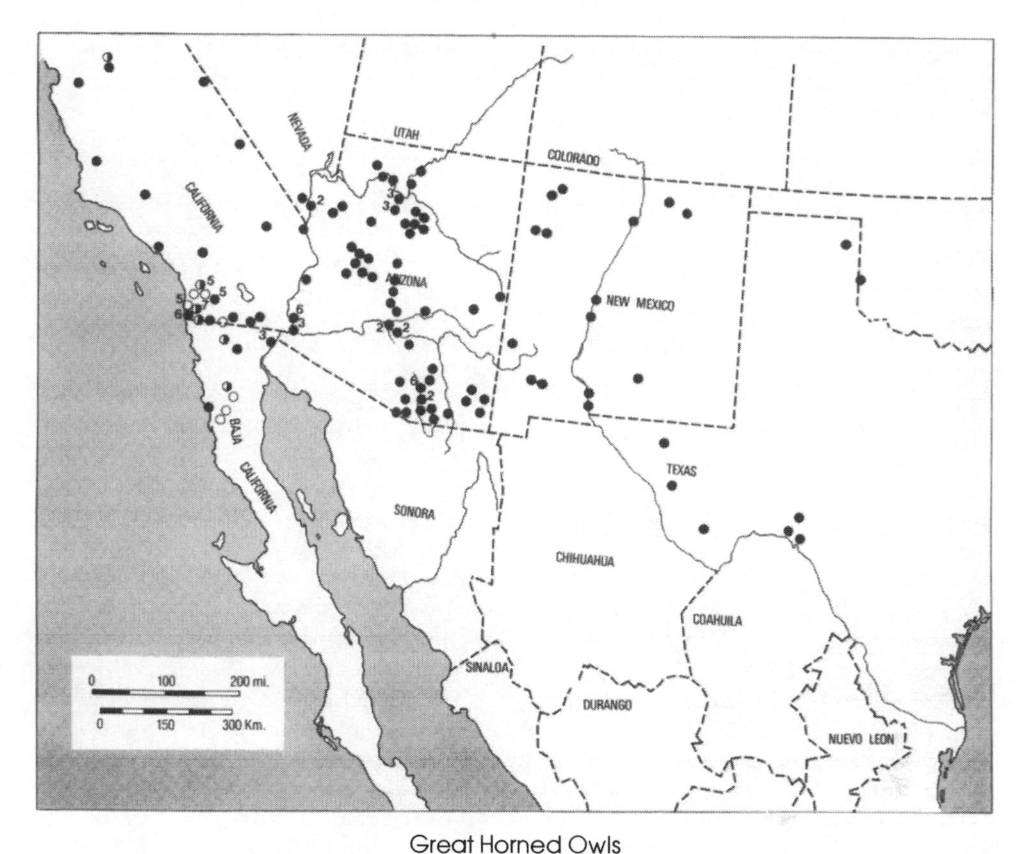

Great Horned Owls

● = Bubo virginianus pallescens ○ = B.v. pacificus
◑ = B.v. pacificus x pallescens

from around San Diego Bay, SD). But typical rich brown examples of *pacificus* occur as far south as northern Baja California (SD, MVZ).

All Arizona specimens except one I consider to be *pallescens*. The ascription of *pacificus* to the state (Oberholser, 1904; A.O.U. Check-list, 1957) is based merely on somewhat darker or more ochraceous specimens (US, GCN, MNA) of *pallescens*. One quite dark bird (MVZ 134581, 17 August 1956, Chiricahua Mountains, Arizona) is indistinguishable from coastal birds except for the umber rather than black areas of the head and shoulders. A truly blackish bird, both above and below, is a post-ovulating female from near Snowflake, Navajo County, Arizona (AMR 4850, 2 August 1974, juniper grassland, G. Fritz). This bird is *not pacificus* and Kenneth C. Parkes (personal communication) considers it *B. v. saturatus* Ridgway, the race breeding along the humid Pacific coast. I find it matches quite closely the blackish, non-brown-washed winter specimens (SD, perhaps *saturatus* × *occidentalis* intergrades?) from Siskiyou County, California, and Oregon. Contrary to the opinion of Phillips et al. (1964), I can find no correlation within the Southwest between coloration or barring of the legs and either altitude or habitat.

B. v. occidentalis Stone breeds in the Great Basin and northern Great Plains. It differs from the two taxa discussed above in larger size and apparently in color. I have reexamined the Arizona specimen (SD) which Huey (1939) believed to represent this form. Though still largely in juvenal plumage, it is a dark gray *pallescens.* Other larger specimens from the state are apparently only missexed females; their wings do not exceed 370 mm. Northern Arizona, southern Nevada, and (*fide* Behle, 1960) southern Utah birds are *pallescens.* The nearest *occidentalis* I have seen are from north-central and northeastern New Mexico (UNM), but others from this area appear to be good *pallescens* (e.g., AMR 810, female, 30 August 1965, Colfax County, New Mexico). Sutton (1967) considers western Oklahoma owls nearest *occidentalis.*

Common Screech-Owl, *Otus asio* (Linnaeus)

Ethnographic: *kú:kul;* also *kú:kvul*

The name is possibly onomatopoetic. This morpheme appears to be the same as *koo-ah-kohld* that Gilman (1909b:147, 150) obtained for Elf and Pygmy (Ferruginous) Owls. He could elicit no name for the Screech Owl: "The Indians would give me no name for this owl; one man said it had a name but he had forgotten it; another lookt [*sic*] puzzled and said he thought it had a name but he had never heard it. They all knew the bird however." Delores Lewis, a Papago from Gu Oidak, called both the Screech and Elf Owl *kú:kul.* Screech Owls are common about Pima houses. Rosita and David Brown raised three young at Santa Cruz village, which were allowed their freedom in old tamarisk trees. The dread of the Great Horned Owl appears not to extend to this or other owl species.

Historic: Gilman (MS) recorded the species as "abundant" and gave (1909b) egg dates from 29 March to 12 April (fresh) with several nests of recently hatched young on 12 April. Nests were from 7 to 25 feet (2 to 7.5 m) from the ground, mostly in Gilded Flicker holes.

Modern: Common. The nests that I have found were in flicker holes in saguaros, whereas Gilman (1909b) mentioned only willows and cottonwoods. In the deep arroyos coming down from the Sierra Estrella, Common Screech-Owls often roost in large ironwood trees, where they are virtually invisible against the gray branches.

Taxonomy: Gilman collected the holotype of *O. a. gilmani* Swarth, 1910 (MVZ 10651, type examined) at Blackwater. This race is considered (Miller and Miller, 1951; also, Marshall *in* Phillips et al., 1964; and Marshall, 1967) a synonym of *O. a. cinerasceus* (Ridgway, 1895). *Cinerasceus,* as currently used, is a race intermediate between two extremes, *O. a. aikeni* (Brewster, 1891) in the east and *O. a. yumanensis* Miller and Miller, 1951 in the west. It is unfortunate that the Huachuca Mountains are the type locality for the older name *cinerasceus* because the owl population there (ARIZ, fall series) is actually intermediate toward *aikeni* in heaviness of ventral markings and in size. Much more typical of the extensive central lowlands (Maricopa, Pinal, Pima counties, and extending far southward into Sonora) is *gilmani* (near topotypes, AMR), which I recognize as a valid race.

Flammulated Screech-Owl, *Otus flammeolus* (Kaup)

Modern: Rare. I secured a male (AMR 4728) 5 May 1975 in a small grove of oaks below Montezuma Sleeping, a peak in the Sierra Estrella. The date is consistent with other lowland spring migrants for Arizona (Phillips et al., 1964; Balda et al., 1975).

Taxonomy: Delacour (1941), Friedmann et al. (1950), Oberholser (1974), and others have recommended the merger of New World *flammeolus* into Old World *scopus* (Linnaeus). However, Marshall (1967) found the two to be biologically distinct species. Often vocalizations are more important than morphology in species recognition in nocturnal birds.

Two races are indistinguishable in the Southwest. The population breeding in the mountains of southeastern Arizona (ARIZ) has heavy ventral shaft streaks, fewer and broader cross bars, and generally more rufous suffusion dorsally and ventrally. Its ventral pattern strongly resembles that of the sympatric *Otus trichopsis* (Wagler). More northern birds, breeding south to the Mogollon Plateau and the Bradshaw Mountains (AMR, MNA, NAU) are finely patterned below (as in *Otus asio gilmani*) and have the rufescent areas greatly reduced or lacking altogether. The confusion regarding which race Kaup described and the uncertainties concerning major trends in variation were discussed by Phillips (1942), Marshall (*in* Phillips et al., 1964), and Marshall (1967). Thus no formal subspecific names are presently assignable. The migrant Sierra Estrella specimen is of the more northern of the two Southwestern races.

Ferruginous Pygmy-Owl, *Glaucidium brasilianum* (Gmelin)

Ethnographic: Gilman (1909b) recorded *koo-ah-kohld* as the name for both Elf and Ferruginous Pygmy-Owls. His informants supplied no name for the Common Screech-Owl, however. My consultants supplied *kú:kul* (the same morpheme in a different orthography) for the Screech-Owl but did not know Elf or Ferruginous Pygmy-Owls or even recognize skins of these species. This confusing shift in name might be caused by (a) *kú:kul* being generic, including Elf, Ferruginous, and Screech without a qualifying term; or (b) *kú:kul* being some 'small owl,' the name being applied to the most prominent species of the area—Ferruginous of Gilman's day at Sacaton and Common Screech presently among Pima villages.

Historic: F. W. Bennett collected a set of four eggs (ARIZ) at Phoenix 3 April 1897. Breninger (1898:128) wrote of the Ferruginous Pygmy-Owl in the general area:

> Among the growth of cottonwood that fringe the Gila and Salt rivers of Arizona this owl is of common occurrence. Nidification in this valley usually takes place about the 20th of April. During the winter months as well as early spring the tawny coloring of the tail and upper parts is very prominent; but with the approach of summer, the entire plumage becomes worn and bleached. In more recent years, and since trees planted by man have become large enough to afford nesting sites for woodpeckers, this Owl has gradually worked its way from the natural growth of timber bordering the rivers to that bordering the banks of irrigating canals until now it can be found in places ten miles from the rivers. I have never known it to use holes in giant cacti as does the little Elf Owl.

Beninger collected at least one specimen (MNA 1300).

Gilman (MS) found the Ferruginous Pygmy-Owl "Fairly common, nesting in deserted nest holes of Gila Woodpeckers, in cottonwoods and willow trees. Have seen the bird in spring, summer, and early fall, but have no record [*sic*] during winter months. Fly about during daytime to some extent." He gives additional notes in his 1909b paper. The lack of winter records was apparently a lapsus on Gilman's part. In 1908 he collected two females (orig. MFG 133 and 152, now SBC) at Blackwater, the first on 25 January, the other on 28 March. The first is in rich, unworn plumage, appropriate for that season with no indication of having been caged. (And Gilman [1909b] specifically mentioned releasing the one Ferruginous he had captive.)

Modern: I have never found this species on the reservation nor did any of my consultants from the western villages even recognize skins of this distinctively colored owl.

Taxonomy: The reservation specimens are the pale *G. b. cactorum* van Rossem, the northwestern extreme of a widespread Neotropical species.

Change: The evident spread of the Ferruginous Pygmy-Owl witnessed by Breninger proved only temporary, as it is no longer found in the Phoenix region. Such a diurnal owl is not likely to be overlooked. The cause of extirpation is enigmatic. In part it is undoubtedly related to the destruction of good gallery forest. But it has not been found during the past twenty-five years (Johnson and Simpson, personal communication) in still wooded portions of the Salt River Valley, as at Blue Point Cottonwoods. And the owl should inhabit saguaro stands (*fide* Phillips et al., 1964; S. M. Russell, personal communication), as well as riparian woodland.

Elf Owl, *Micrathene whitneyi* (J. Cooper)

Ethnographic: See notes on preceding species.

Historic: Breninger (1901) apparently did not work the reservation at night, as he records no owls except the conspicuous and largely diurnal Burrowing Owl. Gilman (MS) found the Elf Owl "Rather common though not often seen as they move about very little during daytime.... None were seen during the winter months, but owing to their nocturnal habits they might be overlooked." He related additional life history in the longer paper (1909b) and took a male and female specimen. He found the bird both in cottonwood and saguaro. The Elf Owl is now known to be migratory (Phillips, 1942), as Gilman suspected.

Modern: Rare transient (?). I searched saguaro stands since my arrival on the reservation without finding evidence of this species. Even using tape recordings of Elf Owls on moonlight nights along the Sierra Estrella bajada (as 14 May 1971 from Santa Cruz village to Hidden Valley) produced negative results. During the day 11 March 1975 I flushed one individual four times in Hummingbird Canyon, west of Santa Cruz village. This is my only observation and I suspect the bird was a transient.

Change: I am unable to account for the apparent disappearance of Elf Owls from the reservation. Phillips et al. (1964:52) considered the species "Arizona's most abundant owl." Perhaps some saguaro stand not yet searched may reveal this owl breeding here in modern times.

Burrowing Owl, *Athene cunicularia* (Molina)

Ethnographic: *kókoho*

The name is onomatopoetic. This ground-nesting owl is well known to the Pima and Papago, but it is apparently not enriched with the mortuary elaborations of the Yuman culture, notwithstanding that pairs frequently live in Pima cemeteries. Gilman (1909b) recorded the Pima taxon as *kau-kau-ha* (same morpheme in slightly different orthography). I elicited the name as *kawkuhwa* from Delores Lewis of Gu Oidak Papago village, Arizona, and *kokawa* from Luciano Noriego of Quitovac Papago village, Sonora.

Historic: Scott (1886) reported a colony near Florence, but considered the species unusual for the area. Breninger (1901:45) recorded "One seen on the desert." Gilman (MS) found Burrowing Owls "resident in small numbers," nesting in badger holes. He mentioned seeing only four birds.

Modern: Common in specialized habitats. Burrowing Owls inhabit open areas affording good visibility and holes, especially badger burrows. Fields left fallow a few years, undisturbed sections of fence rows, and cemeteries are favored, particularly in *Atriplex* or *Ambrosia* communities. The only approximation to colonial nesting I have found was two nests 130 to 135 feet (40 to 41 m) apart at Bapchule, 24 May 1968. The eggs had already hatched, and about one month later young were perched about the nest entrances. I believe the bulk of the population abandons the reservation in winter. I have made special efforts to locate the species at that time, usually without finding it. For instance on 21 October 1967 I searched their usual burrows about Sacaton, Santan, and Bapchule, but located none. My only "winter" records in thirteen years are two pairs seen 4 November 1972 at Vahki and West Casa Blanca, and possibly another seen (and a feather found) 27 January 1970 at Barehand Lane. Thus I do not consider it "probably permanently resident" (Phillips et al., 1964:53).

Taxonomy: The Burrowing Owl, found interruptedly in the grasslands and semi-deserts of the Western Hemisphere, is immediately derived from the Old World genus *Athene* Boie 1822. Eurasia and North America share six species and eight other genera of owls. Most owl genera have a long paleontological history (Brodkorb, 1971), but the Burrowing Owl is relatively recent, appearing first in the Upper Pliocene of Kansas as *Speotyto megalopeza* Ford, a form a bit larger than the modern species. The Burrowing Owl is similar to the three species of Old World *Athene* in behavior (social organization, activity period, hunting methods). They share a similar dorsal and ventral coloration pattern, external morphology (wing formula, cere structure, external ear opening morphology), and an arrangement of carpal tendons peculiar for strigid owls. The crania are virtually identical and the postcranial osteology differs principally in the hind limbs, which are lengthened in the Burrowing Owl. This difference I think is best recognized by maintaining *Speotyto* Gloger 1842 as a subgenus. The only genus that I have found close to this complex is *Micrathene*.

Long-eared Owl, *Asio otus* (Linnaeus)

Modern: Rare. Lucius Kyyitan and I first discovered this species one moonless evening in fields at Sacate 20 February 1968. The specimen (AMR 1973) was a female with light to moderate fat. Most subsequent records were for the fall and winter of 1972. I flushed one bird three times at Sweetwater 16 October in a tamarisk thicket. I found Long-eared Owl feathers (AMR 4217) at Komatke 5 November in another tamarisk grove and found more feathers at Maricopa Colony 23 December along a tamarisk row. These last two owls apparently were victims of a predator, and no bones were found. At Vahki 24 December 1972 I found another Long-eared Owl (AMR 4217 [male], skin and skeleton) that had become entangled some days before in two loose strands of barbed wire. My only summer observation of an apparent Long-eared Owl is one flushed four or five times in a tamarisk grove south of Sacaton 21 June 1970.

Taxonomy: Males are considerably darker than females, both dorsally and ventrally (AMR, SD, ARIZ: 30 skins). A very blackish winter bird recently taken in Oregon (SD, D. F. Payne) raised the question of the separability of a paler western race, *A. o. tuftsi* Godfrey, from the eastern *A. o. wilsonianus* (Lesson). Kenneth C. Parkes (personal communication) compared western birds of various museum ages (including the unusual Oregon skin) with topotypical Pennsylvanian material and was unable to substantiate a western race. He found a tendency for eastern birds to have paler, less orange-buff, legs and flanks.

Short-eared Owl, *Asio flammeus* (Pontoppidan)

Modern: Rare. I found one specimen (dried mummy) 26 or 28 November 1965 hanging on a fence along an alfalfa field on 91st Avenue just north of the Salt River.

It was apparently recent as dermestids had not yet attacked it. I picked up diagnostic feathers of this species along a canal bank at Santan 13 March 1971. Phillips et al. (1964:54) considered this owl "Generally rare winter visitant in open grasslands, marshes, swales and fields of southern and western Arizona."

Caprimulgidae

Nuttall's Poorwill, *Phalaenoptilus nuttallii* (Audubon)

Ethnographic: *kó:logam*

Matthias' grandmother said that *kó:logam* hibernated. This was based on finding apparently dead birds in winter but discovering that they were still alive, according to Matthias, but he never observed this himself. I do not know how widespread this information is, but the distinction between animals that hibernate and those that migrate is clearly conceptualized by the Pima. Nest behavior is described as similar to that of the nighthawk. Papago name the same.

Historic: Gilman (MS) recorded it as "Heard frequently and seen occasionally. Earliest date 22 March and latest 19 October." He took a specimen (MFG 306) 19 October 1913.

Modern: Common during hot months. My extreme dates, substantiated by specimens, are 5 April to 19 September, but no special effort has been made to find them at other times.

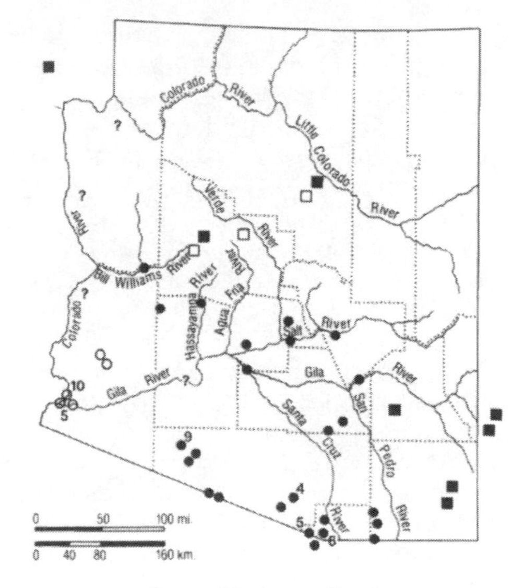

Poorwills (breeding)

■ = Phalaenoptilus nuttallii nuttallii
□ = blackish nuttallii? ● = P. n. adustus
○ = P. n. hueyi ? = breeding race unknown

Taxonomy: Phillips (1959:28) declined to recognize races of poorwills away from the Pacific coastal area, calling the variation pronounced, locally dominant color phases. "But alongside these birds occur, not infrequently, normal individuals, just like ordinary Poor-wills elsewhere. These are now called 'migrants,' yet some were taken in late May, after migration has probably ended (there are no helpful annotations on the labels)." In my opinion, this interpretation of polymorphism is not acceptable. First, the very pale pinkish birds of the lower Colorado, *P. n. hueyi* Dickey, have significantly shorter wings (eight males, nine females, $\bar{X}$ = 135.2, R = 129.6–137.8) than breeding topotypical *P. n. adustus* van Rossem, an intermediate race of eastern parts of Arizona lowlands (eight males, seven females, $\bar{X}$ = 142.8, R = 134.4–145.3). (Males average about 1 mm larger than females in a given population; therefore, the sexes are not segregated in these samples.) Nominate *nuttallii,* breeding in the Great Basin, the Great Plains, and the southwest uplands is even larger, as shown by fifteen birds from eastern Oregon (eleven males, four females, $\bar{X}$ = 143.7, R = 136.0–148.8). Dark, boldly marked birds (SD) taken within the range of *hueyi* are almost invariably larger, often considerably larger, than the pale local birds. Furthermore, though over half the paler lowland birds (SD) are marked as breeding or with enlarging gonads, not one of the larger, darker birds has such label data. I consider these latter to be migrant *P. n. nuttallii.* The large and very dusky coastal race *P. n. californicus* Ridgway also appears during spring and fall migration in areas where it does not breed (e.g., metropolitan San Diego area; Channel Islands, SD), further evidence that at least northern populations are migratory. Wintering specimens, particularly in Mexico, need to be taken to determine the extent of migration.

The variation shown by southwestern populations of poorwill (*hueyi, adustus, nuttallii*) is remarkably parallel to that found in *Otus asio (yumanensis, gilmani, aikeni),* with the smallest, pinkest, most finely marked populations of both being from the lower Colorado River Valley.

The systematic status of the middle Gila poorwills remains to be determined. The only reservation specimen in good plumage (AMR 601, female, 19 September 1964) is the color and pattern of *adustus,* but the small size of *hueyi* (wing 133.8). A very reddish juvenal female (31 July), with apparently all but the outer primary grown, measures only 129.3. Gilman's 19 October specimen (SBC) was unfortunately lost in the mails before it was critically determined.

Lesser Nighthawk, *Chordeiles acutipennis* (Hermann)

Ethnographic; *ñúput;* [pl.] *ñúñuput*

This species is well known to the Pima. Sylvester Matthias relates the following folk natural history. *Ñúñuput* nested in the saltbush-alkaline areas around Komatke. They moved their eggs around the bush to keep them in the shade from morning to afternoon. He never saw them actually move the eggs, but apparently they pushed the eggs with their breast. His grandfather told them this and Sylvester and his cousins had to find out for themselves. *Ñúñuput* were very annoyed at youngsters. The eggs were gray and spotted like a quail's and very hard to see on the alkali. The children would check them throughout the day, and they always moved around the same bush, in the shade, never in the sun; but the *ñúñuput* never made any attempt to hide them or to move them to an entirely new place. The same name is used by the Papago (*ñú:pud*) but not by the Pima Bajo, who call this bird *du:duar.*

Historic: Breninger (1901:45) noted "Several seen flying overhead" in September. Gilman (MS) found them a "Common summer visitant arriving about 24 March and seen as late as 12 October."

Modern: Common in summer. I have no satisfactory dates of arrival for this species. There is a specimen (CGR 41, male) from Casa Grande Ruins 25 April 1938. My late dates are 22 September 1973 at Casa Blanca and 28 October 1972 at Santan, when seven (including an adult male) were feeding over a Pima field. All my late records are far from any bright artificial lights, but over or near fields with flying moths.

Apodidae

[Vaux's Swift, *Chaetura vauxi* (Townsend)]

Historic: Breninger (1901:45) observed "a number seen in company with swallows," in mid or late September. Gilman (MS) considered this species "a rare migrant, identified with certainty but once, 12 May 1909." I have no further records for the species. Apparently Gilman collected no specimen.

White-throated Swift, *Aeronautes saxatalis* (Woodhouse)

Historic: Gilman (MS) found this swift only a "Winter visitant, a few [maximum of seven individuals] being seen the following dates: 27 December, 8 15, 31 January, 10, 20 February. These dates were not all the same year and therefore a few of the birds were seen each year sometime during the three months named."

Modern: Fairly common in winter and spring. The status of this species is puzzling, as I have no observations for July through mid-September, or for October and November. The single fall record (AMR 2364, adult female) was found alive 17 September 1968 at Komatke. In the Sierra Estrella west of Santa Cruz village I saw a few 9 January 1966. Swifts were rather abundant late in the afternoon 16 January 1966, when a male (AMR 913) was taken. At least five birds were repeatedly entering a fissure in a cliff here (nesting?) 7 and 8 April 1973. Two were copulating in the Sierra Estrella 5 May 1975. My lowland records from the northwestern part of the reservation include at least forty (probably over a hundred) on 14 December 1974, twenty-two on 26 December 1973, twelve on 7 February 1982, and at least twenty-eight on 23–24 June 1974. I saw three or four over the lower Salt in June 1977 and 1978.

Taxonomy: Behle (1973) found too much overlap in size to recognize formally the larger *A. s. sclateri* Rogers of the northern interior. The reservation specimens are both small (wing 136.7 and 136+ [ensheathed outer primary]).

Trochilidae

Hummingbirds (all species)

Ethnographic: *vípismal*

The initial morpheme is probably derived from *vi:pim,* 'to suck an object.'

Possibly the terminal morpheme *-mal* is derived from *-mat,* 'child of,' or diminutive, hence, 'little sucker.' But none of my consultants offered this etymology. I found slight variants of this lexeme throughout the northern Piman speaking groups: *ví:pismal* (Kohadtk), *wí:pismal* (Papago), *wí:pisumal* (Quitovac), *ví:psimal* (lowland Pima Bajo). In Northern Tepehuan the name is *vipíshi* (Pennington, 1969). The Pima distinguish no individual species of hummingbirds. *Ví:pismal* is one of the four birds that figure in the Piman creation myth.

Black-chinned Hummingbird, *Archilochus alexandri* (Bourcier and Molsant)

Historic: Breninger (1901:45) recorded "one seen attracted by the open flowers of the morning-glory," in September, but he did not mention the sex. Gilman (MS) did not record seeing this species.

Modern: Rare, except in the *Salix-Populus* community of the Salt River. I took an adult male (AMR 760) at Komatke 21 May 1965. In Hummingbird Canyon on the northeast slope of the Sierra Estrella another adult male was courting a female Costa's Hummingbird 4 March 1973, but no copulation took place. Two male Black-chinned Hummingbirds were present here 7 and 8 April 1973, but they were not courting. These are my only records for the species in dry habitats. I collected an immature female (AMR 2206) at Barehand Lane marsh 13 June 1968. Undoubtedly the species nested in the deciduous trees above the Salt-Gila confluence, though no nests were found. I observed courtship here 27 and 28 March 1976. An adult male and several female Black-chinned Hummingbirds were usually in the main grove during May and June, but most of the hummingbirds that seek the cool riparian shade here during the summer months are young of the year, including males (AMR, JSW) of this species and the next.

Change: I strongly suspect that Gilman's (MS) records for *A. costae* ("Summer visitant, a few found nesting") refer actually to this species. The deciduous trees of the riparian remnants of Gilman's time would more likely have attracted nesting Black-chinned, not the xerophilous Costa's.

Costa's Hummingbird, *Archilochus costae* (Bourcier)

Historic: Gilman (MS) recorded the species as a "summer visitant, a few found nesting." As mentioned in the previous account, I suspect these notes apply rather to *A. alexandri.* Gilman apparently took no substantiating specimen, nor are the diagnostic males mentioned.

Modern: Locally common to abundant, winter and spring. In the Sierra Estrella, west of Santa Cruz village, is a deep arroyo where I first discovered Costa's in numbers January 1965. The uppermost section of the arroyo supports abundant chuparosa, *Justicia* (*=Beloperone*) *californica,* and desert lavender, *Hyptis emoryi,* the two primary local plant foods for this bird. There is considerable variation from year to year in arrival and breeding dates. Adult males arrive sometimes as early as 4 December (8 December 1966 the earliest specimen date: AMR 1265) or as late as 29 January. The bulk of the adult males depart by the end of April, though I observed several exceptionally late males 5 May 1975. Nuptial flights begin as early as 17 December and may continue into early January, apparently with no females present. Females return as early as 8 December (AMR 1266) or as late as 21 January. Either males or females may arrive first. The greatest numbers of both sexes and the most intense courtship are during February. I have found females incubating by 11 February and a nest still under construction (with one egg) as late as 8 April. The earliest I have found young was 12 March. By 16 April 1967 two nests I had under observation had already fledged young, and the young from a third nest were out on a nearby limb begging. I have found no Costa's in this canyon from the end of April to the end of November. But beginning about mid-April and continuing through early May, worn adult females (AMR 4731), young, and occasionally an adult male may be found out on the Sierra Estrella bajada feeding.

By June, birds of the year (AMR, ARIZ, JSW) appear at the Salt-Gila confluence. The latest date for young Costa's in this riparian growth is 23 June (JSW, AMR 4614). We have found an adult female there only once (ARIZ 12184, 19 June 1976).

These reservation dates (both arrivals and nesting) are considerably earlier than known dates for the rest of Arizona. Phillips et al. (1964:62) summarized, "Males migrate commonly through the blossoming ocotillos in late February and March; by the end of May they have virtually disappeared from the deserts." Substantiated arrival dates for other parts of Arizona are 25 January (Tucson and Bill Williams delta) and 9 February 1854 in the Big Sandy Valley. Though Costa's Hummingbirds nest early in Baja California, the earliest egg date reported by Bent (1940) is 24 February. A circumstantial account (attributed to this species but unsupported by specimen evidence) of a nest with young 3 February 1962 was given by Bakus (1962) from eastern San Diego County, California. Grinnell (1904) found males at Palm Springs, California in mid-winter (25 December 1903 – 2 January 1904).

Change: I doubt that there has been any real change in the status of this species. Rather, this particular arroyo in the Sierra Estrella presents a favorable set of circumstances for early reproductive activities.

Anna's Hummingbird, *Archilochus anna* (Lesson)

Historic: Gilman (MS, 1914b) collected an immature male 4 September 1910. I have not relocated this specimen.

Modern: Rare. I have but three observations of this species, all adult males: 26 November 1972 near Barehand Lane (in *Prosopis-Atriplex-Lycium* community), from 25 October to 2 November 1975 at the Salt-Gila confluence (singing vigorously in *Nicotiana glauca*), and 2 April 1978 on the Salt at 91st Avenue (also in tree tobacco).

At least adult male Anna's Hummingbirds were wintering fairly commonly at tree tobacco thickets along the Gila between The Buttes and Cochran in December 1981, but I found none the following summer.

Change: The species now breeds locally in the Phoenix urban area (Salome R. Demaree, personal communication; winter nests examined, ARIZ, but I have not found it breeding on the reservation. Zimmerman (1973) discussed range expansion in this Pacific coastal species.

Rufous Hummingbird, *Selasphorus rufus* (Gmelin)

Historic: Breninger (1901) reported seeing the species in September. Gilman (1914b) secured two immature males 4 September and 1 October 1910.

Modern: Rare spring, uncommon fall transient. I obtained an adult female and an immature female (AMR 2267, 3126) at Barehand Lane 31 July and 1 September 1968. Phillips (personal communication) observed an adult male at Sacaton 9 September 1947. Additional late summer individuals of this genus appear at the Salt-Gila confluence, but field identification to species is unreliable. Adult males, rather certainly this species, have appeared 9 and 26 March and 7 – 8 April. On this last date four or five males were present in the Sierra Estrella canyon used by nesting Costa's Hummingbirds.

Alcedinidae

Belted Kingfisher, *Megaceryle alcyon* (Linnaeus)

Ethnographic: *báifchul;* [pl.] *bábaifchul*

The choral ensemble from Bapchule village is the "Kingfisher Singers." Matthias related that kingfishers nested in bluffs of the Gila River near Komatke. He noted the peculiar head shape and crest. The nests were in holes in banks about 6 feet (2 m) above ground or higher. Several pairs nested in a single cliff. Matthias and other children used to poke sticks in the holes to make the parents fly out. The kingfishers fed on "trout," suckers, catfish, carp, and "sardines." Food was taken by diving. The kingfishers were last seen there in 1928. None of my consultants recognized the Green Kingfisher, *Chloroceryle americana* (Gmelin), but all knew the Belted. The Northern Tepehuan call the Belted Kingfisher *baivukali* (Pennington, 1969; accent not indicated).

Historic: In the early 1880s, W. E. D. Scott collected on the Gila from Florence to Riverside (near Kelvin) and south along the San Pedro River. Scott (1886) considered the kingfisher: "A resident species. Met with at the several points where I collected, but it retires from the mountains in the winter. It is a curious fact that the species is frequently to be found in this region far from water, feeding on the larger insects and lizards. It always seemed strange to meet the bird under 'desert' conditions." Breninger (1901:45) merely mentioned it 'seen about flooded willow ground" in September, without further details. Gilman (MS) saw the kingfisher occasionally from 17 September to 15 April "Along the 'Little River,' which contains a few small fish besides frogs and water insects."

Modern: Rare transient, wintering locally. I saw the species seven times (April, August, September, December, January) in thirteen years. One specimen (AMR 1455) was obtained at Komatke on 18 April. At least one kingfisher usually wintered along the lower Salt, where small exotic fish were abundant until 1980.

Taxonomy: Many recent authors have merged *Megaceryle* Kaup, 1848 in *Ceryle* Boie 1824. However, *C. rudis,* the type species, differs rather considerably in cranial characters (articular at mandible posterior, pterygoid shaft) as well as postcranial characters (coracoid, furcula, bicipital crest of humerus, condyles and tendinal bridge of tibiotarsus) from the two species I compared (*M. torquata* and *M. alcyon*). In most characters where *C. rudis* differs, it agrees instead with states found in *Chloroceryle aenea, C. inda,* and

C. americana. Therefore I am maintaining the genus *Megaceryle.* (A logical and perhaps equally justifiable course would be to lump both *Megaceryle* and *Chloroceryle* Kaup, 1848 in the intermediate *Ceryle.*)

No subspecies of *M. alcyon* are recognized, following Phillips (1962).

Change: The Belted Kingfisher has been extirpated from the reservation as well as the rest of the state (Phillips et al., 1964) as a breeding species. Phillips (1968:139) believed "by the 1930s it certainly did not breed in southern Arizona." The Pima ethnographic account was the only specific location given for kingfisher breeding in Arizona until Steven D. Emslie discovered a nesting attempt in 1980 on the Verde River (dead young salvaged, NAU). Kingfishers wintered on the lower Salt when food was available before floods scoured the channel.

Picidae

Northern Flicker, *Colaptes auratus* (Linnaeus)

Archaeological: One specimen (tarsometatarsus) from Snaketown is in the size range of the small local subspecies *mearnsi.*

Ethnographic: *kúdat*

The Pima lexeme *kúdat* includes both the Red-shafted and Gilded Flickers without distinction. The recognition of their conspecificity is a taxonomic refinement officially adopted only in 1973 (A.O.U. Committee on Classification and Nomenclature).

The Papago name is also *kúdat.* Lowland Pima Bajo call the Gilded Flicker *dú:dga:t.* Pennington (1969:116) recorded a spotted white breasted woodpecker the Northern Tepehuan called *vïpïgi kúratu,* 'red woodpecker.' This is surely the Red-shafted Flicker. These are all cognates for local forms of the same biological species.

Historic: Breninger (1901) reported seeing a number of Gilded Flickers but no Red-shafted Flickers in the latter part of September. Gilman (1915b:159–60) found the Gilded Flicker "abundant throughout this region...found in cottonwood and willow groves as well as wherever giant cactus grows." He found the height of nesting in April (nineteen of twenty-seven occupied nests) and the latest eggs on 17 May. The Red-shafted Flicker he found "the first week in September, the earliest date I have recorded being 4 September. Most of the birds leave for their breeding grounds about the first of April, though I have recorded them as late as 15 April."

Modern: The Gilded Flicker is common throughout the year. There is no evidence for winter movements in this subspecies except for fall-winter congregations in fields with abundant food. At this season the Gilded Flicker is greatly outnumbered by the red montane forms. Red-shafted Flickers are abundant in winter. Arrival dates supported by specimens (AMR) for five years are 22 (twice), 26, 28, and 30 September. Latest spring dates (specimens) for five years are 10 (twice), 12 (twice), and 14 March. The Yellow-shafted Flicker is casual. There is a sight record (5–13 December 1970, Johnson et al., MS) and a specimen (AMR).

Taxonomy: Red-shafted Flickers breeding throughout the humid Pacific coast area are brown-naped, concolor with the back. There is a cline of increasing darkness of dorsum from south to north. Those from southern Alaska to about the northwest corner of California are *C. a. cafer* (Gmelin). Less sooty brown birds breed along the Coast Range to northern Baja California. These are *C. a. collaris* Vigors. The type of *collaris* no longer exists, but the illustration of the type shows a brown-naped bird, as are premigration (late August through early September) topotypes (AMR) from Monterey, California.

Red-shafted Flickers breeding in the interior West (from the east slope of the Cascades and the high Sierra Nevadas) are pale and grayish on the crown, nape, and back. The brown is restricted to a narrow frontal area extending back above the eye as a superciliary. These are *C. a. canescens* Brodkorb, described from southern Idaho (paratypes examined). The holotype of *C. a. martirensis* Grinnell is a typical example of *collaris,* though most of the type series are migrant *canescens,* and the formal description applies to this interior form.

Canescens accounts for the bulk of the migrants into the lowlands of the West. Even within the restricted breeding range of *collaris,* the interior gray bird is the more common wintering bird.

The abundant flicker wintering on the reservation is *canescens.* Outside specimen dates are: 22 September–14 March (forty-seven specimens). Coastal *collaris* is an uncommon winter visitant to south-central Arizona. I refer twelve reservation specimens (AMR) to this race. Outside dates are: 24 September–12 March. *Collaris* is more common in winter in far western Arizona and southern California (AMR, LAM MVZ, SD, UCLA). No reservation specimen is truly representative of dark brown *cafer,* though several approach it (e.g., SD 40833, 17 October 1979).

The Yellow-shafted Flicker is casual in winter, with one specimen (AMR 4533, female, 13 October 1973) taken at the Chandler boundary cottonwoods. It measures: wing chord 169.5, tail 117.6. There is a cline of decreasing size from the northwestern part of the continent to the southeast. The larger birds are *C. a. luteus* Bangs, from which *C. a. borealis* Ridgway is not satisfactorily separable (Parkes, personal communication; Phillips et al. 1964; Short, 1965; Rand, 1944). All Arizona specimens are *luteus.*

The Gilded Flicker, *Colaptes a. mearnsi* Ridgway, breeds commonly throughout the saguaro stands on the reservation.

Observations of "hybrids" (= secondary intergrades between major subspecies groups) require specimen verification. Red-shafted Flickers may have yellow flight feather linings or red nuchal patch (e.g., AMR 3954, Otero County, New Mexico, June 1972).

Change: The only change in the species on the reservation seems to be a several-week earlier departure of the montane form since Gilman's time. In thirteen years I have never found any lingering to early April, when Gilman thought most of them departed. His arrival dates are so exceptionally early for the West in general that they would require substantiation. I suspect they were *mearnsi* intergrades.

Gila Woodpecker, *Melanerpes uropygialis* Baird

Ethnographic: *híkvik*

This species is well known to the Pima and abundant in virtually every yard. It figures in the creation story. This name might be used distributively for other woodpeckers not formally in the Pima lexicon, as for the Yellow-breasted Sapsucker and Acorn Woodpecker. The Piman choral group from Komatke is known as the "Woodpecker Singers." The Papago name may be either *híkiuwij, híkwik,* or *híkvik.* The lowland Pima Bajo use *hí:kvik.*

Historic: Breninger (1901:45) recorded it as "Common; often seen perched on the houses of the Indians." Gilman (1915b) found the height of the nesting season from middle April to mid-May and half-grown young 10 July, indicating a second brood. Most nests were in saguaros, though decayed willow and cottonwood were also used.

Modern: Common. Concentrates in fall and winter in cornfields.

Taxonomy: I do not recognize *M. u. albescens* (van Rossem) of the lower Colorado Valley, though an occasional individual does fit the description assigned to this population (type and type series examined, UCLA). I have collected a series of fall topotypes from Alamo Crossing on the Bill Williams River, type locality of the nominate race. Most of these have distinctly deeper, brighter yellow abdomens than birds from the remainder of the state, but this character appears occasionally in birds from other localities (Komatke, Tucson; AMR, ARIZ). Therefore I include all Gila Woodpeckers of Arizona and southern California in the nominate race.

Acorn Woodpecker, *Melanerpes formicivorous* (Swainson)

Historic: Gilman (MS) saw this species only three times: 22 May 1908, 5 September 1910 (specimen collected), and 7 December 1914.

Modern: Rare winter visitant. I have seen this species on the reservation only twice: 27 November 1964, at Komatke (AMR 640, adult female) and [28 December 1977, mummified male] on the Salt.

Taxonomy: Both specimens are the thin-billed Arizona race *M. f. aculeata* Mearns, separable from the nominate race of Mexico (see Phillips et al., 1964; and Oberholser, 1974). I have been unable to relocate Gilman's specimen for racial identification.

[Lewis' Woodpecker, *Asyndesmus lewis* (Gray)]

Historic: Breninger (1901:45) noted: "Several seen; first time I've seen this species in this valley." Gilman's (1915b, MS) observations were singles seen 6 October and 13 November 1910, at Sacaton.

Modern: Rare. I saw two Lewis' Woodpeckers: 29 September (Komatke) and 16 October (Chandler boundary cottonwoods), both 1972. These birds were in the tops of tall trees in agricultural areas.

Taxonomy: The melanerpine woodpeckers *Melanerpes, Centurus,* and *Asyndesmus* are exceedingly close osteologically and perhaps might all best be merged into the oldest genus, *Melanerpes,* as was done by Mayr and Short (1970). However, *Asyndesmus* has structural feather characteristics and a distinctive corvidlike flight pattern not found to my knowledge in any of the *Melanerpes-"Tripsurus"-Centurus* group, and I believe it has the best claim to separate generic status of this complex.

Yellow-bellied Sapsucker, *Sphyrapicus varius* (Linnaeus)

Historic: Breninger (1901:45) saw a Red-naped Sapsucker "busily pecking into the trunk of a willow" on either 18–19 or 25–26 September. Gilman (1915b:152; MS) found the Red-naped form from 6 October to 17 April "in cottonwoods, willow, and occasionally Arizona ash." Gilman (1915b) recorded the coastal Red-breasted form from 5 October 1910 (specimen lost?), 5 October 1914, and 9 February 1910.

Modern: Common in winter. My specimens of the Red-naped Sapsucker span from 27 September to 12 March, but the September date is exceptionally early. The species becomes common about a month later. I found this woodpecker most frequently in large tamarisk trees, *Tamarix aphylla,* which the Pima formerly planted about their homes (Fig. 3.9). I have no modern record for the Red-breasted group.

Taxonomy: All my specimens (n = 10) and two of Gilman's (Blackwater 1907 and 1909) are referable to *S. v. nuchalis* Baird, the Rocky Mountain or Red-naped form that breeds in Arizona and winters abundantly over the lowlands of the state. One of Gilman's specimens (MFG 117, SBC) superficially appears to lack the red nuchal crest but the nape skin is tucked inside the occiput. I have failed to find the nominate race (more extensively white, less red, and distinguishable in the field), though it occurs in winter in Arizona (Phillips et al., 1964). Of Gilman's several Red-breasted Sapsuckers, I could locate only the 9 February 1919 specimen (MFG 233, SBC) from Sacaton. I compared this with both coastal races (AMR, ARIZ). The colors agree well with the more southern *S. v. daggetti* Grinnell, particularly in the paler, less sooty-washed flanks. There is one other state specimen of this race recognized by Phillips et al. (1964).

Williamson's Sapsucker, *Sphyrapicus thyroideus* (Cassin)

Modern: Rare winter visitant. An adult male (AMR 3081) taken 26 November 1970 at Komatke.

Taxonomy: The specimen is the small-billed *S. t. nataliae* (Malherbe) of the interior, the only race thus far recorded from the state (Phillips et al., 1964). The total culmen measures 25.5.

Ladder-backed Woodpecker, *Dendrocopos scalaris* (Wagler)

Ethnographic: *chúhugam* (presumably from *chú'u,* 'dark,' 'darkness'; *-gam,* attributive or abstractive)

This confiding and distinctive little woodpecker was identified by both David Brown and Joseph Giff. The ethnotaxon is not local. Luciano Noriego, a Papago from Quitovac village, Sonora, also identified the bird as *chúhugam,* noting that it is different from *hikvik,* 'Gila Woodpecker.' Pedro Estrella, lowland Pima Bajo, called it *ú'us túpuidum (ú'us, 'palos,'* 'sticks,' 'branches'). The Tepecano word for darkness is *tutukam* (Mason, 1917).

Historic: Breninger (1901:45) noted it "among the cottonwoods along the river." Gilman's (1915b:151) account of the reservation woodpeckers says, "The Cactus Woodpecker [a former name] may be seen in limited numbers at all times of the year. It is seemingly at home in any location, in the open country working on the various species of cactus (*Opuntia*); in dense mesquite and screw-bean thickets; or in cottonwood and willow groves." He found nests in mesquite, screwbean, ironwood, cottonwood, willow, paloverde, and cholla cactus.

Modern: Common. My observations of the species closely parallel those of Gilman's. It is at home in the mountains, bajadas, floodplains, and riparian woods. At the confluence grove I found the first nest holes in tall willows in April 1978.

Taxonomy: The merger of *Dendrocopos* Koch 1816 into *Picoides* Lacépède 1799 suggested by Delacour (1951) has been gaining recent popularity. Ouellet's (1977) study of this complex is based almost exclusively on color patterns rather than structure and anatomy. While the loss of a toe, by itself, may not be of great evolutionary importance (see Delacour, 1951), the very different bill and cranial structure readily sets the two groups apart. The posterior horns (epibranchial) of the hybrid in *Picoides* reach only to about the midline of the eye rather than coming forward and inserting strongly into the right nostril, as in all *Dendrocopos* species I have seen. Since feet and skull are not parts of a single functional complex, I maintain both genera (see Ridgway, 1914:194, 195, 290, for generic diagnoses).

I concur with Phillips et al. (1964) and Short (1968) in considering *D. s. yumanensis* van Rossem a synonym of *D. s. cactophilus* (Oberholser); some reservation specimens are as extensively white dorsally as birds from the lower Colorado River Valley (AMR).

Tyrannidae

Western Kingbird, *Tyrannus verticalis* Say

Ethnographic: *chíkukmal*

This is a species well known to the Pima, who report it perching on wires and nesting in cottonwoods. Its fondness for butterflies, *hawhakĭmal,* is also noted. The Kingbird's name would appear to be derived from 'butterfly.'

Historic: Breninger (1901:45) noted "A few belated migrants" in September. Gilman (MS) found the species arriving from 20 March to 1 April and observed it as late as 16 September. In June he counted fifty active nests along 9 miles (14.5 km) of newly installed high power poles.

Modern: Common in summer. My specimen records extend from 30 March to 14 October. The bulk of the birds arrive in breeding areas in mid-April and leave in early October. The 30 March specimen (AMR 2058) was taken in the saguaro-paloverde bajada community and no doubt still had far to go as it was very fat. Kingbirds nest at Pima ranches and in large fields where a few cottonwoods are allowed to grow along canal banks. I have observed nesting from mid-May through 31 July. During the summer of 1976 I found but few kingbirds in the leased farms about Co-op Colony and none at all at the Chandler boundary cottonwoods, places where they nested abundantly in previous years.

Cassin's Kingbird, *Tyrannus vociferans* Swainson

Historic: Breninger's (1901:45) September observation of this kingbird as "more numerous than the foregoing species [*T. verticalis*]" is almost certainly an error. Western Kingbirds molting their diagnostic outer pair of rectrices must have been considered Cassin's. Gilman (MS) recorded Cassin's as "seen during spring and fall." His specific dates are: 30–31 March and 20 September to 31

October, apparently all in 1914 at Sacaton. Gilman's March observation is considered one of the three reliable lowland records for the state (*fide* Phillips et al., 1964).

Modern: Rare. I have never seen the Cassin's Kingbird on the reservation. Dickerman collected an adult male (CU) 20 September 1953 on the Salt River just above the Salt-Gila confluence. The fall records by Gilman and Dickerman correspond to what is known for fall migration in the rest of Arizona.

Brown-crested (Wied's Crested) Flycatcher, *Myiarchus tyrannulus* (Müller)

Historic: Gilman (MS) found the species a "summer visitant, not very numerous." He discovered five nests between 10 June and 10 July.

Modern: Rare summer visitant. My only definite records for this species are two mummified specimens. The first (AMR 4846, adult) was found in a horizontal iron pipe at Komatke by Leonard Miller in 1966 or 1967. Also in this pipe was a dried *M. cinerascens.* The pipes were stored on a rack alongside a shop at St. John's Indian School. The second specimen (AMR 4857, immature) was found [13 September 1976] hanging in a young willow thicket at the Salt-Gila confluence. On occasion I have seen a *Myiarchus* in late May or early June that I suspected might be a transient Wied's but secured no specimen to determine the identity.

Change: Gilman's records no doubt were from the remnants of mature riparian trees (cottonwood, willow, ash) affording suitable shady nesting locations for this large, hole-nesting species. Here and in the Salt River Valley (Johnson et al., MS) it eschews the drier areas with apparently suitable saguaro stands and abundant mature ironwood trees, although it nests in such habitats in other parts of Arizona. Crested Flycatchers still breed upstream where the Gila is a living stream (see p. 73).

Ash-throated Flycatcher, *Myiarchus cinerascens* (Lawrence)

Historic: Gilman (MS) called the Ash-throated Flycatcher a "common summer visitant, sometimes seen in winter (16 January)." His nest records were all for May.

Modern: Common breeding in summer, uncommon and local in winter. A pair was nesting by 11 April at Santa Cruz. I have taken juvenals (AMR 2186, 1122) from 25 May to 16 September at Komatke, indicating an extraordinary span for nesting, if the latter were local. Family groups gather at cooler locations (as at the Salt-Gila confluence) during the summer months. Desert areas are sometimes partially to entirely vacated from early June through August. But from 8 to 9 July 1973, when saguaro fruit were ripe in the Sierra Estrella bajada from Santa Cruz to Hidden Valley, pairs of Ash-throated Flycatchers were still common. The previous winter had been exceptionally wet (10.7 inches [272 mm] of rain from October to March). This species winters, sometimes in numbers, in mesquite bosques or along arroyos with large ironwoods (AMR specimens, November through February).

Eastern Phoebe, *Sayornis phoebe* (Latham)

Modern: Casual. I collected an immature male (orig. AMR, now MVZ 150789) at Komatke 19 October 1963. I took another male (SD 41248) on the Salt below Maricopa Colony in the remnants of willow thickets 27 December 1980. The species is considered a "rare fall transient and winter visitor in southern Arizona" (Phillips et al., 1964:83). There are eleven previous Arizona specimens.

Black Phoebe, *Sayornis nigricans* (Swainson)

Historic: Breninger (1901:45) noted "a few seen near water" in September. Gilman (MS) considered the species "resident in small numbers" but discovered only one nest with four eggs under a bridge across the "Little River" 15 May. Monson and Phillips (personal communication) observed one south of Olberg 17 June 1939 and two along a ditch northwest of Sacaton 26 July 1947.

Modern: Uncommon except along the lower Salt River, where fairly common in nonbreeding season. I have only three specimen records of the Black Phoebe away from the rivers, all females from Santa Cruz or Komatke villages: 25 or 26 September 1966 (found dead, entangled in low, sticky *Boerhaavia* plants), 15 February 1964, and 3 March 1966 (one of two or three present). At the diversion dam at Olberg bridge, I saw one adult 12 May 1968 and one immature 23 July 1974. There is an old nest high up in one of the concrete rooms, and the species may nest here. From Barehand Lane I have but three observations: 22 October 1968, 26 November 1972, and 27 December 1980. The species (one or rarely two individuals) occurs most commonly along the lower Salt, with observations (Johnson, Simpson, Rea) from July through February but not for the breeding season (April–June). We have found no young, and we doubt that there are any locations here suitable for nesting.

Change: The absence of the Black Phoebe during the breeding season is directly related to the loss of water near suitable nest sites on the reservation.

Say's Phoebe, *Sayornis saya* (Bonaparte)

Ethnographic: *hé:vel maws* (*hé:vel*, 'the Wind's [female],' *maws* [or] *mos*, 'grandchild' [either sex]; literally, a woman's daughter's child; hence, 'grandchild of the Wind.')

David and Rosita Brown had 'the Wind's grandchild' nesting annually in the bell tower of St. Catherine's Church, Santa Cruz, and were quite familiar with the life history of this confiding flycatcher. The kinship term *maws* or *mos* indicates that the wind, personified, is a woman. Anyone familiar with the graceful, powerful flight of the Say's Phoebe will appreciate the poetic beauty of its Pima name. I long assumed that such a complex ethnotaxon would be of only local usage. However, at Quitovac, Sonora, in May 1982, I asked Papago Luciano Noriego what was the bird nesting in the hole in his house. He responded, *he:wui maws, 'abuela de viento.'*"

Rosita Brown suspected that *he:vel mawsmat* was the ***komalk** mawkum,* or 'flat headed bird,' of legend that was stepped on during the wine feast.

Historic: In September Breninger (1901:45) recorded the Say's Phoebe "occasionally seen about the houses." Gilman (MS) found it a "common summer visitant and on occasion seen in winter months." The only winter date he gave is 15 January. He found nests in April (sets of four eggs) in an old dry well and on the crossbeam of an abandoned adobe house.

Modern: Common in winter, departing after breeding. My specimens extend from 23 October to 23 April and sight records from 30 August (transient?) and 10 September to 31 May and 4 June. Say's Phoebes abandon the reservation soon after breeding in spring. My late dates are 31 May 1966, 26 May 1968, and 19 May 1972. Fall arrival dates are more conspicuous: 4 October 1963, 30 August 1964, 10 September 1966, 29 September 1972, 22 September 1973, and 28 September 1975. Outside these dates I have but three observations of postbreeding vagrants: a juvenal (AMR 2178) 6 June 1968; an adult from 8 to 18 June 1969, both at Komatke; and an immature at Casa Blanca 23 July 1974.

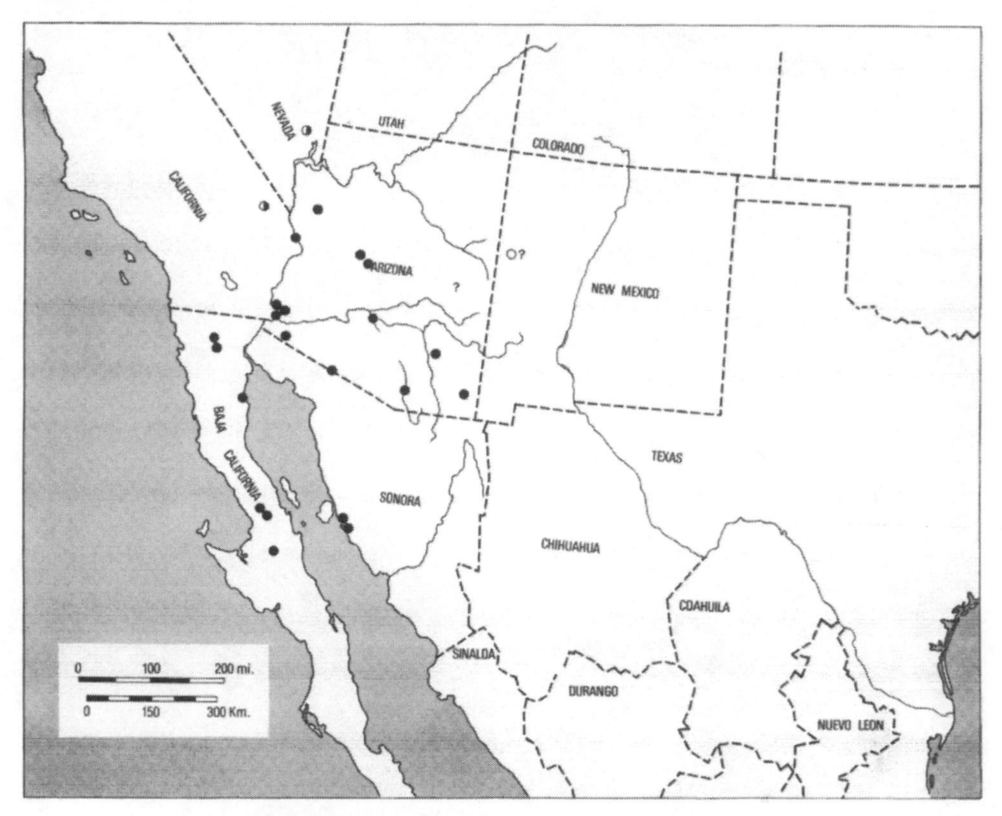

Say's Phoebes (breeding adults and juvenals)

● = Sayornis saya quiescens ○ = S.s. saya ◑ = apparent intermediate

At least sometimes on the middle Gila Say's Phoebes are double brooded. In 1977 David Brown found five eggs in the Santa Cruz nest. Rosita commented that she thought there were supposed to be only four eggs in a clutch. All five young fledged and three more eggs were laid in the same nest. The young were about a week from fledging when I saw them 22 May.

Taxonomy: Three races have been described north of the southern central plateau of Mexico: *S. s. saya* (Bonaparte) of the Rocky Mountains, *S. s. yukonensis* Bishop 1900 of the Yukon Valley, and *S. s. quiescens* Grinnell 1926 of Baja California. The Say's Phoebe presents virtually every taxonomic problem imaginable. There are postmortem color changes (foxing), so that material collected in the past two decades is not comparable with the bulk of museum material, of much greater vintage. The species favors exposed perches in arid habitats so that by late February most specimens are hopelessly faded. Postbreeding populations vacate desert areas while still in abraded plumage. Wintering populations in fresh plumage may be composed of local birds mixed with northern migrants. This leaves early season juvenals as the best starting point for working out geographic variation in the species.

Juvenal specimens from Alaska, eastern Oregon east at least to Denver and south apparently to Zuni, New Mexico,

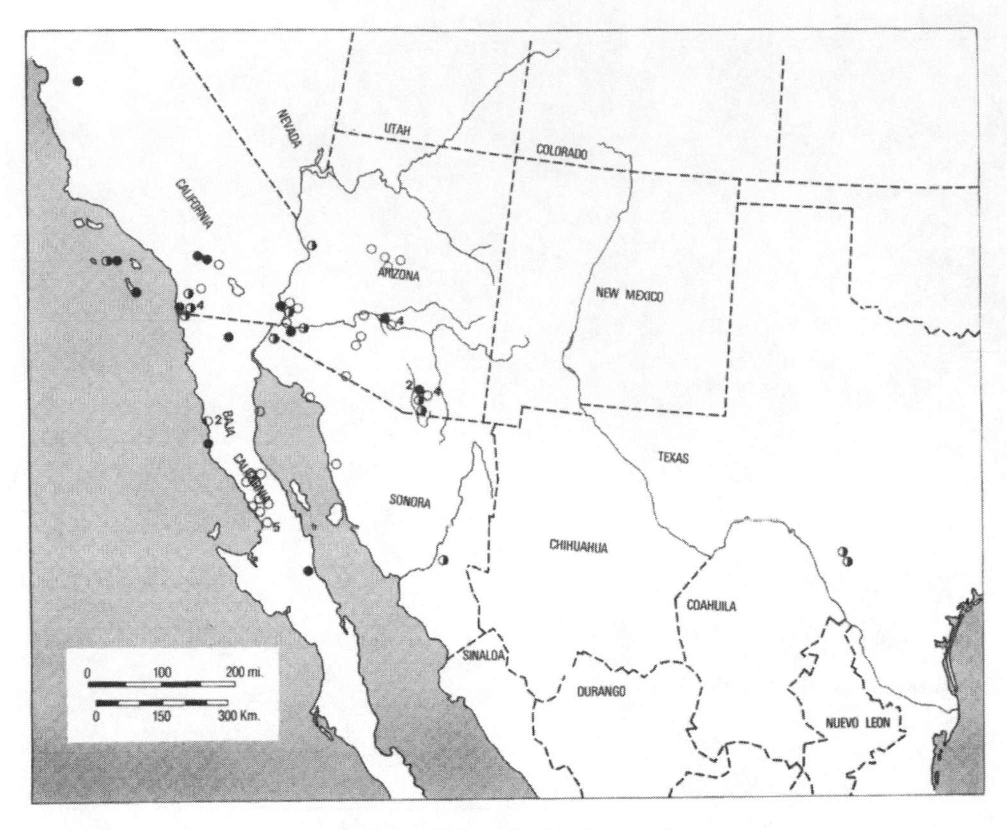

Say's Phoebes (wintering)

● = Sayornis saya saya ○ = S.s. quiescens ◑ = S.s. saya x quiescens

and perhaps the higher elevations of northern Arizona are darker, with distinctly darker crowns and broad, dark gray chest bands that invade most or all the throat. Overall paler birds with nearly concolor crowns and backs and pale gray, narrower chest bands, and whitish throats and chins are from Baja California and at least all the Lower Sonoran portions of Arizona (AMR, ARIZ, DM, SD, UCLA). A fully grown juvenal (already wandering?) taken at Komatke 6 June 1968 is paler on the chest than another taken three days later at Zuni, New Mexico. Adults can be segregated readily into dark and light groups. The pale birds in fall and winter occupy all the southern lowlands (Baja California and at least coastal Sonora north to the Salton Sea), all along the Gila at least to Komatke and north to Camp Verde and Prescott. These are *quiescens.* A few early spring Arizona birds, fortunately still in good plumage and with labels indicating gonads enlarging or already in breeding condition, are definitely *quiescens* (e.g., 2 April at Aravaipa Creek, Pinal County; 27 February near Topock, Mohave County; and a pair taken 3 March at Topawa, Pima County). These four are exceptional, as most individuals by the breeding time are taxonomically useless.

Some dark northern birds occur in winter in the lowlands. Arrival dates (specimens, SD, ARIZ) appear to be: 17 September, Yuma, Arizona; 27 September, El Rosario, Baja California; 9 October, Santa Barbara Island, California (where the species does not breed); 17 October, Laguna Hanson, and 18 October, Santa Rosalía, Baja California; 22 October, Tucson. There is one dark bird from Komatke (4 December 1964, AMR 645). The taxonomy of these dark birds has not been determined satisfactorily. Browning (1976) considered *yukonensis* a synonym of nominate *saya*. This was based on birds he considered to be breeding, without taking into account wear. Juvenals are not discussed even though nine of the eleven specimens in the type series of *yukonensis* are juvenals (Bishop, 1900). I suspect that Bishop may have had more than just foxing involved when he described the nominate Colorado birds as more "scorched." Certainly a broader geographic representation of juvenals, especially unfaded fledglings, will need to be collected to determine how many dark races should be recognized.

One final comment. Bishop (1900), van Rossem (1945:145), and other early authors correctly noted the identity of the pale birds of Arizona and southeastern California with those of Baja California, but this was later interpreted as a northward migration "to lower Colorado River Basin of southeastern California, western Arizona, and western Sonora" (Miller et al., 1957; see also Grinnell and Miller, 1944:255).

Change: If I have interpreted Gilman's data correctly, then the Say's Phoebe has exactly reversed its status from a summer to a winter species on the reservation. My observations are in keeping with the status given in Phillips et al. (1964:84): "A post-breeding migration carries most of the birds out of the southwestern part of the state, where virtually absent in July and August."

Dusky Flycatcher, *Empidonax oberholseri* Phillips

Historic: Gilman (MS) recorded this species "Numerous during spring migration: 15 April–14 May and in between." But this account must include *E. hammondii* (see that species), not mentioned in Gilman's manuscript.

Modern: Rare transient. I have but four specimens of Dusky Flycatcher from the reservation: 18 and 22 April 1967, 6 May 1964, and 1 October 1972. The first is from Barehand Lane, and the rest are from Komatke. Both sexes are represented. One specimen was taken in thick mesquite, one in an *Atriplex-Prosopis* association, and one in tamarisks. Phillips (personal communication) writes, "This species is rare on migration, in valleys of central, western Arizona and southeastern California."

Gray Flycatcher, *Empidonax wrightii* Baird

Historic: Gilman (MS) secured one specimen 12 December, which he listed under the former name, *E. griseus*.

Modern: Uncommon transient and winter visitant. I collected seven Gray Flycatcher specimens between 25 September and 31 December, all at Komatke. I have no winter sight records after the first week of January. There are but three spring specimens from the reservation, 22, 24 and 25 April, the latter from the Salt-Gila confluence (JSW). Both spring and fall specimens are equally divided between males and females. Three were taken in mesquite, two in mesquite fence rows, two in open mesquite scrub, and one in tamarisk.

Taxonomy: I reject Oberholser's (1974) proposal to replace the name *E. wrightii* Baird (1858) with *E. obscurus* (Swainson 1827). Even were this not a *nomen oblitum*, unused in the primary literature for fifty years, still Oberholser's grounds for such a change are extremely tenuous: "...the probability that Swainson had *E. wrightii* in hand....The type...is lost [but] the name still seems applicable."

Hammond's Flycatcher, *Empidonax hammondii* (Xantus)

Historic: Undoubtedly the bulk of Gilman's (MS) records attributed to the Dusky Flycatcher were actually this species.

Modern: Common during spring migration, uncommon in fall. I collected eighteen Hammond's Flycatchers in spring from 3 April to 19 May. The peak of migration (eleven specimens) is from 20 to 25 April. Females appear to predominate in spring (six of eight sexed specimens). I collected this species but twice in fall: 25 September 1965 (AMR 834, immature male) and 4 October 1967 (AMR 1746, female). This is the most ecologically diverse of the five *Empidonaces* migrating through the reservation. Four specimens were taken in ironwood-paloverde association in bajadas, three in floodplain mesquite bosques, and one in willow-cottonwood association.

Least Flycatcher, *Empidonax pusillus* (Swainson)

Modern: Accidental. I collected an odd, solitary *Empidonax* flycatcher on 12 April 1978 along the Salt west of 91st Avenue. It was flipping its tail up, foraging in dry branches midway up young willows. The specimen (SD 40502, first-year female) was identified by Phillips. There are several fall records for extreme western Arizona but no other in spring nor any this far east in Arizona.

Taxonomy: Phillips' cryptic revival of Swainson's name, *Platyrhynchus pusillus* 1827, replacing Baird and Baird's *Tyrannula minima* 1843, appears to be justified. Swainson's bird (or birds?) was from the "maritime parts of Mexico." The only commonly occurring white-throated *Empidonax* in Caribbean Mexico is the Least. Others, though decidedly less common, are the transient Traill's (or Alder) and Acadian Flycatchers, clearly not the bird Swainson had before him. Acting in effect as first revisor (Oberholser, 1918, notwithstanding), Swainson (1831) corrected the generic allocation of *pusillus* to *Tyrannula* and equated a Carlton House, Saskatchewan, specimen with the type (or types?) from eastern Mexico. This he contrasted with *T. querula* Wilson (= *acadica* Gmelin, = *virescens* Vieillot). The wing formula given is an accurate description of the Least Flycatcher. Swainson's *pusillus* was considered a migrant in mid-May, retiring to the north to breed. Baird and Baird (1843) had "no specimen of *T. pusilla,* of Swainson, but [depended] upon comparison with the description in Swainson and Richardson's Zoology of North America, (so favourably known for accuracy)." Their new species, *Tyrannula minima,* was a migrant in Pennsylvania "as after the last of May it is no longer to be seen." They characterized it as a smaller version of *traillii.* In the original description, Swainson said the tail of *pusilla* is even, in the subsequent revision "very slightly emarginate." *Minima* is described as having an even tail also. There seems to me no question that both are descriptions of the same species and that Swainson's name must stand unless it can be demonstrated conclusively to be a *nomen oblitum* (a game some taxonomists play to avoid their supposed fundamental principle, priority).

Western Flycatcher, *Empidonax difficilis* Baird

Historic: Breninger (1901:45) observed a Western Flycatcher in September. Gilman (MS) noted it "Seen during spring migration [and] several 25 September 1910 and one 1 October 1910, Sacaton."

Modern: Fairly common during migrations. I collected thirty-two specimens of the Western Flycatcher. Spring records are from 16 April to 21 June (ten specimens). Fall specimens are from [?29 July] and 3 August to 15 October. The age groups during the fall migration are sharply divided, with little overlap: the last adult was 3 September and the earliest immature 29 August. Females predominate among fall adult migrants (eleven of fifteen specimens), but the sexes of the immatures appear to be equal. The two summer birds are problematic. The 21 June 1970 female (AMR 2975, Sacaton) is very worn but the ovary appeared to be prebreeding. I consider it to be a belated spring migrant. The other is a worn adult headless mummy found [29 July 1969] (Komatke). However, both these dates are within the respective spring and fall extremes given in Phillips et al. (1964).

Of the five migrant *Empidonaces,* the Western shows the greatest habitat specificity while crossing the reservation. The bulk of individuals was found along deep arroyos lined with large ironwood and paloverde trees. A few others were taken in *Salix-Populus* groves or in clusters of *Tamarix aphylla.*

Taxonomy: Western Flycatchers breeding from the Rocky Mountains south to the Sierra Madre Occidental are the large *E. d. hellmayri* Brodkorb. Only four of the reservation specimens (three males, one unsexed) are this race: 16 April and 26 August (adults) and 24 and 29 September (immatures). The remainder are the small *E. d. difficilis,* breeding along the Pacific Coast. My identifications follow Brodkorb (1949).

Willow (Traill's) Flycatcher, *Empidonax traillii* (Audubon)

Historic: Gilman (MS) recorded this species as "Seen in May and September."

Modern: Uncommon transient. I collected seven specimens on the reservation, the adults ranging from 10 August to 2 September, the immatures from 2 to 11 September. The sexes appear to be evenly distributed. No spring records.

Taxonomy: Phillips (1948) revised the species. He identified the seven reservation specimens as *E. t. brewsteri* Oberholser, the race breeding along the Pacific Coast. These birds agree with my series of migrants from San Diego County, California.

The formal decision (A.O.U. Checklist Committee, 1973) recognizing two sibling species within the "Traill's" complex follows primarily from ethological and other studies by Stein (1958, 1963).

Western Wood Pewee, *Contopus sordidulus* (Sclater)

Historic: Breninger (1901:45) recorded the species as "Quite common." Gilman (MS) considered this flycatcher "common during spring and fall migration."

Modern: Fairly common on migration. I collected thirty-five Western Wood Pewees on the reservation. The numbers vary greatly from year to year. In 1966 and 1967 I collected twenty specimens and the species was common, but I averaged only 1.4 birds for the remaining years. Generally the migrations are protracted: 25 April to 21 June and 16 August to 10 October. As is the case with a number of other species, males begin the spring migration earlier: 25 April to 21 June for twelve males in contrast to 29 May to 13 June for five females. The discrepancy in the sex ratio during spring might indicate differential migratory routes, a problem warranting further investigation.

I obtained only one adult pewee during the fall migration—17 August. Seventeen immatures ranged from 16 August to 10 October. The sexes in the fall appear to be equally represented. There are strong parallels in the fall migration pattern between this pewee and the Western and Willow Flycatchers. However, it appears that adult pewees in fall either go south via a different route or fly over the low, hot desert at this season (August) without stopping.

Taxonomy: Burleigh (1960b) recognized four United States subspecies of Western Wood Pewee on the basis of color and size, but I do not agree with his descriptions (see also Browning, 1977). Two well-marked races (Austin and Rea, 1976:405) are distinguishable north of Mexico. *C. s. veliei* Coues is paler and grayer. It breeds throughout most of the United States' range of the species. *C. s. saturatus* Bishop is darker, more olive, and with a blackish crown and a yellowish wash on the throat and belly. It breeds in the humid Northwest. In comparing races of this pewee it is critical not only to segregate the sexes and age classes, which are very different, but to have specimens of comparable museum age. Foxing (postmortem color change) enhances the browns at the expense of olives in approximately four years.

Both races migrate through the reservation. Spring specimens of *saturatus* extend from 25 April to 21 June and *veliei* from 30 April to 13 June. There are three times as many *saturatus* (thirteen of seventeen) in spring. The one fall adult (AMR 1662, male) is *veliei*. The two races appear to be approximately equal among immature fall migrants (ten *saturatus* and seven *veliei*). But the peak migration of immature *saturatus* appears to be somewhat earlier. Ranges for *saturatus* are 16 August to 24 September and for *veliei* 29 August to 10 October.

Olive-sided Flycatcher, *Contopus borealis* (Swainson)

Historic: Breninger (1901:45) recorded "a number seen here only as migrants," while calling the Western Wood Pewee "quite common." Gilman (MS) did not mention the species.

Modern: Irregular, uncommon migrant. I have seen this species but five times on the reservation: in 1967, 1968, and 1972. The three spring specimens are females taken 29 April 1968, and 28 and 30 May 1967. I took another female (SD) on the Gila at Cochran 11 May 1982. A male (CGR 52) was taken 2 miles (3 km) northeast of Casa Grande Ruins 3 May 1938. The fall specimens are immatures: a male on 10 September 1967 and females on 29 September 1968 and 1972.

Taxonomy: There is no real basis for maintaining the Olive-sided Flycatcher in the monotypic genus *Nuttallornis*. Osteologically and behaviorally it is a *Contopus* (Austin and Rea, 1976:405; see also Phillips et al., 1964; Mayr and Short, 1970; and others). The name *Contopus mesoleucus* (W. Deppe) is used by Phillips et al. (1964), but no reason is given.

Bangs and Penard (1921) described a western race, *C. b. majorinus,* based primarily on size (type locality Los Angeles County, California), but this was never officially recognized (A.O.U. Check-list 1931, 1957). In a carefully documented review of the species, Todd (1963) pointed out that *majorinus* is indeed a valid subspecies but with a range much more restricted than given in the original description. Breeding birds of southern California and northern Baja California are *majorinus*. Notwithstanding Todd's revision, Oberholser (1974) considered *majorinus* (1921) a junior synonym of *borealis* (Swainson, 1831 [= February 1832], Carlton House, Saskatchewan), thus necessitating the revival of the next older name, *cooperi* (Nuttall, 1832, Boston) for the eastern race (see also van Rossem, 1934). Oberholser's action is unwarranted, as Todd's appended measurements clearly demonstrate. Phillips (personal communication) considers the Arizona breeding birds to be *majorinus,* but the few sexed breeding birds I have been able to locate (JSW, LLH, MNA, DEL) appear too small for this race.

Three of the spring reservation specimens are *majorinus* and one is *borealis,* as is the Cochran bird; the three fall specimens are all *borealis*.

Vermilion Flycatcher, *Pyrocephalus rubinus* (Boddaert)

Historic: Breninger (1901:45) reported "several seen, all females." Gilman (MS) described this flycatcher as "resident in fair numbers... nesting takes place in April, May, and June, and in one case two broods were raised in the same nest, one in April and the other in June. Three eggs formed the set in each nest found... usually from 20–40 feet [6–12 m] from the ground though one was only six and one half feet [1 m]." His extreme nesting dates are 15 April to 24 June. The species may have bred locally as late as 1938 (CGR 67, adult male, 20 May, 2 miles [3 km] north of Casa Grande Ruins, giving courtship flight).

Modern: Fairly common irregular early winter visitant. Only males have been collected or observed: 14 October to 27 January. Individual adults often maintain territories about Pima farms, as at Joseph Giff's (Komatke) from 18 October to 8 December 1967 and again 31 October 1969 to 27 January 1970. An individual appearing there 26 September 1968 was apparently only transient and was never seen again. There are perhaps not more than several per winter and in seven winters I found none. Outside these dates I have but two records: 21 March 1964 (AMR 465, immature male, Komatke) and 20 June 1968 (Barehand Lane marsh, adult male chased away by Ash-throated Flycatcher and not seen again).

Change: The Vermilion Flycatcher has radically changed its status on the reservation from a breeding bird (supposedly resident) to an early winter resident, absent from February through September. The cause of this dramatic change is enigmatic, as the species still nests elsewhere in mesquite (as near Blue Point Cottonwoods, Taylor and Hanson, 1970). Phillips et al. (1964:92) summarized its status as, "Common to abundant summer resident in mesquites, willows, and cottonwoods (always near water in the lower, western valleys), in southern and central Arizona, but rather local along the Salt and Colorado River Valleys." Lack of open water near nesting localities seems to be the factor limiting breeding. Vermilion Flycatchers still breed at *charcos (represos)* with mesquite trees on the Papago Reservation.

Alaudidae

Horned Lark, *Eremophila alpestris* (Linnaeus)

Historic: Gilman (MS) recorded Horned Larks in "small flocks at various times," mentioning specific dates of 26 October to 10 January. It appears that the species was not regular at that time and surely was not breeding.

Modern: My first observation of Horned Larks on the reservation was 29 September 1972, when a flock of twenty-five flew over Barehand Lane in the Co-op Colony area, where I had worked regularly, summer and winter, for a number of years. In November we collected six specimens. By December the species had become abundant. The winter of 1972–1973 was atypically wet (272 mm from October to March).

At Komatke 1 to 2 June 1973 I found a pair in dry grass and *Atriplex* behind Sally Pablo's house. The male was giving flight song and the female behaving territorially as if with eggs or young, but I failed to find a nest. On 18 June 1974 I saw one male at the edge of a dirt road near Barehand Lane and other males (but no juvenals) all along Lower Buckeye Road, north of the Salt River. A. R. Phillips and I saw many Horned Larks near Co-op Colony and Barehand Lane 9 May 1974. I returned 12 May and collected two males in breeding condition and a female with an egg in her oviduct. Another female was gathering nesting materials. By 8 June juvenals were being fed by males at the Chandler boundary north of Santan. The following day I collected a juvenal and adult male together in Co-op Colony. By 10 March 1975 males here were numerous, paired, and singing. I took a male with enlarged cloaca. These birds used alfalfa fields with broad, bare borders and roads. Nesting success must depend upon irrigation times, as the fields are flooded periodically. I collected one female and three males (ARIZ 12130–12133) 22 June 1975, but saw no juvenals in Co-op at that time. By 28 August 1975 Horned Larks were actually abundant in the fields, flying about conspicuously. This was evidently an influx of northern migrants but no specimens were collected to demonstrate this conclusively. At least two pairs were courting continuously near the Komatke cemetery on 20 May 1979.

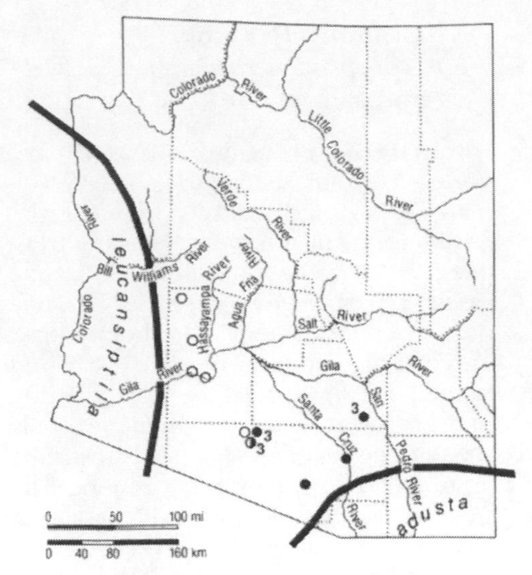

Horned Larks to end of 1940s

○ = Eremophila alpestris leucansiptila
● = E.a. adusta ◑ = E.a. adusta x leucansiptila
Heavy lines enclose historic range.

Taxonomy: My study of Horned Lark subspecies has been confined to the three Arizona breeding races: *E. a. occidentalis* (McCall type locality near Santa Fe, New Mexico); *E. a. adusta* (Dwight, Fort Huachuca, Arizona); *E. a. leucansiptila* (Oberholser, Yuma, Arizona). All determinations of breeding birds, both recent and historical, are my own. All breeding season specimens were washed. Identification of winter birds is principally by Phillips, who (in Phillips et al., 1964) has published the most recent revision of western subspecies.

Two quite different races are involved in the invasion of the reservation (and the Phoenix area in general) as breeding birds. *Leucansiptila* breeds in the low hot deserts of southeastern California, adjacent far western Arizona, and northwestern Sonora. It is grayish (canescent), with a vinaceous tinge to the nape, and pale, buffy chest and flanks in the females. *Adusta,* the "scorched" Horned Lark, breeds in the elevated grasslands of extreme southeastern Arizona. It is altogether a warmer colored bird, with buffier ground color, browner dorsal streaks, and salmon nape, with darker flanks, chest, and shoulders in females. Even juvenals of the two (AMR, ARIZ, ARP, UCLA) are readily distinguishable. Of twenty breeding Horned Larks taken on or adjacent to the reservation, five were *leucansiptila,* four were *adusta,* and eleven were intermediates (with the *leucansiptila* influence predominating).

Three freshly molted birds taken on 29 September were still local (*leucansiptila, adusta,* and an intergrade). Migrants and winter visitants appear on the reservation at least by mid-October. These include one *occidentalis* (not typical; approaching next race?); several *E. a. "leucolaema"* Coues, a local variant of *occidentalis* breeding in the western Great Plains and Rockies; and two female *E. a. enthymia* (Oberholser), an uncommon to rare wintering race from the Northern Great Plains.

Change: The historical ranges of the two sedentary races, *adusta* and *leucansiptila,* were widely separated (Behle, 1942; Phillips, 1946), with no Horned Larks inhabiting the intermediate areas. Originally van Rossem (1936) found *leucansiptila* (UCLA) only in Yuma and Mohave counties, Arizona. He later found

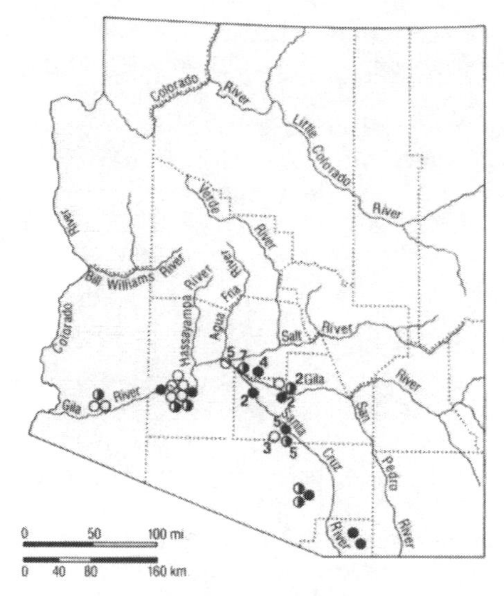

Horned Larks in 1970s

○ = Eremophila alpestris leucansiptila
● = E.a. adusta ◑ = E.a. adusta x leucansiptila

it (1947) in western Maricopa County, Arizona, at an abandoned airstrip overgrown with grass. Phillips and Hargrave (Monson and Phillips, 1941) took mid-February specimens (ARP) in agricultural fields west of Gila Bend, Maricopa County. Likewise, *adusta* of the southeastern grasslands seems to have undergone some range extensions. Van Rossem found it west to the Altar Valley, east of the Baboquivaris (UCLA), and Brown took a specimen (ARIZ 2758) in January 1885 near Tucson. This appears to bracket the original western and northern limit of this form. In 1940 Monson took breeding *adusta* specimens (GM, ARIZ, including juvenal) 7 miles (11 km) northwest of Oracle, Pinal County, Arizona, where it had not been found previously (Monson and Phillips, 1941). Van Rossem and Phillips found Horned Larks in a tobosa (*Hilaria mutica*) meadow near Ventana Ranch, northwestern Papago Indian Reservation. These were recorded first as *adusta* (Sutton and Phillips, 1942), but later thought to be nearer *leucansiptila* (Phillips et al., 1964). I have reexamined the Ventana Ranch specimens (ARP, UCLA) and identify them as follows: from 1940 one *adusta* and one intermediate; from 1947 one *leucansiptila,* two intermediate, and two *adusta* approaching *leucansiptila;* from 1952 one *adusta,* one intermediate, and three *adusta* approaching *leucansiptila.* This situation suggests to me continued recruitment of parental forms to this presumably ancient grassy island.

It is unfortunate that the Horned Lark invasion of the Phoenix region was not periodically documented with breeding specimens. The earliest specimen is a juvenal (UCLA 32908) taken by van Rossem 8 May 1945, 3 miles (5 km) northwest of Hassayampa (ca. 40 miles [65 km] west of Phoenix). Housholder took another juvenal (DEL 19204) in Paradise Valley 15 miles (24 km) northwest of Phoenix 10 May 1958. Both are *leucansiptila.* Johnson collected a Phoenix adult male (MNA 3476) 14 June 1959 that is *adusta* approaching *leucansiptila.* Thomas D. Burleigh took an intermediate female (DEL 46301) 21 March 1966 and a juvenal *adusta* (DEL 46285) 10 May 1966, both at Scottsdale. The next breeding specimens are those I collected from 1973 to 1976. Both parental races and intermediates are represented, as noted above. *Adusta* now strongly influences at least some individuals even in extreme western Maricopa County (e.g. SD 40599, 2 May 1978, Agua Caliente Road, a very "scorched" specimen).

In the past three decades *leucansiptila* has extended its breeding range eastward by at least 80 miles (130 km) and *adusta* northwestward by over 100 miles (160 km). The two factors that appear to be responsible for these range extensions are unusually wet winters with subsequent grass growth and the recent extensive agricultural development of the floodplains throughout the Sonoran Desert.

There have been changes even in wintering status. Gilman found the birds wintering irregularly. Between 1963 and 1971 I found no wintering larks. But since fall 1972, some individuals of northern races as well as the breeding populations have wintered locally.

Hirundinidae

Violet-green Swallow, *Tachycineta thalassina* (Swainson)

Historic: Gilman (MS) reported the Violet-green Swallow "seen occasionally in early spring," giving specific dates of 22 February, 22 March, and 15 April.

Modern: Uncommon (or rare) transient. I observed three Violet-green Swallows several times during 29 October 1972 at Barehand Lane marsh and possibly one in this area 28 August 1975. Two specimens (CGR 53, 54; males, 2 miles [3 km] NE of Casa Grande Ruins) were taken 3 May 1938.

Taxonomy: The two specimens are *T. t. lepida* Mearns. Their wings measure 109.2 and 113.0.

Tree Swallow, *Tachycineta bicolor* (Vieillot)

Historic: Gilman (MS) noted on "15, 16, and 17 February there were many of these swallows flying up and down the 'Little River' and a specimen was secured."

Modern: Uncommon migrant. My only specimens are immatures taken 4 and 17 August 1967 (Gila Crossing and Barehand Lane). This species returns exceptionally early (see Phillips *in* Phillips et al., 1964:96); a flock appeared at Barehand Lane 6 January 1973 (possibly with some Violet-green Swallows), but I failed to secure a specimen. Both adult and immature Tree Swallows were again common over Barehand Land 16–17 July 1972 and also over the Salt-Gila confluence 5 August 1972 (*fide* Johnson and Simpson). Tree Swallows surely do not winter or breed at either place.

Taxonomy: Brodkorb (1957) united *Iridoprocne* Coues (1878) with *Tachycineta* Cabanis (1851) on the basis of osteological characters of the humerus, carpometacarpus, and coracoid. Phillips et al. (1964) also lumped the genera. I agree with this merger. The entire family Hirundinidae needs a worldwide generic revision based on ecology, nest construction, plumage, and, most importantly, anatomy.

Oberholser (1974) advocated separation of California Tree Swallows as *T. b. vespertina* (Cooper) on the basis of smaller size. Though this might prove a usefully recognizable subspecies, I have seen too few California specimens of known breeding status to evaluate the taxon. The two reservation specimens, though immature, are very small and would be referable to *vespertina.*

Bank Swallow, *Riparia riparia* (Linnaeus)

Modern: Rare migrant. I have two specimens (AMR 2297, 4653) and two further observations of the Bank Swallow extending from 14 to 29 August. An additional partial skeleton was removed from the crop of a Turkey Vulture (small summering race, *Cathartes a. aura)* found long dead [28 November 1969] (Rea, 1973). The Bank Swallow was unrecorded by Gilman.

Taxonomy: New World Bank Swallows for the most part average considerably smaller than those from Europe. Oberholser (1974) advocated formal recognition of the New World populations as *R. r. maximiliani* (Stejneger).

Rough-winged Swallow, *Stelgidopteryx ruficollis* (Vieillot)

Ethnographic: See Barn Swallow

Historic: Breninger (1901:46) noted, "A number were seen near the river" in September. Gilman (MS) found the Rough-wing Swallow a "Summer visitant, breeding along the Gila River. Seen as early as 28 February, remaining until the middle of September." He noted 107 perched on some wires 12 September. Nests were placed from 3 to 8 feet (1 to 2.4 m) above the bottom of the bank and from 1 to 4 feet (0.3 to 1.2 m) from the top, all in mid-May. The birds were loosely colonial, though no nests were closer than 50 feet (15 m).

Modern: Common on migrations. Migration dates (specimens) are 29 March to 25 April and 1 to 29 August. Birds seen occasionally in June and July suggest the possibility that a few may still breed on the main part of the reservation, but I have failed to locate nests. A small entirely black swallow seen 23 June 1975 at the Salt-Gila confluence proved in the hand to be a stained Rough-wing (AMR 4743, adult female). It was not in breeding condition.

On 25 June 1977 Rough-wings were common on the Salt at 91st Avenue. During early April the following year groups of fifteen to twenty pale-looking birds were flying about holes in a cliff cut out by the previous winter's floods. Later in the month they appeared to be nesting. The first week of June I took a postbreeding adult and a juvenal still growing its outer primaries from several dozen in the young willow trees nearby. I found none here in 1981.

Taxonomy: Phillips et al. (1964) merged *Stelgidopteryx* (1858) in *Riparia* (1817). Mayr and Short (1970:62) noted simply that "Its relationships are uncertain." Brodkorb (1957, 1968) kept both genera. I think the osteology of *ruficollis* and *riparia* is sufficiently distinct to merit retention of both genera pending a thorough study of all the species in question.

Two subspecies of Rough-wing Swallow are currently recognized (Brodkorb, 1942; A.O.U. Check-list, 1957; Phillips *in* Phillips et al., 1964) from northern Mexico and the United States: darker *S. r. serripennis* (Audubon) and paler, more southern *S. r. psammochroa* Griscom. Both breed in the Southwest. I find them good subspecies, most readily identified in fall immatures, with their distinctively colored secondary edgings. In identifying reservation migrants, I used unfoxed adult and immature specimens recently taken in early summer from Chino Valley, Arizona (*psammochroa*), and Zuni, New Mexico (*serripennis*). I have taken three spring migrant *serripennis* on the reservation and two others farther down the Gila at Gillespie Dam. The only spring female (AMR 2067 10 April, local?) is *psammochroa*. This subspecies nests in southern and western Arizona. An adult and a scarcely volant juvenal (SD) taken in early June 1978 on the Salt at 91st Avenue are *psammochroa*. Four fall immatures (two males, two females) are likewise this pale race. The only fall adult (AMR 4060, female, 1 August 1972, Barehand Lane) is a rather dark example of *serripennis*.

Change: Phillips et al. (1964:97) called the Rough-winged Swallow a "common summer resident in dirt banks of streams throughout Sonoran zones of state." Gilman found it breeding evidently in numbers. Its virtual extirpation from most of the reservation as a breeding species is attributable to lack of suitable nesting banks near water. Even on migration this swallow keeps near water (irrigation ponds and ditches, Barehand Lane, Salt River). The species (*psammochroa*) still nests in suitable locations in the Salt River Valley (Johnson et al., MS).

Barn Swallow, *Hirundo rustica* (Linnaeus)

Ethnographic: *hévaichut* [or] *hévachut; gídval,* [pl.] *gígidval*

Consultants used both of these names for the Barn Swallow, but the first was indicated more often and I suspect that the second Pima taxon actually refers to the Rough-winged Swallow, the other species formerly nesting on the reservation; see Purple Martin account.

In 1968 Matthias described *hevachut* as follows: "It migrates but comes back each summer. Seen in the fields when someone is plowing. No longer seen here [Komatke]. Nested down in wells." Joseph Giff's description of "So blue that it shines, nests in wells, and has a forked tail" could apply to no other species. The Browns of Santa Cruz confirmed that *hevaichut* used to nest in a well at St. Catherine's Church, where they were caretakers. Matthias glossed it 'bluebird' as did Russell (1908:345, 367).

Historic: Breninger (1901) noted the species in September, and Gilman (MS) wrote, "The following are dates of occurrences of this swallow: 27 April, 11, 16 May, 1 June, 6, 7 September, and 1 October when 12 were seen." But he did not find it breeding.

Modern: Abundant during migrations. This is the most conspicuous daytime migrant in the area. The peak of fall migration is 22 September to 20 October, with adults as well as immatures sometimes passing by the hundreds for hours, usually tapering off at mid-day. Spring migration is in early May.

Change: The Barn Swallow disappeared from the reservation as a breeding species with the loss of surface water and the lowering of the water table, with consequent loss of shallow wells.

Cliff Swallow, *Hirundo lunifrons* Say

Historic: In September Breninger (1901:46) noted, "A number seen circling about the fields." Gilman (MS) recorded it as "Seen occasionally in migration on dates as follows: 11 April, 4 May, 9 July, 7 September." There is no ethnographic account of the Cliff Swallow, nor did Pima consultants know anything of its distinctively shaped mud nests.

Modern: Rare in spring, fairly common in fall migrations. Cliff Swallows are seen usually at open water (Barehand Lane, Salt River, pumping pond at Co-op). My spring record is for one individual seen on the Salt River above 91st Avenue on 14 March 1971. Fall records extend from 23 June (at least fifty, probably all immatures) to 28 August, with specimen dates from 16 July to 28 August.

Taxonomy: The genera *Hirundo* Linnaeus (1758) and *Petrochelidon* Cabanis (1851) are clearly congeneric on osteological evidence. In fact, bones of young Barn and Cliff Swallows from Pleistocene caves are difficult or impossible to identify to species (Rea, personal observation). Phillips et al. (1964) and Phillips (1973) united the genera.

By no stretch of the imagination can Vieillot's description of *Hirundo pyrrhonota,* with its blackish lower belly and a russet brown forehead, be construed to apply to our northern Cliff Swallows (*contra* Hellmayr, 1935; A.O.U. Checklist, 1957; Oberholser, 1974; and others), which have both the forehead and belly white. Ridgway (1904:50) justifiably disposed of this name as doubtfully applicable and used *H. lunifrons* Say 1823 (type from Arkansas River, Colorado). This name, long in general use (A.O.U. Checklist, 1931), should stand unless *H. albifrons* Rafinesque, 1822 proves acceptable.

Cliff Swallows are greatly in need of a taxonomic revision covering the entire range of the species. The ranges as given in the A.O.U. Check-list (1957) are unsatisfactory. Confusion is caused in part by preoccupation with a single variable, forehead color, that not only varies individually within populations (even within the chestnut-fronted populations) but is greatly affected by abrasion and soiling once nest construction begins. Phillips' revision (*in* Phillips et al., 1964) is only tentative.

Of the eight reservation specimens, seven are referable to *H. l. tachina* (Oberholser), the race breeding across the major portion of the Southwest, from at least southern California (San Diego County, specimens AMR), most of

Arizona and New Mexico (AMR, ARP, LLH, ARIZ), to south-central Texas. One specimen (AMR 2298, 14 August 1968, wing chord 110.6+) is either *H. l. aprophata* (Oberholser), a Great Basin subspecies, or *H. l. hypopolia* (Oberholser), the largest and northernmost population (determinations by Phillips and Rea). In the most recent revision of the western races, Behle (1976) considered *aprophata* a synonym of *hypopolia*.

Purple Martin, *Progne subis* (Linnaeus)

Ethnographic: Russell (1908) gave a number of names for 'swallows' from various songs and legends, including *kikitâvalĭ* and *ñiñitâvalĭ* (p. 292) and *tcotcok kiñitâvald* (p. 293), glossed 'black swallow.' I suspect that this last (it is in plural form) might refer to the Purple Martin, but I found no Pima consultant who knew the species or recognized the morpheme. Martins may once have bred near the Pima villages. They are still common and well known in Papago country, where they are called *gi:gidwal.*

Historic: Gilman (MS) recorded the Purple Martin "Seen but once, 13 May, when several perched on an upstairs porch railing at Sacaton."

Modern: Rare transient. I saw this species only once: adult male in Santa Cruz village 31 August 1967. Johnson and Simpson took an adult female (JSW, MNA 4040) at the Salt-Gila confluence 30 May 1971. A small desert race of Purple Martins breeds across the Papago country in saguaros of southern Arizona but curiously is absent from the Pima country and all of Maricopa County. The nearest breeding locations, at least formerly, are at Florence and to the north (Phillips et al., 1964:101). On each summer trip I have seen at least one adult male in the elevated country between Florence and Cochran.

Taxonomy: Behle (1968) described a Rocky Mountain race, *P. s. arboricola,* which is as large as the eastern nominate race but paler in females and young. The reservation female is this race (identified by Phillips and Rea).

Corvidae

Steller's Jay, *Cyanocitta stelleri* (Gmelin)

Ethnographic: (?) *unmú'uhik*

Gila River consultants relate that the Pima singers from the Salt River Reservation have a song about a bird that is blue, comes from the cold country, and "lives in the spruce, but nobody knows what it is." As this conspicuous jay is seen in the desert at least irregularly in winter, perhaps this is *unmú'uhik,* which consultants believed to be some "blue jay."

Historic: Gilman (1911a, MS) observed that "a few spent the winter of 1910–1911 at Sacaton and Santan." He had seventeen records from 17 October to 1 March. The maximum number of jays was a flock of seven seen 27 October.

Modern: Rare winter visitant. Cyril Magooshboy collected one specimen (AMR 1399, immature female) in the tamarisks lining the football field at St. John's Indian School, Komatke, 23 February 1967. Steven Giff, hunting with me at Barehand Lane, saw another 29 October 1972, but the bird disappeared into the dense saltcedars before I could see it.

Taxonomy: Two subspecies of interior Steller's Jays are recognizable, and both occur in the Southwest. *C. s. macrolopha* Baird is the darker backed, more deeply blue bird breeding south to southern Nevada (Austin and Rea, 1976:406) and northwestern New Mexico (Phillips, 1950b). *C. s. diademata* (Bonaparte) is a grayer backed, altogether paler blue bird breeding over most of Arizona, (AMR, ARIZ, ARP, LLH) south through the Sierra Madre Occidental. *C. s. browni* Phillips (type locality Santa Catalina Mountains, Pima County; incorrectly cited by Blake *in* Mayr and Greenway, 1962) is a synonym of *diademata fide* Phillips et al. (1964). The one reservation specimen is *diademata.*

Scrub Jay, *Aphelocoma coerulescens* (Bosc)

Archaeological: McKusick identified a Scrub Jay (left ulna, complete) from Snaketown.

Historic: Breninger (1901:45) found Scrub Jays in September "Fairly common; have never seen so many in this valley before; winter visitants." Gilman (MS) recorded the species "Winter visitant in small numbers each season." He gave extreme dates from 17 September to 21 April, but "rarely after 1 March." The maximum number given is fifteen on 9 January 1915 at Sacaton and three on 18 November at Santan.

Modern: Rare winter visitant, seen by me in three out of thirteen winters. My first observation of the Scrub Jay on the reservation was four individuals on the Salt River below 91st Avenue (in young cottonwoods) and one at Barehand Lane (in saltcedars) from 13 to 15 February 1971. On the lower Salt near the reservation Johnson and Simpson (personal communication) saw four on 2 September 1972. Another was in a dense mesquite fence row at Sacaton 30 September 1972. I collected one specimen (AMR 4826, immature male) at the Salt-Gila confluence 31 December 1975.

Taxonomy: Two color races are distinguishable in the interior West and Southwest. Phillips (1964) argued that the type of *A. c. woodhouseii* (Baird 1858) is a pale bird, perhaps collected at Fort Webster rather than Fort Thorn, New Mexico. If this is the case, *A. c. nevadae* Pitelka (1945) is a synonym. Phillips renamed the darker, more eastern jay *A. c. suttoni* (type locality 22 miles [35 km] south of Pueblo, Colorado). I have examined the types of *suttoni* but have not seen Baird's type, on which Phillips' action hinges.

Oberholser (1974) argued for the retention of the spelling *woodhousii* (plate xliii) over *woodhouseii* (pp. 584–585) because of anteriority in Baird's original. *Contra* Oberholser, this printer's error need not be honored; the spelling is emended according to Article 33 of the International Code of Zoological Nomenclature. The situation is parallel in part to Oberholser's (1920, 1930, 1937) upholding of the suppressed lapsus *Toxostoma "dorsale"* over *T. crissale* and others.

The reservation specimen is the paler subspecies, *nevadae, sensu* A.O.U. (1957) (= *woodhouseii, sensu* Phillips).

Change: It is evident from Gilman's account, at least (1901 must have been a flight year), that the Scrub Jay formerly appeared with regularity and in numbers along this section of the Gila. Phillips et al. (1964:104) indicated that "In most winters a few descend to Lower Sonoran brush, orchards, and trees mostly along streams of central Arizona." The spectacular decline of Scrub Jays as winter visitants to the reservation area I attribute to the destruction of riparian growth. The species appears more regularly in less disturbed areas with surface water, as at Blue Point Cottonwoods (Johnson and Simpson, MS).

[Piñon Jay, *Gymnorhinus cyanocephalus* Wied]

Historic: In September Breninger (1901:45) "First heard, then saw three fly from the trees; first record for this part of Arizona." Gilman (MS) saw a flock of seven at Sacaton in April (year?) and one along a field fence at Santan 25 September 1914. I have no further records for the species.

Taxonomy: Much is made of the "crow-like" behavior (Phillips et al., 1964, and others) of the Piñon Jay, and it is traditionally placed after the genus *Corvus* in lists (A.O.U. Check-list, 1957; Amadon, 1944; Mayr and Short,

1970; and others). I have examined the osteology of most Old and New World corvid genera and find that *Gymnorhinus* shows no affinities to *Corvus* or to any Old World genus but is indeed nearest the New World jay evolutionary line, *contra* Hardy (1969). Therefore I place the genus here, following *Cyanocitta* and *Aphelocoma* which it closely resembles in postcranial osteology. Ligon (1974) reached a similar conclusion, calling *Gymnorhinus* a specialized and perhaps early offshoot of the lineage known as the New World jays.

Common Raven, *Corvus corax* Linnaeus

Archaeological: This species was reported from the early Snaketown excavations (Haury *in* Gladwin et al., 1937) and from the recent work there. I have examined the latter bone (Snaketown #6043, left carpometacarpus) and have confirmed the identification. Additionally, I have identified an ulna and two femora from the Las Colinas Site, Phoenix.

Ethnographic: *schuk ú'uhik (schuk,* 'black'; *ú'uhik,* 'bird' [life form]).

I have rechecked this name independently with various consultants with the same result: *schuk ú'uhik* is a solitary (at most paired) bird, distinguishable from the *hávaiñ* (Common Crow), which flocks. Sylvester Matthias identified *schúk u'uhik* in the field in April 1978 and May 1982.

In Papago country, where two raven species occur, both are called *háwañi* or *háwañ*. More southern Piman speakers use an entirely different word for '*cuervo*': *kókon* or *kóko'on* (lowland Pima Bajo), *kóko:n* (Tepecano; Mason 1917), *kokóñi* (Northern Tepehuan; Pennington, 1978). This lexeme is apparently unknown among the northernmost Pimans.

Historic: Breninger (1901:45) reported that "A pair flew down into a field near my camp soon after daylight; the only ones seen" in September. Gilman (MS) recorded the Common Raven as "Seen occasionally in winter and spring. Two spent most of one winter around a slaughter house and became quite tame." He gave late dates of 17 and 27 May.

Modern: Rare. I have seen the species but three times on the reservation, in each case two fully adult birds together: at Joseph Giff's fields, Komatke, 13 March 1971, and the next day at Barehand Lane; in the Sierra Estrella 4 March 1973; and on the Gila River bridge between Gila Crossing and Santa Cruz 11 April 1978.

Specimen evidence rests on a hackle feather from an adult (AMR, uncatalogued) found at a *tinaja* (tank) in the Sierra Estrella 7 February 1968 (Dempsey Kanteena, A. Rea and class).

Taxonomy: The general failure to segregate first year from fully adult specimens when making size comparisons in *Corvus corax* has led to taxomonic confusion. At least three races can be distinguished in North America (AMR, ARIZ, ARP, LLH, MNA, SBM, SD, SWAC, UCLA, UNLV). *C.c. principalis* Ridgway, 1887 of the far northwestern (and northern?) part of the continent, is large and heavy billed. This race is almost certainly indistinguishable from *C.c. kamschaticus* Dybowski, 1831 of eastern Siberia, a name already abundantly supplied with synonyms. *Contra* Meinertzhagen (1926), the Siberian-Alaskan population is fully separable from *C.c. tibetanus* Hodgson, 1849 on wing measurements, as Meinertzhagen's own table shows. The smallest North American birds, in all dimensions, are *C.c. clarionensis* Rothschild and Hartert from Baja California and most of California. *C.c. sinuatus* Wagler, from the remainder of the continent, is rather larger, approaching "*principalis*" in size, but more slightly built, with a thinner bill. Skeletons suggest it is probably a much lighter weight bird, but too few weights are available from the far north. Arizona upland ravens are all *sinuatus*. Lowland specimens are few. A femur from the Las

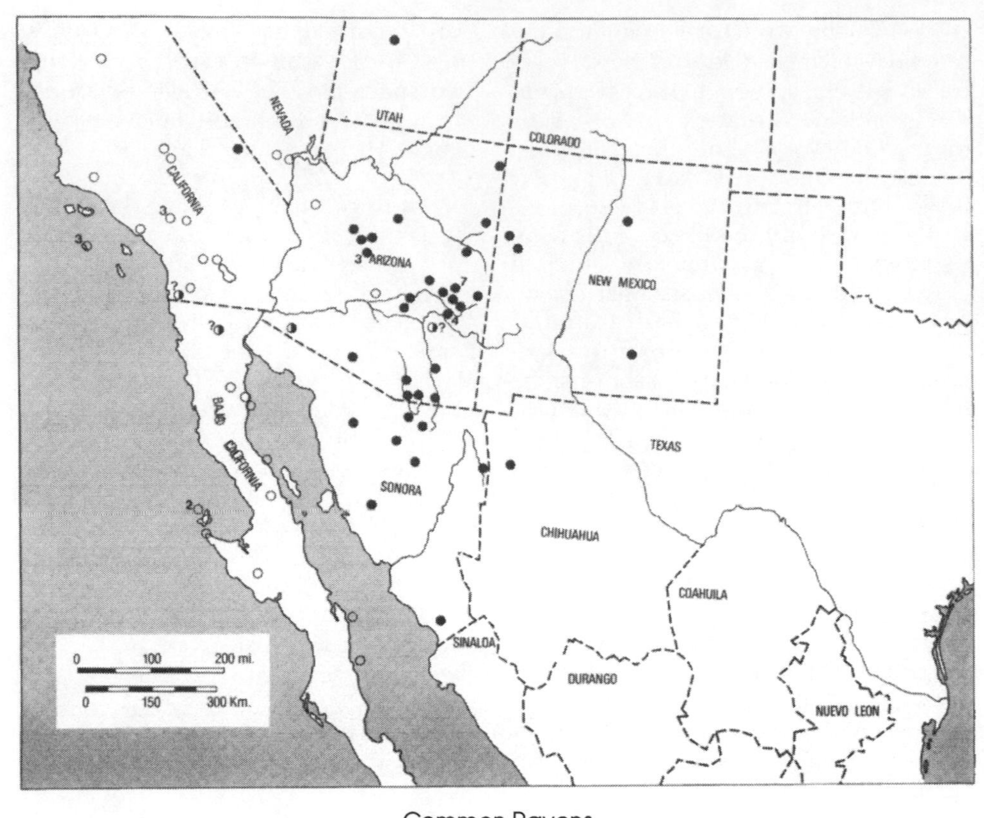

Common Ravens

○ = Corvus corax clarionensis ● = C.c. sinuatus ◑ = ♀ in size overlap range

Colinas site is within the *sinuatus* size range (Rea, 1981b). An adult male (AMR 2534) found dead by S. R. Demaree on North Central Avenue, Phoenix, is as small as *clarionensis* of the coast. While it is possible that this was someone's free-ranging pet (there were no indications of confinement), there are other suggestions that the small race occurs eastward into Arizona. A partial skeleton (SD) collected by Alan Ferg 26 miles [42 km] southwest of Seligman, Yavapai County, is too small for *sinuatus.* (It differs in characters and is larger than *C. cryptoleucus.)* An adult female from Tinajas Altas, Yuma County (SD, 8 March 1937) is in the range of overlap in all dimensions. Specimens from the coast and deserts of California and southern Nevada are just as small as those from Clarion Island, Mexico. The only apparent *sinuatus* vagrant I have seen is an adult male (UNLV) taken 4 December 1965, at Saratoga Springs, Death Valley.

[Common Crow, *Corvus brachyrhynchos* Brehm]

Ethnographic: *hávaiñ*

The Pima clearly distinguish "crows" from Common Ravens. Crows were flocking and colonial nesting birds. The north end of the Sierra Estrella is still called *hávaiñ kosh,* 'crows' nest.' This name is an old one; it was mentioned a century ago (1876–1877) in the Gila Crossing calendar stick: "There was an Apache village called Hâvany Kâs at the junction of the Gila and Salt Rivers while a truce existed between the Pimas and Apaches"

(Russell, 1908:55). The Pima say there were abundant *hávaiñ* there when there was water. Crows were especially troublesome in the fall when the Pima harvested their corn. Matthias relates that his parents described flocks of *hávaiñ* until ca. 1918. Russell (1908:345) glosses *hávaiñ* in a speech as 'raven.'

Historic: Russell (1908:92) wrote, "Not with the withering drought alone has the Gileño [Pima] planter to contend but also with the myriads of crows that are extravagantly fond of a corn diet, and with the numerous squirrels and gophers that thrive apace where protected in a measure from the coyotes, which are themselves a menace to the fields." Russell conducted his field work in 1901–1902. But scarcely a decade later Gilman (MS) recorded crows "Seen but twice, when three were seen at one time in April and travelling toward the northeast. Two were seen another year in the spring."

Modern: No observations.

Change: There is no conclusive evidence that the flocking *Corvus* species appearing in fall and winter was the Common Crow rather than the White-necked Raven, *Corvus cryptoleucus* Couch. It would have been about as far for this raven (which gathers in enormous fall flocks) to wander west as for the crow to move south in order to take advantage of Pima agriculture. But the colonial nesting in cottonwoods at the Salt-Gila confluence suggests the crow as the biologically correct gloss of *hávaiñ*. Neither species occurs today.

[Clark's Nutcracker, *Nucifraga columbiana* (Wilson)]

Historic: Gilman (1911a:35) recorded this species at Sacaton. "The most incongruous combination was a Clarke Nutcracker perched on a Deglet Noor date tree the morning of 17 October [1910]. He was quite tame and though an instinct demanded his acquisition as an avian record for this locality I refrained and he departed in peace about noon." This is the only record for the reservation. Nutcrackers invade Arizona lowlands periodically (Phillips et al., 1964), and there is even a specimen (AMR) from as far south as coastal Sonora.

Laniidae

Loggerhead Shrike, *Lanius ludovicianus* Linnaeus

Ethnographic: *gáw'uguf* [no plural form]

The food-storing habits of shrikes are well known to all the old Pima I have talked with. Matthias relates this story: "Once a *s'vekí vamat* ['red snake,' i.e., *Masticophus flagellum*] went up a mesquite after a *gáw'uguf*. When the snake got tired the *gáw'uguf* pecked its eyes out and the racer fell to the ground dead." The name is common to both Pima and Papago.

Historic: Breninger (1901) found shrikes common in September. Gilman (MS) considered them "Common resident," recording nests from 8 March to 28 May.

Modern: Common. Unlike the Phainopepla and Say's Phoebe, there is no evidence of a postbreeding exodus. I have taken juvenals in June, July, and early August, and an adult in *Ambrosia-Atriplex* association 3 August. On this date in the open desert shrikes were everywhere, usually in pairs, and wary. A drive along the Sierra Estrella bajada from Santa Cruz south to the end of the mountains 8–9 July 1973 showed the species well spaced but abundant.

Taxonomy: I collected sixteen specimens from the reservation, nine of these being in fresh fall or winter plumage. Most of the latter are pale *L. l. excubitorides* Swainson (which includes *L. l. sonoriensis* Miller, in large part, and *L. l. nevadensis* Miller, *fide* Phillips), breeding in the interior West. Juvenals were taken from 3 June to 2 August (AMR). Only one specimen (AMR 1153, 16 October 1966) appears to be *L. l. gambeli* Ridgway, the darker Pacific and interior Northwest form, though Phillips et al. (1964:141) considered the two races "about equally common" in southern Arizona.

Paridae

Bridled Titmouse, *Parus wollweberi* (Bonaparte)

Historic: Gilman (MS) recorded the Bridled Titmouse as "Seen in limited numbers during October, November, and December 1909 and again same months of 1910." He had seven observations between 3 October and 21 October 1910, the greatest number of individuals being six on 16 October. In 1915 he observed four on 8 February. All records are for Sacaton. Gilman collected two specimens 10 November 1909.

Taxonomy: Gilman's existing November specimen (SBC, orig. MFG 211) is an excellent example of the still undescribed northern subspecies (Rea, in press) that breeds north of the Gila River below the Mogollon Rim and wanders occasionally in winter into the Salt River Valley.

Change: Phillips et al. (1964:110) called this normally Upper Sonoran woodland species a "Regular winter visitant to willow-cottonwood association along larger streams in southern and central Arizona, west to Tempe and Sacaton." The complete absence of the species here in recent times is directly attributable to the destruction of the riparian woodlands along the Gila and lower Salt rivers. No doubt the former gallery forests were avenues for this tit to wander down into the lowlands.

Aegithalidae

Bushtit, *Aegithalos minimus* (Townsend)

Modern: Casual. I collected two adult males (AMR 4828, 4829) from a flock of perhaps five or six at the Salt-Gila confluence 31 December 1975. They were feeding in saltcedars and willows. I saw the flock here again 27 March 1976. I have no other observations of this species on the reservation.

On the Gila River, just upstream from the Buttes, I found a flock of Bushtits 29 December 1981. They were foraging in saltcedars and in a *Celtis-Acacia-Prosopis* thicket. On 11 May the flock was still there, and I took an adult female and two fledged young (SD).

Taxonomy: There is no morphological or anatomical justification for generically separating the Bushtits of North America from the various species of long-tailed tits of Eurasia, which differ significantly only in pattern of coloration (see discussion in Ridgway 1904:424–425). Their nests are remarkably similar. I have had field experience only with *caudatus* and *minimus*. While these two species are probably the extremes of form and coloration, their behavior and voice immediately suggest they are congeners. *Aegithalos* Hermann, 1804, has priority over *Psaltriparus* Bonaparte, 1850. Gadow (1883:54) gave a diagnosis of the inclusive genus. I have seen no skeleton of the closely related Javan tit *Psaltria* Temminck, 1836, which differs externally primarily in having a longer outer primary; see Baird (1858:395–396) for other differences. I would include both in the family Aegithalidae. A similar conclusion was reached by Snow (1967a).

Pacific coastal Bushtits are smoky brown-headed. Interior Bushtits are clear gray-headed and gray-backed, with brown auriculars west of the continental divide. (For a discussion of polymorphism and auricular colors see Phillips et al., 1964:111–113; and Raitt, 1967.) The oldest name for the southwestern birds is *A. m. plumbeus* (Baird 1854), with type locality on the Little Colorado River, Arizona. If a smaller interior race is recognized, *A. m. lloydi* (Sennett, January 1888) is the next oldest available name in the proper range, with *A. m. santaritae* (Ridgway, October 1888 and *A. m. cecaumentorum* (Thayer and Bangs 1906) being synonyms. For a different conclusion (recognizing three species) see Oberholser (1974). Bushtits are in need of a revision based on carefully sexed birds. The two reservation males are clear gray-headed and small (wing chord 50.4 and 48+; tail 55.8 and 58[+]), hence are tentatively referred to southern *lloydi,* the subspecies breeding in adjacent Upper Sonoran areas of Arizona. The three Bushtits taken in May near The Buttes are also small (ad. female wing chord 48.2[+]; tail 53.7+; juvenal male and female wing 48.6 and 47.8, respectively.

Remizidae

Verdin, *Auriparus flaviceps* (Sundevall)

Ethnographic: *gí:sop;* [pl.] *gí:gi:sop*

The *gí:sop,* builder of a large, enduring, and conspicuous nest, figures importantly in Piman stories, song, and speeches, but unfortunately is almost never identified in English translations. The name is common to Papago and lowland Pima Bajo.

Historic: Gilman (MS) found Verdins:

> Common resident, nesting usually in mesquite, screwbean, zizyphus [graythorn] and catsclaw trees or shrubs. The birds build winter nests in which to roost, not more than one bird to each nest though big enough for two or three and they seem so fond of the nests that frequently they retire to roost before sunset.... March 23 was the earliest date of finding eggs and 26 June the latest. Most of the nests were found in March and April, it being the exception to find any later nestings.

Modern: Common. Reservation Verdins breed in virtually all habitats affording a suitable location for the nest. This may be along arroyos in the bajadas, in the floodplains, and even occasionally in

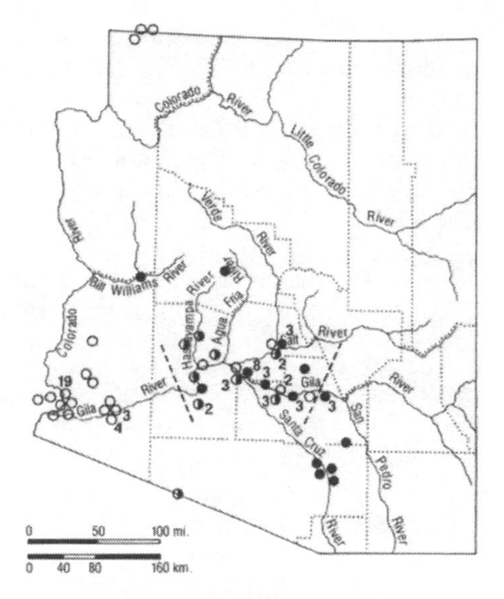

Verdins (fall specimens)

● = Auriparus flaviceps ornatus
○ = A.f. acaciarum ◑ = intermediate
Dashed lines show intergrade zone on Middle Gila drainage.

Populus-Salix communities or the usually sterile saltcedars. There is no evidence for any movements except for gatherings in the postbreeding season at damp and shady locations (as at Barehand Lane, 22 June 1975, and at the Salt-Gila confluence).

Taxonomy: Though currently usually placed in the family Paridae, *Auriparus* shows no close affinities to the tits. Rather, its nearest relative, based on nesting, external morphology, and osteology, is the African genus *Anthoscopus* (Rea and Austin, personal observation), and we include both in the family Remizidae as was done also by Snow (1976b). Remizidids, aegithalids, sittidids, certhiidids, and true parids are probably of equal taxonomic rank. In order to reflect this relationship in taxonomy, the only alternative to keeping each of these five as a family is to reduce all to subfamilies in the family Sittidae, which has priority. In keeping with the taxonomy of the remainder of the Passeriformes, I favor full family status for each.

None of these is anatomically close to the Corvidae as the current sequence suggests.

As presently conceived, Verdin populations in the United States consist of a pale race, *A. f. acaciarum* Grinnell in the extreme West (desert area of southern California and adjacent western Arizona), and a slightly darker race, *A. f. ornatus* Lawrence, across the remainder of the range to the lower Rio Grande.

Because the Verdin lives in highly xeric, exposed habitats, its plumage is subject to considerable abrasion and wear. To be taxonomically useful, specimens must be in fresh plumage (September to early December). I assembled a series of eighty-six unworn specimens (AMR, ARIZ, UNLV, UT) from California, Arizona, Nevada, Utah, New Mexico, Texas, Sonora, and Coahuila. The specimens indicate that the reservation is an area of intergradation of the two northern races: five specimens are *acaciarum,* fourteen are *ornatus,* and five are intermediates.

Sittidae

Red-breasted Nuthatch, *Sitta canadensis* Linnaeus

Historic: Gilman (MS, 1911a) observed "a few" (maximum of two) Red-breasted Nuthatches from 5 October to 12 November. All six records were at Sacaton in 1910. "One day I noticed one fly several times from a tree trunk, warbler-like, and snap up worms hanging at the ends of webs" (Gilman 1911a:35).

Modern: Uncommon. I observed this nuthatch (never more than two individuals) in four winters (1963, 1972, 1973, 1975) out of thirteen. The earliest record (AMR 463) is one in bright fall plumage found [15 September 1963] impaled and already dried on a paloverde spine at the south end of the Sierra Estrella. My latest date is 9 December.

Taxonomy: Burleigh (1960a) described a race of Red-breasted Nuthatch, but Phillips et al. (1964), Greenway (1967), Banks (1970), and Mayr and Short (1970), found the species to be monotypic.

Certhiidae

Brown Creeper, *Certhia familiaris* Linnaeus

Modern: Uncommon winter visitant. I discovered creepers (never more than two individuals) in five of thirteen winters. Extreme dates (observations with specimens) extend from 30 September to 1 February. Unaccountably, Gilman (MS) did not record this species. The six specimens are equally divided as to sex, but five are immatures.

Taxonomy: Creeper populations form a number of well-marked subspecies in North America. Most races were reviewed by Phillips *in* Phillips et al. (1964). However, I do not agree in not recognizing *C. f. leucosticta* van Rossem of southern Nevada (AMR and type series, UCLA, consulted), which I find distinctive (see Austin and Rea, 1976:406). The race breeding throughout the Rocky Mountains south to the border mountains of Arizona is *C. f. montana* Ridgway (type from Mt. Graham, Arizona, 23 September 1874). This form is long-billed, has a white superciliary, a brown and white back with no buff, and an ochraceous-tawny rump. Two reservation specimens (AMR 4145, 30 September 1972; AMR 4658, 25 October 1974) are reasonably typical *montana.* Presently included within the range of *montana* (A.O.U. Check-list, 1957) is an apparently undescribed race, which migrates to Arizona (AMR 4807, 1 November 1975, Komatke; other specimens ARP and UCLA). *C. f. americana* Bonaparte breeds in most of the eastern United States and across the wooded portions of much of Canada (*fide* Phillips). It is short-billed, with warm, buffy-tinged upper parts (superciliary, back, rump). Three reservation specimens are *americana* (AMR 1250, 27

November 1966; AMR 4824, 31 December 1975; AMR 423 [now ARP] 1 February 1964). Parkes and Phillips assisted me with these identifications. However, J. Dan Webster (personal communication), who recently revised the species, considers these first two within the range of *montana* variation.

Troglodytidae

[Winter Wren, *Troglodytes troglodytes* (Linnaeus)]

Modern: Rare winter visitant. I observed the Winter Wren but once on the reservation, 17 March 1973 near Bapchule. This bird appeared three times at very close range in a tangle of debris and tumbleweed (*Salsola iberica*) under a dense tamarisk fallen over a dirt irrigation ditch with running water. When I returned 26 March, a bulldozer was parked on the site and the tamarisk thicket, cottonwoods, and clump of mesquite were gone, as well as the Winter Wren!

House Wren, *Troglodytes domesticus* (Wilson)

Historic: Breninger (1901:46) noted the House Wren "Several seen at intervals in brush fences," in mid or late September. Gilman (MS) recorded it "Seen in small numbers each winter: November, December, and January."

Modern: Common winter visitant. Normal lowland records for the reservation extend from 19 September to 25 April (specimens). But on 1 September 1980 I found three individuals in flood debris on the Salt at 91st Avenue. I took a late male (SD 40530) here 11 May 1978. In an oak grove atop the Sierra Estrella there were still several House Wrens present 5 May 1975 and one the following day. Phillips et al. (1964:117) wrote "Normal occurrence in southern Arizona lowlands is from the last of August to early May. Extreme dates [without specimens] are 16 August. . . to 24 May."

Taxonomy: *T. d. parkmanii* Audubon is the grayish House Wren breeding across western North America to the mountains of southeastern Arizona. All the reservation winter specimens are this race.

Oberholser (1974) has shown that the name *Sylvia domestica* Wilson, 1808, has priority over the currently used *Troglodytes aedon* Vieillot "1807" (= 1809). Wilson's name cannot be considered a *nomen oblitum* since Oberholser (1934) and Huey (1942) have used it within the last fifty years. Unless Wilson's name is officially suppressed by the International Commission of Zoological Nomenclature, *Troglodytes domesticus* (Wilson) must be considered the correct name for this species. (Name suppression is a way of avoiding conformity to the taxonomic rules—analogous to having four strikes in a ball game when convenient.)

Bewick's Wren, *Troglodytes bewickii* Audubon

Historic: Breninger (1901) recorded only one, in September. Gilman (MS) reported Bewick's Wrens as "Seen occasionally during the winter season." His specific dates are 25 September, 2, 7, 8, and 18 October, 9 January, and again in March (day not specified).

Modern: Uncommon in winter. I saw this wren but five times on the reservation and collected three specimens from 16 October to 4 March.

Taxonomy: The Troglodytidae is one of the best-defined oscine families, but the genera are generously over-split. The Holarctic species *Troglodytes troglodytes* was once divided into two different genera! The accepted "generic" diagnoses (see Ridgway, 1904:548, 569, 595) serve only to distinguish single species. The best external morphological characters are the configuration of the nasal fossa and the operculum (if these have not been mutilated by the preparator). The nominal genus *Thryomanes* Sclater, 1862, is best merged into *Troglodytes* Vieillot, 1807, for the taxa *troglodytes,* *"aedon" (=domesticus), sissoni, tanneri,* and *bewickii. Bewickii* on the one hand and *troglodytes* on the other represent the extremes, but the remaining three taxa group closely. The osteology is compatible with this interpretation.

The breeding race of the interior West, *T. b. eremophilus* (Oberholser), is large and grayish. Two of the specimens are *eremophilus,* but a third (AMR 1897, 28 December 1967) is darker, perhaps approaching *T. b. "atrestus"* (Oberholser), a synonym of *T. b. drymoecus* (Oberholser). Identifications are by Phillips and Rea. (See Rea, in press b.)

Cactus Wren, *Campylorhynchus brunneicapillus* (Lafresnaye)

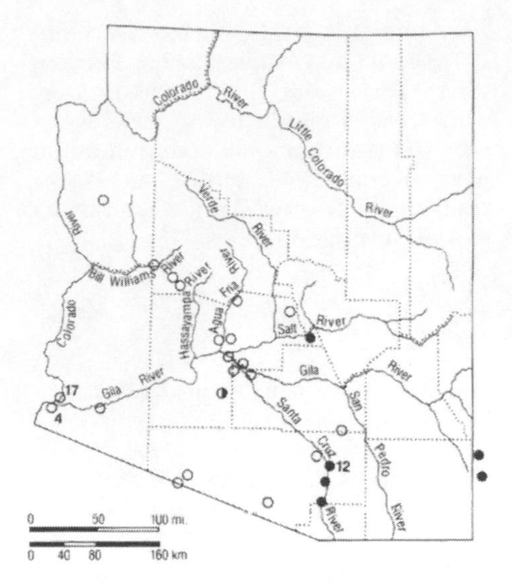

Cactus Wrens (fall specimens)

○ = Campylorhynchus brunneicapillus anthonyi
● = C.b. couesi ◑ = intermediate

Ethnographic: *háwkut;* [pl.] *háwhawkut*

This name is common to Riverine Pima and Papago.

Historic: Unaccountably, Breninger (1901) failed to mention this ubiquitous species. Gilman (MS) found the "earliest eggs 15 February and the latest 8 July, April being the most frequent." He noted that the young remained housed for twenty-one days.

Modern: Abundant. As with the Verdin, the Cactus Wren is a characteristic species of the reservation, found at Pima dooryards (particularly if there is a *vahtto* or 'arbor'), in fence rows and mesquite bosques, and well up into the bajadas and even into the mountains.

Taxonomy: The Cactus Wrens of Arizona are currently considered (A.O.U. Check-list, 1957; Phillips et al., 1964) to be the race *C. b. couesi* Sharpe (type locality Laredo, lower Rio Grande, Texas). This is incorrect. Three September topotypes of *couesi* (AMR) are quite different from Arizona specimens in having brownish-black heads and brown backs with very heavy throat spotting to the ramal area. Western and central Arizona birds have a dark rufous head and contrasting pale warm sandy dorsum, with less dense chest and throat spotting, and with usually unmarked chin. These are *C. b. anthonyi* (Mearns, type locality Adonde Siding, near present Tacna, Yuma County, Arizona, one fall topotype, AMR). Selander (1964) likewise recognized *anthonyi* but synonymized *couesi* with *C. b. guttatus* Gould. Phillips (personal communication) suggests that *guttatus* represents a different population (type locality given simply as "Mexico") and that at that date, 1837, Gould's specimen was most probably collected in or just north of the Valley of Mexico, *contra* Hellmayr (1934) who restricted the type locality to Tamaulipas. Oberholser (1974:993) agreed with the separation of *anthonyi,* noting that the wrens of "east Trans-Pecos Texas verge somewhat toward *couesi*...." Birds from the Tucson Valley and east through southern New Mexico and western Texas I find nearest *couesi* (ARIZ, UNM, US).

Long-billed Marsh Wren, *Cistothorus palustris* (Wilson)

Historic: Breninger (1901) did not record this species. Gilman (MS) noted it as "Seen three times in a tule marsh during spring migration."

Modern: Locally common in winter. Dates of the ten reservation specimens span from 24 September to 12 March, with sight records of several still present 26 March 1973 (when Barehand Lane marsh was ditched and dry) and one 28 March 1976 (Salt-Gila confluence). I saw an early migrant 19 September 1968 in Komatke in a damp, weedy edge of a cornfield. All other marsh wrens were in *Typha* or *Scirpus* thickets at Barehand Lane or along the Salt River. Wintering marsh wrens became numerous on the Salt River near the Maricopa Colony. The winter floods of 1966–1967 washed out most of the suitable habitat, and it was several years before the wrens returned.

Taxonomy: The merger of *Telmatodytes* Cabanis into *Cistothorus* Cabanis follows Brodkorb (1957) and others (see A.O.U. Check-list Committee, 1976). *Cistothorus* has page priority.

On the basis of my examination of over 230 specimens in fresh plumage (AMR, CM, DEL, JSW, MVZ, SD, UW), there are four recognizable races in the West: two in the interior that are larger and paler and two smaller coastal ones that are quite dark. *Cistothorus palustris paludicola* Baird of the Pacific Northwest is dark but dull, with reduced amounts in blackish in the crown and interscapulars. It appears to be sedentary. *Cistothorus p. aestuarinus* (Swarth) of the southern coast is also dark but is more black on the crown, nape, and back, and more reddish-brown on the rump and flanks. The breeding population of the lower Colorado River and the middle and lower Gila is somewhat larger, but otherwise appears to be indistinguishable from California *aestuarinus.* There is a series of freshly molted mid-August and winter specimens (DEL) taken from Palo Verde Marsh (1955–1956) about 23 miles [37 km] downstream from the Salt-Gila confluence, but I have never seen this form on the reservation. *Cistothorus p. plesius* Oberholser of the Rocky Mountains, large, pale, but bright, especially on the rump, accounts for the bulk of Arizona wintering birds. Reservation specimens range from 24 September to 12 March. *Cistothorus p. pulverius* (Aldrich) of the Great Basin is large and pale, but dull, especially on the crown, nape, and rump, with the blackish parts softly colored. It has been taken four times on the reservation: 29 September to 10 December. One or two of these may be intermediate toward *plesius.*

Change: Marsh wrens once bred at a *Scirpus* marsh near Palo Verde, not far from the reservation (Simpson and Werner, 1958; eggs and skins secured). They doubtless must have bred in the formerly extensive marshes at and below the Salt–Gila–Agua Fria confluences described by Charles H. Lowe (personal communication). As the *Scirpus-Typha* marsh at Barehand Lane expanded from year to year, I looked for marsh wrens to remain in spring, but they departed each March. Apparently the dark resident race is completely extirpated from the middle Gila and the wintering races *plesius* and *pulverius* do not colonize redeveloping potential breeding habitats this far south.

Parenthetically, in all Gilman's writings on reservation birds and vegetation, the above mention of "a tule marsh" is his sole reference to emergent vegetation.

Cañon Wren, *Catherpes mexicanus* (Swainson)

Ethnographic: According to some, this is probably the 'flat-headed bird' whose head was stepped on at a wine feast. The incident occurs in both Pima and Papago story and song. This wren is probably called *vavas* also (see Rock Wren account).

Historic: Breninger (1901:46) in September reported "One heard and finally seen; usually found in rocky places." Presumably this record is from the lowlands. Gilman (MS) recorded Cañon Wrens as "Fairly common resident." He found a nest at Sacaton in April.

Modern: Fairly common in mountains and deep arroyos, rare in floodplains. The Cañon Wren is not encountered as frequently as the Rock Wren, perhaps because of more narrow habitat requirements. In the Sierra Estrella four young were out of the nest and being fed 7 and 8 April 1973. But the nesting season may be prolonged. I took a juvenal (AMR 4639) 23 July 1974 and another (AMR 1081, in prebasic molt) 15 August 1966. Lowland winter records are few. I saw one Cañon Wren climbing on the old grammar school walls at Komatke 25 October 1965 and another in Komatke from 20 October to 11 November 1973.

Taxonomy: Paynter (1960) and Mayr and Short (1970) favored merging *Catherpes* (1858) into *Salpinctes* (1847). The postcranial osteology of these wren genera appears to me to be very similar. The two species share a similar dorsal plumage pattern. *Catherpes* has a distinctive foot structure with the nails strongly compressed laterally and incised on each surface with deep pairs of grooves, largely lacking in *Salpinctes.* I tentatively keep *Catherpes,* pending detailed study of *fasciatus, guttatus,* and *maculatus,* the Central American taxa in the *Salpinctes* group.

Reservation specimens are the smaller, paler *C. m. conspersus* Ridgway.

Rock Wren, *Salpinctes obsoletus* (Say)

Ethnographic: *vávas*

This bird is well known. Vávas was the name of the deceased father of one of my Pima consultants; hence, it was a word taboo in her presence. *Vávas* occurs in a number of songs. That this species is intended rather than the sympatric Cañon Wren is clear from a song, *Kómakĭ vávas,* 'gray wren.' But Matthias believes both species are included in the term *vávas.*

Delores Lewis, a Papago from Gu Oidak, called this wren *wa:wusigim,* which he said came from *waw,* 'cliff' or 'rock.' He thought both the Cañon and Rock Wren were included in this ethnotaxon. The Pima name *vávas* is probably derived from *vaf* or *vav,* 'cliff.'

Historic: Gilman (MS) found the Rock Wren a "Fairly common resident."

Modern: Common in upper bajadas and mountains; uncommon and irregular in winter in lowlands. I have not studied this wren to discover if there is any year-to-year change of status as suggested by Phillips et al. (1964:120). It appears always to be present in the Sierra Estrella. Here young were already out of the nest 7 April 1973. I have seen young as late as 31 May 1966. The species still seemed to be present in numbers 15 August 1966, when I collected a specimen (AMR 1082, adult male) in heavy molt. At midday on 7 June 1978 I found three Rock Wrens, two poorwills, and an Ash-throated Flycatcher taking refuge from the intense heat in a mine shaft on the south slope of the South Mountains. The Rock Wren is more likely than the Cañon Wren to be found wintering in the floodplains, but not every winter. Extreme dates for these lowland birds are 17 October to 10 March and apparently once on 3 September.

Mimidae

Northern (Common) Mockingbird, *Mimus polyglottos* (Linnaeus)

Archaeological: Two bones of mockingbird were identified by McKusick from Snaketown.

Ethnographic: *shu:g* (identical lexeme in Pima Baja and Papago)

A well-known Pima taxon, this bird figures importantly in stories and song. A child who is slow in acquiring speech is fed the meat of the mockingbird either roasted or boiled.

Historic: Breninger (1901) noted "A family seen on a brush fence." Gilman (MS) found mockingbirds "Fairly common resident." His nest and egg dates extend from 2 April to 7 July, with most birds nesting during April and May.

Modern: Common. Mockingbirds are partial to more mesic habitats so are most frequently seen in *Lycium,* overgrown fence rows, and mesquite bosques. I cannot confirm a postbreeding exodus as suggested by Phillips et al. (1964) for other desert areas. However, some spectacular aggregations occur. At the Riggs Road pond I counted sixty-three mockingbirds in four small cottonwoods on 21 June 1973, and by 10 July fully 150 birds were there. By this time but a few inches of water remained. At Barehand Lane, in contrast, the greatest number encountered was six on 23 December 1972.

Taxonomy: I have compared fall and winter mockingbirds (AMR, ARIZ) from eastern and western states and can find no differences to support the separation of *M. p. leucopterus* (Vigors), a conclusion reached also by Phillips (1961), Phillips et al. (1964), and Mayr and Short (1970). Age differences are important, birds of the year being paler and browner than adults.

If *gilvus* (Vieillot) proves to be conspecific with *polyglottos,* Common Mockingbird would be the appropriate English name.

Thrashers, *Toxostoma* spp.

Ethnographic: Four thrasher species are common or fairly common residents breeding in or near the Pima villages: Curve-billed, Bendire's, Crissal, and LeConte's. The first two and the last two are morphologically similar pairs. I have obtained five Pima taxa for thrashers, but their application to biological species is uncertain. Possible correspondences between Piman and Linnaean taxonomies are suggested below, the first (Bendire's) being almost certain.

bet kúishnam (*bet,* 'excrement,' *kúishnam,* 'it stepped on'; rendered less euphemistically by consultants). This is the common, confiding thrasher with spots, hence must be the Bendire's Thrasher.

ku:d vik [no gloss obtained]. A spotted species, probably the Curve-billed Thrasher.

kúli huch (*kúli,* 'old man['s]'; *huch,* 'nail'). The name is apparently in reference to the bird's bill. The bird is described as being colored like an Abert's Towhee; hence, this must be the Crissal Thrasher.

chuf chíñgam (chuf, 'long'; *chiñ,* 'mouth,' 'bill'; '*-gam*' attributive or possession marker, hence, 'it has a long bill'). This is either the Le Conte's or Crissal Thrasher.

kómakĭ ú'uhik (kómakĭ, 'gray'; *ú'uhik,* 'bird' [life form]). This is most likely the Le Conte's Thrasher.

Bendire's Thrasher, *Toxostoma bendirei* (Coues)

Ethnographic: See "Thrashers." Probably *bet kúishnam.*

Ruth Giff considered this thrasher a nuisance in fields because it digs up newly sprouting plants.

Historic: Breninger (1901:46) noted Bendire's "on the desert." Gilman (MS) recorded it as an "Abundant resident." Gilman (1909a) kept notes on 112 thrasher nests found in the Sacaton area in 1908; thirty-nine of these were Bendire's. Egg dates (Gilman MS) for seven years extended from 19 February to 18 July (fresh). He found two broods not unusual and once noted three. "Bendire in particular seems scarce during the latter part of September and during October and November, but is occasionally seen during all that time" (Gilman 1909a:50).

Modern: Abundant on the floodplains. The species is partial to native Pima ranches, where *Lycium, Prosopis,* and *Zizyphus* bushes are left growing along fence rows. Here nearly every row has one or more pairs of Bendire's nesting by early May. I have not witnessed any emigration from the reservation as described by Gilman (1909a) and Phillips et al. (1964). Ten of twenty-two reservation specimens were collected between October and January, when the birds were numerous. All seven specimens taken 19 and 22 February 1968 had enlarged gonads approaching breeding size, and an already volant juvenal (AMR 2062) was taken 6 April 1969.

At least for the Southwestern populations there is a misconception in field guides concerning eye color differences between Bendire's and Curve-billed Thrashers. I recorded the following eye colors of twenty Bendire's Thrasher specimens: four bright medium yellow, fifteen bright chrome yellow, and one orange-yellow. Of eleven Arizona Curve-billed Thrashers, I found five with bright chrome yellow, five with light orange-yellow, and one with orange. One had one eye pale orange and the other deep yellow. (None had orange-red eyes.) Thus eye color would not distinguish these species 83 percent of the time.

Taxonomy: Van Rossem (1942) considered the species polytypic, but Phillips (1962) cautioned against racial recognition, since Bendire's Thrashers migrate first, then molt.

Change: Gilman (MS) predicted:

> The Bendire Thrasher is one bird that from all indications takes kindly to settlement. These birds nest near houses, on which they perch to sing, come into the yards, and seem fearless if not molested. If their natural shelter is cleared up they take kindly to artificial or planted growth and I believe will persist in the face of civilization. All of this, of course, provided they receive some measure of protection and encouragement.

But Gilman did not visualize the huge mechanized farms now found on some parts of the reservation, covering thousands of acres without a single bush left for escape or nest of the Bendire's Thrasher (see Fig. 3.11).

Curve-billed Thrasher, *Toxostoma curvirostre* (Swainson)

Ethnographic: See "Thrashers." Probably *kud vik.* The Pima name is probably derived from the verb *ku:d,* 'to take a firebrand from one fire to light another fire' (see Mathiot, n.d.:12) in allusion to the diffuse or smoky breast markings which distinguish the Curve-billed from the more contrastingly spotted Bendire's Thrasher. In the field, Papago Delores Lewis identified this bird as *kulwichgam,* 'beak' 'it has,' commenting that "it can sing any kind of notes and eats everything we plant; gives us lots of trouble."

Historic: Breninger (1901) observed several and Gilman (MS) called it "very numerous." Of 112 thrasher nests under observation in 1908, Gilman (1909a) found twenty-seven of Curve-billed. His egg dates extended from 7 February to 17 July.

Modern: Common. This species is normally found out in the dissected bajada country in *Cereus-Cercidium-Olneya-Opuntia* communities. But it occasionally wanders into the *Prosopis*-dominated floodplains in late summer and fall, as at Santan (AMR 1792, fat) and Komatke.

Taxonomy: The race breeding and resident in south-central Arizona is *T. c. palmeri* (Coues) (see A.O.U. Check-list, 1957; Phillips et al., 1964). Reservation specimens were compared with topotypes of *palmeri* from Tucson, Arizona.

Le Conte's Thrasher, *Toxostoma lecontei* Lawrence

Ethnographic: See "Thrashers."

Historic: Breninger (1901:46) observed one "skulking among the desert bushes." Gilman (MS) called it an "Uncommon resident." Of 112 thrasher nests under observation in 1908, only one was the Le Conte's (Gilman 1909a:50–51). All five nests that he found on the reservation were in April. About Sacaton Le Conte's "were in or near the dry sand washes away from the river."

Modern: Fairly common but local. Le Conte's are confined to areas of sand dunes with occasional low *Prosopis* and *Lycium* and the *Ambrosia-Atriplex* flats where these are extensive and undisturbed. These normally rare thrashers may occur here in numbers. The morning of 24 July 1964, following a rain, I observed six together in a graythorn (= *Zizyphus*) bush singing softly. In early mornings of August 1967 they were "singing everywhere," usually in groups of threes, fours, and fives, just south of Gila Crossing.

Phillips et al. (1964:124) write of Le Conte's:

> Intolerant of man and his activities, it has retreated from the newly farmed areas of central Arizona. Its movements are little understood and its former seasonal status in central Arizona will probably never be known.... From October through December it is not definitely recorded in Arizona north of the Gila River, nor east of Yuma County. Presumably it returns in late January.

However, we now know that the species does not leave the reservation during this period (AMR specimens; field notes, Demaree and Rea). Even after sunset on the cold evening of 6 January 1973, three individuals responded to my tape-recorded Le Conte's songs.

Taxonomy: Phillips (1946) has reviewed this species. He finds my specimens (August through November) to be the nominate race. Phillips restricted the type locality to the region between Tacna and Mohawk, Yuma County, Arizona. But from what we now know of the winter distribution of Le Conte's, the type could as well have been taken anywhere downriver from the Pima villages.

Change: Most Le Conte's Thrashers are found today in a relatively small area of undisturbed *Ambrosia-Atriplex-Lycium-Prosopis* desert from the villages of Sacate west to Gila Crossing. As the tribe permits more and more mechanized farming and other nontribal activities onto the reservation, this rare thrasher is bound to disappear.

Crissal Thrasher, *Toxostoma crissale* Henry

Ethnographic: See "Thrashers." Probably *kuli huch*.

Historic: Breninger (1901) simply noted several seen. Gilman (MS) called Crissal an "Abundant resident, frequenting the dense mesquite thickets not far from the river." Of 112 thrasher nests under observation in 1908, Gilman (1909a:51) found forty-five of this species. All years combined, Gilman (MS) recorded one nest in January, fourteen in February, twenty-eight in March, thirteen in April, ten in May, eight in June, and one in July. He found fresh eggs 24 January and called the species "the earliest nester of any of the birds noted." Among Gilman's unpublished observations, these bear repeating: "It was noticed that the nests occupied in the earlier cooler part of the season were much better made and lined than those later on when it was hot. In May, June, and July they were very flimsy structures."

Modern: Common in the mesquite bosques. Crissal Thrashers are more numerous than sometimes suspected. In their favored habitat they call at the very first light of dawn, even in winter, but may remain elusive for the remainder of the day. Abandoned Pima fields, which quickly revert to dense mesquite stands, are favored by Crissals. Just south of Gila Crossing one can stand on a sand dune and see and hear Crissal, Le Conte's, Bendire's, and sometimes a mockingbird, each in its preferred microhabitat.

Taxonomy: I concur with Friedmann (1950), Delacour and Mayr (1945), Phillips (1962), Phillips et al. (1964), and other authors in rejecting the name *Toxostoma "dorsale"* (a printing error) and in keeping *Toxostoma crissale* (see

Intern. Comm. Zool. Nomenclature, 1963, Article 32a [ii] and 32b; and especially Hubbard, 1976).

Where Crissal Thrasher races intergrade along the Gila drainage is still unknown. A series of freshly molted skins from the reservation represents the darker, more eastern nominate race, indistinguishable from the Tucson population (forty-four fall specimens examined). However, an adult from near Tacna, Yuma County (LBSU 3900, 17 September 1969) is just as dark as comparable reservation birds. Phillips et al. (1964) believed that the pale *T. c. coloradense* van Rossem might range up the Gila as far as Arlington, Maricopa County, but my fall specimens from there are all nominate *crissale*. The darkest specimen I have seen (ARIZ 1937, 31 December 1956), as well as the darkest and brownest reported by van Rossem (1946), were taken in New Mexico. Phillips (personal communication) also took a dark specimen (DEL) in that state. Fresh material might demonstrate two races within the range of nominate *crissale,* as presently conceived.

Sage Thrasher, *Oreoscoptes montanus* (Townsend)

Historic: Gilman (MS) found Sage Thrashers a "Common winter visitant." In his thrasher paper, Gilman (1909a:50) wrote, "The Sage Thrasher is here only for the winter, and was first noted 30 November. The last seen was 30 March. They were not numerous at any time and occurred any place from river to hills." His manuscript gives 9 October 1914 as the early date for the species.

Modern: Rare in winter, fairly common to common during spring migration. I know of but two modern winter records for the reservation: near the Salt-Gila confluence 5 December 1970 (JSW, now MNA 5603, female, no fat) and south of Gila Crossing 14 January 1967 (AMR 1256, adult male).

Phillips et al. (1964:126) say of the Sage Thrasher: "Movements poorly understood.... It is rather erratic in its occurrence in southern Arizona." On the reservation there is a conspicuous March flight. Specimens (AMR) extend from 12 March (three different years) to 27 March. Usually they become actually abundant in mid-March, with sometimes a dozen per field (as in 1971 and 1972). Numbers diminish in early April. I recorded them "abundant" on the Sierra Estrella bajada 7 April 1973 but found fewer there the following day, my latest observation of the species. Simpson took a belated migrant (JSW, now MNA 5602, female, fat) near Gila Crossing 4 May 1970.

Taxonomy: The Sage Thrasher is apparently monotypic. It is closely related to (Mayr and Short [1970:68] believed "doubtfully separable from") *Toxostoma.* Of the seven thrasher species I compared osteologically, *montanus* seems nearest to *bendirei.* However, *Oreoscoptes* possesses sufficiently distinctive characters, both osteological (personal observation) and external (Ridgway, 1907) that I do not consider it congeneric with *Toxostoma.*

Turdidae

Northern Robin, *Turdus migratorius* Linnaeus

Ethnographic: *básho svek;* [pl.] *bápsho svepekĭ* (*basho,* 'chest'; *svek,* 'red')

Historic: Gilman (MS) called this robin a "Regular winter visitant though in small numbers." His extreme dates are 26 October 1910 to 12 April 1914, both at Sacaton.

Modern: Erratic in winter: uncommon to common (or absent). In most winters it is uncommon, a few individuals being seen here and there. But the winter of 1972–1973 robins were actually common across the reservation. I found none during the winters of 1965–1966, 1968–1969, 1970–1971, and 1975–1976.

Taxonomy: Six reservation specimens are the paler, larger *T. m. propinquus* Ridgway, breeding in the West exclusive of the humid Northwest. One specimen (AMR 4339, adult male) taken 21 January 1973 is blacker above and rich and dark below, with more white in the tail edges. It is apparently approaching eastern nominate *migratorius*. Identifications are by Hubbard, Parkes, and Phillips.

Varied Thrush, *Zoothera naevia* (Gmelin)

Modern: Casual. I took one specimen (AMR 4249, female) at Komatke 26 November 1972.

Taxonomy: According to Phillips et al. (1964), *Hesperocichla* Baird is the correct name for this genus rather than *Ixoreus* Bonaparte, currently in use (A.O.U. Check-list, 1957). The nomenclatural problem hinges on whether the type specimen or the type species is the basis for the name. Regardless, Ripley (1952, and *in* Deignan, Paynter, and Ripley, 1964) has made a strong case for including both the Varied Thrush and the Aztec Thrush, *Ridgwayia* Stejneger, in the Old World genus *Zoothera* Vigors. Some authors agree (Mayr and Short, 1970). Keith (1968) pointed out the parallel tendency to winter vagrancy between the Varied Thrush and several northern Asian species. *Ridgwayia pinicola, Hesperocichla naevia, Zoothera (Oreocichla) mollissima,* and *Z. (Geokichla) citrina* are virtually identical in osteology and certainly should be considered congeneric. *Turdus* (both Old and New World species examined) is quite similar, differing slightly in details of the pectoral girdle and in heavier build. But *Myiophoneus,* generally placed near these in world lists, differs radically in overall skeletal structure.

Two races of Varied Thrush can be identified on the basis of females. Based on recent specimens, I have identified the reservation specimen (as well as another I took at Roll, Yuma County, Arizona, 18 or 19 November 1972, destroyed, University of Arizona) as the pale interior race *Z. n. meruloides* (Swainson). There are four previous specimens (mid-winter to spring) for Arizona (Phillips et al., 1964; Carothers and Haldeman, 1967; Carothers et al., 1973).

Hermit Thrush, *Catharus guttatus* (Pallas)

Historic: The "Willow Thrush" Breninger (1901) reported in September undoubtedly was either the Hermit or Swainson's Thrush. Gilman (MS, 1914b) collected two specimens, 7 March 1908 and 18 September 1910, and reported another seen 2 October. I have not relocated his specimens for racial determination.

Modern: Fairly common. I collected eighteen Hermit Thrushes on the reservation. Migrants pass through from 10 September through October (and perhaps to 2 November). Spring migration is in March and April, extending (on the top of Sierra Estrella in *Ribes quercetorum*) to 5 and 6 May. They are more numerous in spring. During migration Hermit Thrushes may be anywhere in moist, shady microhabitats. Some thrushes winter from November through February. These wintering birds are pale and medium-sized and have little or no fat deposit. Barehand Lane was a favored wintering area as it afforded moist open spaces beneath mesquites.

Taxonomy: The restriction of the genus *Hylocichla* Baird to the species *mustelina* (Gmelin) proposed by Dilger (1956a, 1956b) and followed by various recent authors, has been officially adopted (A.O.U. Check-list Committee, 1973). The remaining four species form a species group (Mayr and Short, 1970) in the genus *Catharus.* Hermit Thrushes have been reviewed by Phillips (1961; also *in* Phillips et al., 1964) as well as by Aldrich (1968). Most of the reservation specimens were identified by Phillips.

C. g. auduboni (Baird) is the largest race, breeding in the Southwest and northward in the Great Basin and Rocky Mountains; (*C. g. polionotus* [Grinnell] is indistinguishable; see Austin and Rea, 1976). I took two fall specimens (AMR 1117, 10 September 1966; and 2396, 28 September 1968) and one in spring (AMR 4726, 5 May 1975). All three were immatures. Phillips et al. (1964:130) describe *auduboni* as "rarely appearing in the [Arizona] lowlands." The earliest fall specimen previously was 15 September 1948 from Tucson (ARP).

The next largest race, also grayish, is *C. g. sequoiensis* (Belding). I have identified one specimen (AMR 4808, adult female, 2 November 1975) as this race, which breeds in the Sierra Nevada of California.

Phillips identified one specimen (AMR 4152, adult male, 1 October 1972) as *C. g. sleveni* (Grinnell), breeding in coastal California. He identified another small thrush (AMR 1439, immature female, 16 April 1967) as *C. g. jewetti* Phillips, which breeds in the Olympic Mountains of Washington. He now considers (personal communication) *jewetti* probably inseparable from *sleveni* because of too much overlap.

Three specimens (AMR 1765, 21 October 1967; 538, 25 April 1964; 4747, 5 May 1975) are *C. g. verecundus* (Osgood) (= *C. g. "nanus"* of Aldrich, 1968). This race breeds from southeast Alaska and coastal islands south to northern British Columbia. Both spring specimens are late records for the Arizona lowlands. I refer three very dark, nonreddish backed specimens (all males) to *C. g. vaccinius* (Cumming) (if recognizable): 28 October 1977 (SD), 9 December 1972, 14 March 1971 (both AMR). These appear too dark and sooty to be included in *verecundus. Vaccinius* breeds on Vancouver Island.

Six specimens were identified (Phillips and Rea) as nominate *guttatus,* breeding from the Alaskan Peninsula south at least to south-central British Columbia (and perhaps the west slope of the Cascades, *fide* Phillips). These birds are small, grayish-backed, with contrasting deep reddish tails. Five are winter specimens and one (AMR 3162, male, 14 March 1971, heavy fat) is a migrant. Additional Hermit Thrushes seen in winter appeared to be of this race.

I have identified one specimen (AMR 4809, immature female, 2 November 1975) as the rich, brown-flanked *C. g. nanus* (Audubon) (= *C. g. faxoni* Bangs and Penard of the A.O.U. Check-list, 1957). This distinctive subspecies breeds in the eastern part of the continent. My specimen compares well with recent fall and winter skins from Maryland, Virginia, and Louisiana (ARIZ) and with the only other Arizona specimen (GM 392, 29 November 1953, also an immature female). Reasons for the transfer of Audubon's name *nanus* from a western to an eastern race were given by Phillips (1961). This was discussed earlier by Osgood (1901). The identification of the name *nanus* is based primarily on the colors of Audubon's (1841) plate 147 and the fact that all but one in Audubon's series were eastern specimens. Parkes (personal communication) considers the change unfortunate but probably justifiable.

Swainson's Thrush, *Catharus ustulatus* (Nuttall)

Historic: Gilman (MS) recorded the Swainson's Thrush as "Seen occasionally," giving September and October as the only months.

Modern: Rare transient. I observed this thrush only in the fall of 1968 and found one dead in spring 1976. In a net set in the old convent gardens, Komatke, I took five specimens from 13 September to 1 October and found another there dried on [15 October] (but after 2 October). We collected a seventh specimen 5 October at Sacaton. Five of these were immatures.

The only spring record (AMR 4845, adult female) was found dead in a field beneath a powerline near Co-op Colony, 16 May 1976. It had caught its head between its primaries apparently while preening and was unable to extricate itself.

Taxonomy: Phillips (*in* Phillips et al., 1964) has reviewed the races. He identified six of the fall birds as *C. u. ustulatus* (two perhaps being intermediate toward *C. u. oedicus* Oberholser). One is *oedicus* (AMR 2405, adult male, 1 October). The only previous fall record of this race for Arizona is Monson's 5 September 1956 specimen (ARIZ 5081) from the Castle Dome Mountains in the extreme western part of the state. "Normally, *oedicus* does not stop in fall north of southern Sonora" (Phillips et al., 1964:131). The one spring specimen appears to me to be *oedicus*. Both *ustulatus* and *oedicus* are Pacific coastal races, *oedicus* being the more southern.

Western Bluebird, *Sialia mexicana* Swainson

Historic: Gilman (MS) considered the Western Bluebird a "Regular winter visitant, feeding largely on the berries of mistletoe and seeking food in cultivated fields." The only specific date that he mentioned is for ten birds seen at Sacaton 25 November 1910.

Modern: Irregular. Uncommon some winters. I have found the species only about Komatke, and then only during three of thirteen winters. Actual dates are: 15 December 1963 to 1 February 1964 (four specimens taken), eight individuals on 28 December 1967, and three or four on 26 December 1972. All were in mature mesquite trees in open areas. Phillips et al. (1964:132) said Western Bluebirds "Winter...somewhat irregularly in farmlands and on the desert wherever mistletoe occurs."

Taxonomy: Breeding birds (AMR, JSW, ARIZ) of the Bradshaw Mountains near Prescott, Yavapai County, Arizona, not far from the type locality of *S. m. bairdi* Ridgway, show the full range of variation in amount of chestnut in the backs of males. Thus I do not recognize *bairdi* (1894) as a race distinct from *S. m. occidentalis* Townsend (1837). The problems involved in attempting to recognize these northern races were discussed scrupulously by Ridgway (1894). Although breeding specimens of the northern Baja California populations, *S. m. anabelae* Anthony, are indeed quite purple, this is entirely an artifact of feather wear; fall and winter topotypes (SD) are indistinguishable from Oregon *occidentalis* (Jewett Collection, SD). In fact, some Sierra San Pedro Mártir birds have more chestnut above and below than some Oregon *occidentalis*. The alleged size differences all overlap and the three western "races" cannot be distinguished off their breeding grounds.

Change: It would seem from Gilman's characterization of the species as a regular winter visitant that there has been a change of status and considerable diminution in numbers of Western Bluebirds on the reservation.

Mountain Bluebird, *Sialia currucoides* (Bechstein)

Historic: Gilman (MS) recorded the Mountain Bluebird as "Frequently seen in cultivated fields during winter."

Modern: Erratic. Abundant or absent in winter. I found them during only two winters, with extreme dates: 23 November 1972 to 3 March 1973 and 16 November 1974 to 10 March 1975 (only one left). A favored place for Mountain Bluebirds is along fences where flocks of sheep are pastured on permanent alfalfa. When the sheep are moved, the bluebirds leave also. Several hundred may accumulate at pastures, as at Casa Blanca 20 January 1973. But forty-nine were in the open *Atriplex-Lycium* flats about Sally Pablo's house 23–26 November 1972. Phillips et al. (1964:132) said it "Winters...commonly in most years south of its breeding range in farmlands, grasslands, and open berry-bearing woods and brush."

Change: I tend to doubt that any real change has occurred.

Townsend's Solitaire, *Myadestes townsendi* (Audubon)

Historic: Gilman (1911a) found one solitaire in a date grove at Sacaton 23 October 1910.

Modern: Uncommon. I saw Townsend's Solitaires in three years out of thirteen. Two specimens (AMR 2365 and 2388, both immature females) were netted in the old convent garden, Komatke, 17 and 26 September 1968. I saw one individual here 12 November 1963. Two solitaires spent the winter at a cliff in the Sierra Estrella (observed 7 January to 8 April 1973), and E. Linwood Smith found one in this range 5 November 1975. Phillips et al. (1964:133) gave its status as "Winters...rarely in the better-wooded Lower Sonoran valleys and canyons, and even at times in the desert mountains of extreme southwestern Arizona...."

Taxonomy: The reservation specimens are the expected nominate race.

Sylviidae

Black-tailed Gnatcatcher, *Polioptila melanura* Lawrence

Ethnographic: *schuk máw'awkum gí:sop (schuk,* 'black'; *máw'awkum,* 'head' or 'hair' + possession marker; *gí:sop,* 'verdin')

The lullaby *"Schuk máw'akum gí:sop,"* sung to Pima infants in the *naki,* 'cradle,' apparently is based on reference to the deep, finely constructed nest of the gnatcatcher. The Riverine Pima apply the name *gí:sop,* without a modifier, only to the Verdin. However, Delores Lewis, a Papago, called the gnatcatcher itself simply *gí:sop.* This is perhaps a personal ethnotaxonomic variant.

Historic: Heermann (1859) collected this gnatcatcher "from a hedge surrounding the cultivated fields of the Pimos Indians...." Breninger's (1901:46) note of "Western Gnatcatchers, *Polioptila caerulea obscura*—Seen among the mesquites" doubtless applies to this species. Gilman (MS) recorded it (under the name *Polioptila plumbea*) as:

> Abundant resident, being equally at home in the mesquite thickets and in the open desert country. Frequently victimized by the Dwarf Cowbird. The nests were placed in the numerous spiny shrubs and at an average height of 4 1/2 feet and extremes were 2 1/2 and 9 feet. The earliest nesting date was 26 March and the latest 31 May. Of 20 nests noted, 15 were in April, 2 in March, and 3 in May.

Modern: Common. The species stays in pairs, year round, in mesquite bosques, fence rows about rancherias, and in bajadas and their arroyos.

Though Phillips et al. (1964:133–34) considered the Blue-gray Gnatcatcher, *Polioptila caerulea* (Linnaeus), to be a "Common winter resident of wooded river valleys from the Tucson area and the Salt and Big Sandy Rivers southward...common transient in western deserts north of the Gila River," I have failed to discover any, though I have searched every winter. Its absence as a wintering bird is no doubt due to the lack of suitable riparian communities with understory. Johnson and Simpson (MS) have taken it in better wooded areas on the lower Verde and Salt rivers, as at Blue Point Cottonwoods.

Taxonomy: The Black-tailed Gnatcatcher of the Sonoran Desert, *Polioptila m. lucida* van Rossem, with white bellies and broadly white-tipped outer rectrices, sounds to me so entirely different from the gray-breasted, tan-washed flanked bird with narrowly white-edged outer rectrices that lives in the southern coastal chaparral that I consider the latter a full species, *Polioptila californica* Brewster, the California Gnatcatcher (see Hoffmann, 1927). Ranges of the two may meet or overlap at several points.

Ruby-crowned Kinglet, *Regulus calendula* (Linnaeus)

Historic: Breninger (1901:46) recorded "Several seen; evidently just down from the north." This was either 18–19 or 25–26 September. Gilman (MS) found it a "Common winter visitant." His latest date (and latest lowland Arizona record) is 15 May 1914 at Sacaton.

Modern: Abundant in winter. My extreme dates are 4 October to 8 April (male, singing). In a mesquite bosque at Arlington (28 miles or 45 km west of the reservation) I took a female (AMR 1014) 7 May 1966. In winter this species and the Audubon's Warbler are the most numerous insectivorous birds on the reservation. Kinglets even frequent the saltcedar thickets usually avoided by other birds. Apparent arrival dates for four years are: 4 13, 19, and 25 October. However, by this last date the species was already common at the Salt-Gila confluence.

Taxonomy: Browning (1979) reviewed the species, concluding that both *R. c. cinerascens* Grinnell and *R. c. arizonensis* Phillips could not satisfactorily be separated from nominate *calendula,* as Hubbard and Crossin (1974) also had done. Although Browning considered the small, dark, richly colored *R. c. grinnelli* Palmer, breeding from southern Alaska to Vancouver Island, to be resident, there are typical examples of *grinnelli* taken from mid-October on, wintering south to at least Humboldt County, California (AMR, ARP, SD). All reservation specimens are the larger, grayer nominate race.

Motacillidae

Water Pipit, *Anthus spinoletta* (Linnaeus)

Historic: Gilman (MS) recorded this pipit as "Seen occasionally in late fall and early spring."

Modern: Fairly common to abundant in winter. In most years and at most places on the reservation, a pipit or two may be seen at shallow irrigation ponds. Greater numbers are attracted to the agricultural areas about Laveen and Chandler, particularly in mid-winter. However, this species was actually abundant at the west end of the reservation along dirt roads and fallow fields 31 December 1975 and 26 December 1976.

Taxonomy: North American races of this species wintering in Arizona have been reviewed by Phillips (*in* Phillips et al., 1964). I have made no attempt to investigate the races and their relative numbers wintering principally at the edges of the reservation. The single reservation specimen (AMR 2874, male, 29 March 1970) was identified by Phillips as *A. s. geophilus* Oberholser, the most abundant form wintering in southern Arizona.

Bombycillidae

Cedar Waxwing, *Bombycilla cedrorum* Vieillot

Historic: Gilman (MS) recorded "One seen at Sacaton 1 June." Apparently this was his only record.

Modern: Uncommon and erratic, certain winters. My extreme dates for the reservation are 1 September to 28 December. Individuals usually spend only a few days at one location, as one adult and an immature in a garden at Komatke 10–14 October 1965. I observed waxwings in but seven of sixteen winters: 1963, 1965, 1966, 1967, 1972, 1975, 1980.

Phainopepla, *Phainopepla nitens* (Swainson)

Ethnographic: *kwígam (kwi,* 'mesquite'; *-gam,* attributive or ownership marker)

The Phainopepla is called the 'mesquite dweller.' Joseph Giff explained the etymology of this name as 'pertaining to the mesquite,' 'mesquite owner' or 'mesquite, its home is there.' Several Papago and lowland Pima Bajo did not recognize this ethnotaxon, but I could not elicit an alternative name, although the bird itself was familiar to them.

Historic: In mid or late September Breninger (1901:46) called the Phainopepla, "Less numerous than usual with this species; only one male noted." Gilman (MS) treated the species at some length:

> Abundant for the greater part of the year. Winters in considerable numbers, feeding largely during that season on mistletoe berries. Nesting begins in February, the earliest date being 14 February when a nest with two eggs slightly incubated was found. This nest was close up under a mistletoe in a mesquite and was well protected and hidden. Six May was the latest date of finding eggs. Of the nests noted 80% were in mesquites and 20% in palo verdes, and of the nests in the mesquites 80% were in or just under clumps of mistletoe.... Along toward the latter part of June the birds leave and do not return until about the first of September. Probably they seek higher altitudes for the summer to escape the great heat.

His earliest dates for returns were 23 and 28 August 1914 at Sacaton.

Modern: Fairly common to common part of the year. This is a complex species. I can offer only a rough outline toward understanding the reservation Phainopeplas, and even this is possibly an oversimplification. They appear in three different habitats apparently in three different roles (see below). Whether or not these are even the same populations is a matter of conjecture (see Phillips *in* Phillips et al., 1964:139–140). No birds were banded.

1. *Dry mesquite floodplain.* (Mature trees with growths of mistletoe, particularly where trees survive along fence rows.) The species arrives here usually in early October. My earliest record is two at Sacaton (one male, possibly one female) 30 September 1972, but none that day at Vahki, Casa Blanca, or Santan. The usual winter densities build up in November. Breeding occurs from mid-February through March. A female taken 13 February 1965 had an enlarged oviduct and developed broodpatch; one taken 24 February 1967 had two ruptured follicles. Nests are placed in dense clumps of mistletoe and both sexes are defensive about the nest tree. My late dates (Komatke) in dry *Prosopis* are males and females 19 April 1964 and a single pair 30 April 1966; I saw three juvenals flying over a mesic ranch at Komatke 6 June 1978, when young were abundant at Barehand Lane (see below, "Marshes with *Sambucus*").

2. *Mountains above 2,000 feet (610 m).* In the steep upper part of the Sierra Estrella, above the last junipers, Phainopeplas were conspicuous 5 May 1975. I found a nest with two large young being fed *Lycium* berries. This was the only time that I was in the more mesic upper part of the mountains at that date, so I cannot say how regular nesting is.

3. *Marshes with* Sambucus. Beginning mid to late May Phainopeplas accumulate in elderberry trees at marshy areas as at Barehand Lane and points along the Salt River. Densities are greater than in dry *Prosopis.* By early June many of the birds are volant young (perhaps not all locally raised?) and I have seen stub-tailed juvenals still being fed as late as 22 June 1975. There is great year-to-year variation in numbers. At Barehand Lane on 3 June 1978 about a hundred gray individuals (most appearing to be panting juvenals) were in the abundant elderberries. Only three adult males were present. At the open marsh adults and young were abundant over the open water and emergent vegetation on 16–17 July 1971, but on 30 July 1972 I failed to find a single one there. At least by August Phainopeplas vacate even these wet areas. They return in numbers only the following

summer. I worked the marsh from 8:00 to 18:00 hours 29 October 1972 and found but one female. Three were present here 24 November 1972 and one 23 December 1972. On 26 April 1981 I found two or three adult males but no females or young.

Taxonomy: Reservation specimens are the expected northern race *P. n. lepida* Van Tyne.

Change: Presuming that Gilman included both xeric *Prosopis* birds and mesic *Sambucus* birds (he mentions elderberry trees nowhere in the MS), then there has been no change in Phainopepla distribution in the past seventy-five years. Breninger explicitly noted the rarity of the species in September.

Sturnidae

Common Starling, *Sturnus vulgaris* Linnaeus

Modern: Common. Starlings were first observed in Arizona in 1946 and found breeding in the Phoenix area in 1954 (Phillips et al., 1964: 141–142). I first saw the species at Komatke 4 October 1963. The following spring starlings began investigating woodpecker holes in the old convent dormitory at Komatke. I collected ten specimens from this first colony between mid-March and mid-April. Kessel (1953) suggested that breeding range extension is accomplished by colonization by migrant and wandering first-year and nonbreeding second-year birds. Seventy percent of the new Komatke colony proved to be adult (age determinations by David E. Davis). Starlings now breed regularly in saguaro stands as well as in the more urban parts of the reservation, as at Sacaton. Hundreds formerly joined winter flocks of thousands of icterids roosting at Barehand Lane before the marsh was destroyed.

Change: European starlings were introduced into New York in 1890 by direct human agency. From there they spread across the continent. They may now compete with other hole-nesting species.

Vireonidae

Hutton's Vireo, *Vireo huttoni* Cassin

Modern: Rare. I have three late winter to early spring specimen records: 13 February 1965, 14 February 1966 (both Komatke), and 12 March 1966 (Barehand Lane). All were in bare trees. I observed a fourth bird at very close range at the Salt-Gila confluence 2 November 1975 in a young willow thicket. This vireo was not recorded by Gilman. Phillips (personal communication) notes, "The species breeds in Mazatzal Mts. and Sierra Ancha, but these are northeast [of this study area]; rare so far west."

Taxonomy: The three specimens are the pale *V. h. stephensi* Brewster, breeding from southern Arizona southward. It is duller, less greenish than nominate *huttoni,* resident in the Pacific states.

Bell's Vireo, *Vireo bellii* Audubon

Historic: Gilman (MS) recorded the Bell's Vireo as a "Sparse summer visitant along the Gila River, nesting in dense mesquite thickets; a few old nests seen and one in April."

Modern: Rare and local. When this species establishes a breeding territory it is not easily overlooked because of its loud "question-and-answer" song. My first reservation record was a pair in the dense mesquite bosque at Komatke from May to at least 25 September 1965 (still singing). I heard none the following summer. At the St. John's sewage leaching beds (dense young *Prosopis* and *Atriplex canescens*) a pair again was present through the spring and summer of 1967. I documented the species by tape recording. I discovered another territorial bird at Barehand Lane that summer. Both males sang until at least 3 August. The following year the St. John's male sang continuously (same territory as previous year) from 7 April to 4 August. I saw one individual at Barehand Lane 26 May to 20 June 1968, but I heard no territorial song. The sewage pond at St. John's was bulldozed sometime after 1968 to create a large vegetationless pond. I discovered a territorial Bell's Vireo near Gila Crossing 1–3 September 1973. This was along an irrigation ditch with thick *Prosopis* and dense undergrowth. There was one territorial male in a fence row of Don Thomas' fields, Komatke, 5 June 1978, and one in Felix Enos' fields 10–12 May 1982 (see Figs. 3.6, 3.7, 3.8). At the Salt-Gila confluence I heard one or two territorial males 23 June 1974 and 1975 but not in 1976. At the Olberg Bridge spillway on the Gila one was singing 23 July 1974 in mesquite mixed with cattails and saltcedars. In the Salt River at 91st Avenue another was singing in a tall young willow thicket 1 September 1980 and 26 April 1981.

Taxonomy: A single specimen (AMR 946, 24–25 March 1966, Komatke) was identified by Phillips as *V. b. arizonae* Ridgway. It was singing in bare mesquites at St. John's Indian School since the previous day.

Arizonae breeds commonly farther east on the Gila, between The Buttes and Cochran and undoubtedly upstream where the riparian community is intact. A male from Cochran (SD 40521, 8 May 1978) is too bright and too yellow on the flanks for the local race. It was identified independently by Parkes and Phillips as a vagrant *V. b. medius* Oberholser of the Rio Grande, previously unknown from Arizona.

Change: Bell's Vireos formerly bred in the Phoenix region until at least 1889, but later were considered absent from that area and downriver to Yuma and the Colorado River (Phillips et al., 1964; see map p. 142). These authors considered parasitism to be the cause of the decline. The presence of Bell's Vireos today at the western end of the reservation apparently represents a comeback. I believe habitat change to be an important factor. On the reservation I found the species only in quite mesic (usually semi-marshy) microhabitats with very dense vegetation.

Solitary Vireo, *Vireo solitarius* (Wilson)

Historic; Gilman (MS), under *cassinii,* noted that it "Appears during migration in spring and fall." He did not mention the Plumbeous Vireo, *V. s. plumbeus* Coues, readily identifiable in the field.

Modern: Uncommon transient. I have seen this species but nine times on the reservation: 21 April to 15 May and again 29 August, 1 October, and 27 November (confluence).

Taxonomy: *V. s. cassinii* Xantus breeds along the Pacific states. Seven reservation specimens are this smaller, brightly colored race. *V. s. plumbeus,* breeding at higher elevations in the Southwest and Great Basin, is represented by two specimens: AMR 749, 8 May 1965, and AMR 4786, 29 August 1975 (the only female of the nine secured). The August Plumbeous, taken at the Salt-Gila confluence, is a notably early record for the lowlands.

Philadelphia Vireo, *Vireo philadelphicus* (Cassin)

Modern: Accidental. I took a single adult male (MVZ 150790) in a tamarisk grove at Komatke 12 October 1963. The previous three state specimens were immatures taken in October and November.

Warbling Vireo, *Vireo gilvus* (Vieillot)

Historic: Gilman (MS) considered the Warbling Vireo "Fairly common during spring and fall migrations."

Modern: Uncommon spring, fairly common fall transient. I collected two spring specimens and saw an additional bird between 14 March and 6 April. During fall migration the species appears as early as 28 July (old convent garden, Komatke), with specimens from 4 August to 26 September.

Taxonomy: The eleven reservation specimens were identified by Phillips. All but one are the small *V. g. swainsonii* Baird, which breeds along the Pacific States (exclusive of the Great Basin and Southwest, *contra* A.O.U. Check-list, 1957). One specimen (AMR 1678, immature male, 31 August 1967) is considered *V. g. "leucopolius"* Oberholser (possibly approaching *brewsteri* (Ridgway)), an interior race. Browning (1974) would formally recognize *leucopolius* with *V. g. petrorus* Oberholser as a synonym.

Parulidae

[?Black-and-white Warbler, *Mniotilta varia* (Linnaeus)]

Modern: In the thick vegetation *(Salix* and *Arundo)* above the Salt-Gila confluence on 28 October 1977 I briefly saw at close range a warbler that perched creeperlike on a willow. While it may have been a Black-and-white Warbler, which has been taken in Arizona, undoubtedly most of the supposed sightings attributed to these species are actually Black-throated Gray Warblers, which now winter on the lower Salt.

Prothonotary Warbler, *Protonotaria citrea* (Boddaert)

Modern: Accidental. On 27 December 1977 I took an adult female (SD 40982) at the Salt-Gila confluence. It was in young willows at the edge of the river with very dense understory. The bird was foraging head down in dried leaves just above the water. There are three previous specimen records (Austin et al., 1972) for Arizona.

Tennessee Warbler, *Vermivora peregrina* (Wilson)

Modern: Accidental. Two reservation specimens. I found the first (AMR 3899, female, probable adult, 20 May 1972) at Komatke under a mesquite tree where it had been beheaded and dropped by a shrike. The second (AMR 4827, immature female, 31 December 1975) was in willows at the Salt-Gila confluence. There is one previous specimen for Arizona (Phillips et al., 1964). I saw a probable third individual on the Salt west of 91st Avenue on 28 December 1977.

Orange-crowned Warbler, *Vermivora celata* (Say)

Historic: Gilman (MS) found the species a "Common migrant, being seen at various times in spring and fall... noted as follows: 7 March, 25 April, 4, 10 September, 6, 8 October, and 4 November."

Modern: Common in fall migration, wintering locally; uncommon in spring migration. My only spring specimen is 12 March 1966 at Barehand Lane (wintering or spring migrant?). The other spring record is on 2 April 1978, when numbers were seen in cottonwood and willow thickets on the lower Salt. Fall migration dates are from 2 September to 16 November (if not wintering). Of nineteen specimens taken, fifteen were males and two unsexed. At least in recent years the species has wintered in numbers where there are dense vegetation (willows and saltcedars) and water, as at the Salt-Gila confluence and formerly at Barehand Lane.

Taxonomy: *V. c. lutescens* (Ridgway) of the Pacific states and *V. c. orestera* Oberholser of the Rocky Mountains are equally common during fall migration and among wintering birds. *Orestera* is more abundant in winter (nine of thirteen specimens taken in December and January). Although *lutescens* was previously known wintering in Arizona from but a single specimen (Phillips et al., 1964) I have taken four typical male specimens (AMR, SD) between 23 December and 6 January. The one spring specimen (AMR 937) is *orestera*. All the migrants on 2 April appeared to be of this race also.

Change: The change in wintering status of Orange-crowned Warblers is directly related to permanent water with dense, low vegetation, at least partially broadleaved.

Nashville Warbler, *Vermivora ruficapilla* (Wilson)

Historic: Gilman (MS) found this warbler "Fairly common during migration. One secured 10 September [Sacaton]."

Modern: Rare. I collected two specimens and saw the species a third time, between 31 August to 5 October. These data contrast with the status given by Phillips et al. (1964:148).

> Fairly common transient throughout Arizona. Here the spring and fall paths are complementary: in spring, mostly the southwestern and central lowlands; fall, mostly in the mountains of the east and northwest. Both migrations end abruptly and it is unusual to see the bird after the first few days of May and after the end of September.

Taxonomy: Both recent specimens (AMR 1677, 2408) are the bright *V. r. ridgwayi* van Rossem, breeding in the Pacific states and the Great Basin. I have been unable to relocate Gilman's specimen for subspecific identification.

[Virginia's Warbler, *Vermivora virginiae* (Baird)]

Modern: Rare. I saw this species only once, 24 July 1974, at the Salt-Gila confluence. Though I was unable to obtain a specimen for verification, the sight record at that date is probable.

Taxonomy: Phillips et al. (1964) considered *virginiae* a subspecies of *V. ruficapilla*. Mayr and Short (1970) suggested "Field studies are needed in southwestern Idaho and northern Utah where *virginiae* and *ruficapilla* may meet." Johnson (1976) suggested two potential areas of parapatry or limited sympatry in southern California, but not in the north.

Lucy's Warbler, *Vermivora luciae* (J. Cooper)

Historic: Gilman (MS) found Lucy's Warbler a "Common summer visitant, appearing the latter part of March and not seen later than 4 September. Nesting took place in April and May and complete sets were 3, 4, and 5 eggs. Mesquite thickets were their favorite haunts presumably there being a greater number of insects for food to be found." Three mid-May specimens (CGR, 10 – 20 May, one male, two females) collected at Casa Grande Ruins suggest that Lucy's Warblers bred there until at least 1930.

Modern: Uncommon. Known only as a spring migrant 31 March to 17 April in 1966, 1968, 1969, and 1971 (in mesquite bosques from Komatke to Santa Cruz). The birds would sing but were not territorial and never stayed. No more than two birds were ever seen or heard in any year and never longer than one day. I heard no Lucy's Warblers at Barehand Lane during three days' netting in mid-April 1971. The following summer, however, I found at least four families there and collected two volant juvenals 16 July 1972. The species was present until at least 1 August 1972. This was my first indication of breeding on the reservation. The warblers probably bred there the following two summers. At the Salt-Gila confluence Lucy's Warblers were actually abundant 23 – 24 June 1974, but I failed to find a single one the following summer. Again in June 1978 there were three singing pairs at the confluence. I heard a territorial male in the saltcedars on the Gila bed near Komatke 5 June 1975 and another singing at Komatke 15 May 1976 and 12 May 1982.

Change: Lucy's Warblers were absent from most of the Phoenix area (Phillips et al., 1964; see map, p. 149). During my first eight summers on the reservation I found none breeding. As Barehand Lane Marsh developed (old mesquite trees were drowned and *Sambucus* and saltcedars formed thickets about emergent vegetation), the species evidently found conditions suitable for nesting (1972). At the Salt-Gila confluence in 1974 the species appeared to be well established. During the following year many trees, especially *Salix,* were burned or destroyed by natural means, and the main grove became much more open. The warblers did not nest there in 1975. And the Barehand Lane habitat was completely destroyed. The recent recolonization about Komatke has been in places I worked regularly since 1963. Gilman (1909c:166) noted that at Blackwater and Sacaton the Lucy's Warblers "are most numerous in groves of mesquites not far from water, tho this may be for the fact that more trees and other cover are found not far from the river."

Northern Parula Warbler, *Parula americana* (Linnaeus)

Modern: Accidental. I collected one specimen (AMR 2973, male) in a tamarisk grove near Sacaton 21 June 1970. The bird was not in breeding condition (testes 2.8 × 3.5 mm). Spring specimens have been taken in Arizona as late as 25 May (Phillips et al., 1964).

Yellow Warbler, *Dendroica petechia* (Linnaeus)

Historic: Gilman (MS) found the Yellow Warbler "Common during September and April, but none seen during the breeding season." Specific dates given in his notes for Sacaton are 17 and 25 September and "several in mesquite" 25 August 1914.

Modern: Irregular. Uncommon spring migrant; fairly common fall transient. Spring dates (single birds collected in 1964, 1965, 1967; and two in 1978) extend from 29 April to 16 May. Extreme dates for fall migrants (nine specimens, all immatures: 1966, 1967, 1968) are 10 August to 11 September. All but one of the fall specimens were taken atop tall, thin tamarisk trees at St. John's Indian School, Komatke. In late August Yellow Warblers move more regularly through the willows and cottonwoods of the lower Salt.

Taxonomy: Phillips has identified the reservation series as *D. p. morcomi* Coale, which breeds in the interior West south to northern Arizona, where it intergrades with *D. p. sonorana* Brewster (see Phillips et al., 1964; map, p. 150). One or two of the reservation specimens are approaching *sonorana.* Three breeding *D. p. sonorana* (7–8 May 1978, SD) from Cochran on the Gila are at the palest, lightest yellow extreme of over two dozen lower Colorado River *sonorana (SD).*

Change: Surely, Yellow Warblers must have bred all along the Gila in aboriginal times but the riparian timber by Gilman's time was sufficiently devastated that the bird was extirpated as a breeding species. I have watched for the bird to colonize the Salt-Gila confluence stands, but apparently the cottonwoods there were not yet tall enough to attract the species (seldom even as migrants). Mature cottonwoods occur in rows at the Chandler-reservation boundary, but there is no understory there. The nearest breeding habitats known today are at the mouth of the Verde and at The Buttes–Cochran area. Otherwise, Yellow Warblers are now absent from the Gila from the mouth of the San Pedro River to Yuma (see map, Phillips et al., 1964:150).

Yellow-rumped Warbler, *Dendroica coronata* (Linnaeus)

Historic: Breninger (1901) in September recorded one "Myrtle" Warbler and found "*auduboni*—the commonest of the warblers; seen gleaning insects among the branches of the cottonwoods." Gilman (MS) recorded the species (under *auduboni*) as "Abundant during spring and fall but none seen in winter season." The latest fall date he gave is 22 October.

Modern: Common to abundant in winter; common spring transient. Extreme dates are 13 October to 22 May. This warbler and the Ruby-crowned Kinglet are the most abundant wintering insectivorous species in taller trees (willows, tamarisks, exotics) across the reservation. Young willow groves along the lower Salt are favored (Fig. 3.18); in December 1979 I recorded "myriads everywhere!" April and May specimens are transients that appear where the species does not winter (alfalfa fields, marshes, weedy fence rows, young mesquite thickets).

Taxonomy: Most reservation specimens (identified primarily by Phillips) are *D. c. auduboni* (Townsend), the common migrant across Arizona (Phillips et al., 1964). It breeds from British Columbia south along the Pacific. Outside dates for *auduboni* are 30 October to 17 May. Three mid-winter specimens are males. Most of the April and May birds appear to be brightly colored males.

Several specimens (AMR 684, 13 February 1965; 1007, 4 May 1966; and probably AMR 4597, 22 May 1974; and SD 40552, 11 May 1978) are the larger *D. c. memorabilis* Oberholser, breeding from the Sierra Nevada and southern Rocky Mountains south to the Southwest. These tend to have darker faces.

There is one brown-backed specimen of Myrtle Warbler (SD 41512, female, 28 December 1977) taken on the lower Salt in saltcedars and willows. The bird's distinctive call note attracted my attention. The specimen has been identified by K. C. Parkes as *D. c. "hooveri"* McGregor, which breeds from north central Alaska to northern British Columbia and northwestern Alberta. This form "is a rare winter visitant to Lower Sonoran Zone rivers and farms in western and southern Arizona, mostly in the west...." (Phillips et al., 1964). I have seen no other Myrtle Warblers on the reservation. Breninger's observation of this form was most likely a misidentified female of the coastal race of "Audubon's" Warbler, some of which show little or no yellow in the throat.

Change: Though a change in winter status seems incomprehensible, Gilman pointed out "none seen in winter," whereas the species winters commonly today wherever there is suitable habitat. Several other warblers have changed their wintering status within the century.

Black-throated Gray Warbler, *Dendroica nigrescens* (Townsend)

Historic: Breninger (1901) reported one in September. Gilman (MS) considered it "Fairly common in April and May and again in September and early October." His early fall date was 29 August.

Modern: Fairly common transient, wintering locally. Specimen dates for migrants are 30 March to 20 April and 31 August to 21 October. But I have seen the species in spring as late as 12 and 19 May. It began wintering (1973 through 1977) at the Salt-Gila confluence, where it was partial to *Salix*. I found only one here in December 1979 and none the following December.

Taxonomy: Four spring, four fall, and four mid-winter specimens are the small nominate race, breeding in the Pacific states south to Baja California (Sierra San Pedro Mártir and Sierra Juárez, SD). *D. n. halseii* (Giraud), breeding in the Southwest and the Rocky Mountains (see Oberholser, 1930, 1974; Parkes, 1953; Phillips et al., 1964), is a rare lowland migrant in the Southwest.

Change: The recent increase in wintering Black-throated Gray Warblers appears to be directly related to the maturation of the willow stands along the lower Salt. As the broadleafed riparian habitat was eliminated in "flood-control" measures, the warblers ceased to winter (1980).

Townsend's Warbler, *Dendroica townsendi* (Townsend)

Historic: Gilman (MS) found Townsend's Warblers "in small numbers in September and October."

Modern: Common migrant, spring and fall. Males predominate (twelve of fourteen sexed specimens). Spring records (specimens) are from 12 April to 20 May. At this time these warblers are migrating through the blooming mesquite bosques and are more numerous than in fall, when specimen records extend from 2 September (AMR 1686, immature [female]) to 26 November, with a sight record for 30 August. The late November record is for two males (AMR 4244, 4245) taken in tamarisk trees at Komatke. If not wintering, these are the latest fall migrants for Arizona. The species began wintering casually in the willow and saltcedar thickets above the Salt-Gila confluence (December 1974) and by December 1977 was actually common (specimens, SD).

Change: The wintering of Townsend's Warblers is a result of the development of the *Salix-Populus* community on the lower Salt.

Hermit Warbler, *Dendroica occidentalis* (Townsend)

Historic: In September Breninger (1901:46) recorded "One seen; to be sure of no mistake I rode within ten or 12 feet of the bird." Gilman (MS) found it only in September (one for 25 September 1910 given in his notes).

Modern: Rare fall, uncommon spring migrant. Seven spring specimen records extend from 11 April to 22 May. I saw the bird but three times in fall, 26 August, 17 and 25 September (two females and a male, respectively, AMR, SD). The sexes appear equally distributed. As with the Townsend's Warbler, spring migration of the Hermit is principally through the mesquite bosques. In 1977 Hermit Warblers remained in willow thickets at the Salt-Gila confluence well into winter (SD, 27, 28 November, 27 December).

[Grace's Warbler, *Dendroica graciae* (Baird)]

Modern: Twice I saw a warbler which I believed to be this species: 20 April 1967 in the wash at Santa Cruz (behind the church) and 26 September 1968 in the old convent gardens, Komatke. Both eluded collecting, and lowland records of the species are not probable.

Chestnut-sided Warbler, *Dendroica pensylvanica*

Modern: Casual. I found this warbler once, on 28 December 1977 in young willow growth along the Salt River near Maricopa Colony (SD 40983, immature [female]). The species is casual in Arizona, with several previous fall and winter specimens.

[Ovenbird, *Seiurus aurocapillus*]

Modern: Accidental. I saw an Ovenbird on 28 October 1977 running along under low tamarisks and young willows in the eutrophic sludge at the edge of the lower Salt. The crown, bright green back, and white spotted underparts were conspicuous, but the specimen was not obtained for verification.

MacGillivray's Warbler, *Oporornis tolmiei* (Townsend)

Historic: Gilman (MS) recorded this warbler "Common in May and September," with a 16 August date in his notes.

Modern: Common spring and fall transient. At least twelve of the fourteen specimens are males. This may be actual differential migration, but I think, instead, it reflects the fact that MacGillivray's Warblers are usually concealed in low vegetation and must be called up to be collected, the males being more responsive. The true sex ratio could be verified by collecting mist-netted birds. The species appears equally common in spring and fall. Outside dates are 15 April to 11 May and 26 August to 14 September.

Taxonomy: I have not investigated races in this warbler. Mayr and Short (1970) considered *tolmiei* conspecific with *O. philadelphia* (Wilson), as suggested by Phillips (*in* Phillips et al., 1964). Mayr and Short (1970) considered races in the *tolmiei* group too weak for recognition. The A.O.U. Check-list (1957) admits but two subspecies, large *O. t. monticola* Phillips of the interior and nominate *tolmiei* for the remainder of the species' range. Phillips (1947), Phillips (*in* Phillips et al., 1964), and Oberholser (1974) recognized four races. Phillips (personal communication) now considers *O. t. intermedia* Phillips of interior British Columbia south to the Sierra Nevada not satisfactorily separable from nominate *tolmiei.* He has identified eight of the reservation specimens as *O. t. tolmiei* (including two "*intermedia*") and two as *O. t. monticola* of the Great Basin and southern Rocky Mountains. One dull specimen (11 May) is *O. t. austinsmithi* Phillips, breeding in the interior north of *monticola.* This distribution again follows the migration pattern shown by *Empidonax difficilis, Empidonax traillii, Vireo solitarius, Vireo gilvus, Dendroica coronata, Dendroica nigrescens,* and *Wilsonia pusilla,* in which the Pacific coastal subspecies is the common lowland Arizona migrant, and the more interior forms (sometimes breeding in Arizona mountains) are poorly or not at all represented.

Common Yellowthroat, *Geothlypis trichas* (Linnaeus)

Historic: Breninger (1901:46) reported "Several seen near water" in September. Gilman (MS) recorded it "Seen in spring, 27 March to 28 May and 4 September to 4 October. One male seen 9 July." He did not find the species breeding.

Modern: Uncommon migrant, breeding locally; rare and irregular in winter (1975, 1977). The usual spring migration is in March (1965, 1966, 1973; birds at nonbreeding localities). I saw a male at the Salt-Gila confluence 31 December 1975 (associated, curiously, with Orange-crowned Warblers, Black-throated Gray Warblers, Ruby-crowned Kinglets, and a Tennessee Warbler). My first indication of the species at Barehand Lane was an immature male seen various times 29 October 1972. Though the marsh at the east end was dry on 26 March 1973, the entire area was well flooded in June and the individual *Typha* and *Scirpus* clumps had extended to become contiguous. Between three and five male yellowthroats were holding territories on 18 June 1973. I collected a volant juvenal here 27 June 1973 and another near another *Typha* patch west of the road. Males were territorial at a number of points along the entire stream. In 1974, although Barehand Lane was ditched and the cattails were beginning to turn yellow, male yellowthroats were territorial and a juvenal was out of the nest by 9 June. But they did not breed here in 1975 and 1976.

The development of emergent vegetation has a longer history on the Salt. A male was still territorial on a weedy island above 91st Avenue 2 August 1969. At the confluence males were territorial in May and June and juvenals out of the nest by 23 June 1974. A lateral canal overgrown with cattails and willows (originally the confluence pond overflow) still had yellowthroat pairs 7 May 1975, but by summer the vegetation had dried and the birds were gone.

I found three territorial males on the Gila at the Olberg bridge 23 July 1974. They were in *Typha* growing among *Prosopis* below the spillway.

Taxonomy: Breeding birds are *G. t. chryseola* van Rossem, 1930. (In concept, *G. t. arizonicola* Oberholser, 1948, is a synonym of *chryseola,* but the type, *fide* A. R. Phillips, is a migrant *G. t. occidentalis* Brewster, 1883, taken 5 May at Fort Verde, Arizona.) It arrives at least by early May, when territories are established. *Chryseola* has remained as late as 29 September 1975 at the Salt-Gila confluence. Transients, identified by Phillips, include *G. t. occidentalis* (MFG 290, male, 1 October 1910, Sacaton; AMR 4403, female, 26 March 1973, Bapchule) and *G. t. campicola* Behle and Aldrich (AMR 716, male, 27 March 1965, Komatke). *Occidentalis* breeds in the Great Basin and the Colorado River drainage, exclusive of the Gila River drainage and is a common migrant in Arizona. *Campicola* breeds in the interior Northwest and western Great Plains. Phillips et al. (1964:158) consider it "a very rare migrant through Arizona, with only nine specimens, some of which may be faded females of other races." There are two winter specimens (SD), both taken on the lower Salt near Maricopa Colony on 29 December 1977. Phillips suspects that the dull female, collected by Michael Enzler, may be *G. t. yukonicola* Godfrey, breeding in southern Yukon and northwest British Columbia. K. C. Parkes identified the male as *scirpicola* approaching *occidentalis.* Phillips considers the type of *G. t. scirpicola* Grinnel (20 March, Los Angeles County, California) to be indistinguishable from the types of *occidentalis.*

Change: Breeding yellowthroats require emergent vegetation *(Scirpus* or *Typha)* of some extent. The species was no longer breeding in Gilman's day in the Sacaton area. When the marsh vegetation at Barehand Lane finally reached sufficient development (1973), yellowthroats colonized extensively. They continued to breed even as the vegetation was dying in 1974. The species survives in the 1980s only in spots along the Salt River between 91st and 115th avenues.

Wilson's Warbler, *Wilsonia pusilla* (Wilson)

Historic: Breninger (1901:46) noted "Only a few seen," in September. Gilman (MS) noted this warbler "Very numerous during migrations. They appeared as early as 14 March and were in evidence during April and May, appearing again during entire September." Additional dates given in his notes are 16 August 1914 and 16 October 1910 (Sacaton).

Modern: Common transient, spring and fall. Extreme dates are 10 March to 19 May (still common, Komatke, 1972) and 14 August to 26 September. This warbler is most commonly found migrating through *Prosopis* and other deciduous trees.

Taxonomy: Though this warbler is common, I collected only fourteen specimens on the reservation. *W. p. pileolata* (Pallas), breeding in the interior West, is represented by four specimens (22 April–15 May; and 26 September), all females but one. Dates for nine migrant *W. p. chryseola* Ridgway, breeding in the Pacific states, are from 10 March to 15 May (AMR 4740) and 29 August to 26 September (all males but one). Identifications are by Phillips and Rea. Phillips considers the mid-May male to be atypical; migrants of the Pacific race are usually gone from Arizona by mid-April. The only winter specimen of Wilson's Warbler for Arizona is an immature female *chryseola* (SD 41531) taken 28 December 1977 on the lower Salt near Maricopa Colony.

American Redstart, *Setophaga ruticilla* (Linnaeus)

Modern: Casual. Simpson took an adult male (JSW, uncatalogued) at the Salt-Gila confluence 27 July 1974. It was with two females or immature males. I saw what appeared to be another individual 29 December 1977 on the Salt west of 91st Avenue.

Yellow-breasted Chat, *Icteria virens* (Linnaeus)

Historic: Breninger (1901) did not find the chat in September. Gilman (MS) reported it only "Seen occasionally in spring and fall," but did not find it breeding.

Modern: Uncommon locally in summer. At Barehand Lane the first chat (territorial male) appeared in June 1973. Here the following summer were three singing males between 75th and 83rd avenues, and I collected one (AMR 4596). The vegetation was cleared from Barehand Lane in 1975, and chats disappeared. On 26 April 1981 I once again saw a nuptial flight over the newly redeveloped marsh near mesquites (Fig. 3.13b). At the Salt-Gila confluence from three to five males held territories (extreme dates 15 May to 4 August) from 1972 to 1975 (Johnson, Simpson, Rea). There was one territorial male in June 1977 but none during the next four summers following floods.

Taxonomy: This species is doubtfully a parulid and perhaps should be compared osteologically with thraupids (see Eisenmann, 1962). I have found no temperate genus of Parulidae or Thraupidae that resembles the distinctive osteology (notably coracoid and humerus) of *Icteria*. The distinctive wing molt (Phillips, 1974) may provide an additional clue to relationships.

Arizona chats are currently considered to be the grayer western race *I. v. auricollis* (W. Deppe) (A.O.U. Check-list, 1957; Phillips et al., 1964). This race has the tail longer than the wing. Breeding birds (May) from the interior (lower Colorado and Gila drainages) are so much grayer and less olive-green dorsally than the birds (May, June, early July) from the Pacific drainage (Oregon, California, Baja California) that it is difficult to imagine that the differences are entirely due to greater fading in desert birds. Four Cochran males (SD, 7–11 May) are especially more slaty (dorsum, flanks) rather than greenish or brownish. I have seen few specimens in fall plumage, the coastal birds being browner dorsally and on the flanks than those from the interior. The Pacific birds are *I. v. longicauda* Lawrence, those of the interior supposedly *I. v. auricollis*. A thorough examination of trends of variation is necessary to determine the taxonomic status of the chats breeding in Sonoran Desert riparian habitat.

Change: Chats are absent across the reservation, though mesquite thickets abound (Maricopa Colony, Co-op Colony, Komatke, Santa Cruz, Vahki, Sacaton, Sacaton Flats). Phillips et al. (1964) considered the species a very common, sometimes abundant, summer resident in dense Lower Sonoran vegetation. R. Crossin (personal communication) finds chats breeding in mesquite bosques of the San Pedro drainage far from surface water. Both places on the reservation where the chat recently colonized (1972 or earlier on the Salt, 1973 at Barehand Lane) were marshy habitats with deciduous trees (*Populus-Salix* or *Sambucus-Prosopis*) as well as emergent aquatic vegetation.

Thraupidae

Western Tanager, *Piranga ludoviciana* (Wilson)

Historic: Gilman (MS) considered the Western Tanager only an "Uncommon migrant. Seen the last of April, the 28th, and in May as late as the 21st. Appearing again 29 and 31 July (females) and 3 (male) and 18 August."

Modern: Fairly common in migrations. Extreme dates are 29 April to 16 May and again 16 July to 20 September. My three spring records are adult males. On the return, adult males are seen in July and August, but only immatures (specimens) in September. More collecting may demonstrate that the adults are migrating here ahead of the immatures, as with certain *Empidonaces.*

Taxonomy: Oberholser (1974) very briefly described a color race *Piranga ludoviciana zephrica* from the southern Rocky Mountains (type locality Santa Rita Mountains, Arizona). Though the proposed race "averages" larger, there is virtually complete overlap in measurements with the nominate race, as Oberholser divides the species. Though I suspect that this tanager might eventually prove to be polytypic, I recommend that no races be recognized until the important sexual and age variations are carefully analyzed in relation to geographic distribution.

Summer Tanager, *Piranga rubra* (Linnaeus)

Ethnographic: *hé'et kávult (hé'et,* 'red ocher'; *kawult,* 'something round')

This bird is described as being like the cardinal, but with a smooth head (hence, apparently, the second part of the name). A brilliant red ocher was powdered and used for facial paint and for coloring arrows. The word *hé'et* is apparently derived from the cognate *e'et,* 'blood' (see Barn Owl account). Piman *kávult* became *caborca* in Spanish, whence the town in northern Sonora with the characteristic rounded knoll. The Pima relate that this tanager was a common bird about the villages "as long as the cottonwoods lasted."

Historic: In September Breninger (1901:46) noted "A female seen." Gilman (MS) recorded the Summer Tanager "Seen only in June and July, and then but a few. Juvenals taken would indicate breeding birds not far away, even if not in the immediate neighborhood." I have been unable to relocate these juvenals. On 29 June 1910 Gilman took an adult male (SBC, orig. MFG 270) at Sacaton.

Modern: Rare migrant and local breeding species. I have seen migrants but twice (24 July 1967, adult male at Santa Cruz; AMR 4527, 23 September 1973, immature female, Komatke) and one bird in winter (AMR 358, now ARP, adult female, 22 November 1963, Komatke). Starting in 1974 a pair of Summer Tanagers bred in the cottonwoods at the Salt-Gila confluence until that grove was destroyed. They were feeding three young 27 July 1974 (Simpson, personal communication). Outside dates for this pair are 7 May (male singing) and 15 May (female present) to 4 August. Apparently a second pair nested in 1977 in the cottonwood grove on the Salt River near 107th Avenue.

Taxonomy: Racial characteristics in this species have been discussed by Phillips et al. (1964), Phillips (1966), and Rea (1970, 1972). The breeding race in most of Arizona is *P. r. cooperi* Ridgway (two specimens, Cochran). The November specimen is nominate *rubra,* breeding in the eastern United States, and the September bird is an intermediate *rubra* x *cooperi.* Gilman's June bird is *cooperi.*

Change: The Summer Tanager requires tall riparian growth (*Salix* or *Populus*) for breeding. Apparently in Gilman's day the species was nearly extirpated from the dry upper parts of the reservation, but it persisted later at the western end of the reservation (ethnographic data). The birds breeding on the lower Salt represented a recent recolonization.

Emberizidae

Northern Cardinal, *Cardinalis cardinalis* (Linnaeus)

Ethnographic: *sipok;* [pl.] *sispok*

The *sipok* is a species well known to the Pima. It occurs commonly about their houses and *vatto* ('ramada') and in fence rows that are allowed to accumulate mesquite and graythorn. Cardinals entered the quail traps but apparently were not eaten. The Papago and Pima Bajo lexeme is the same, *sí:pok* and *sípo:k,* respectively.

Historic: Curiously, Breninger (1901) did not mention seeing this conspicuous species. Gilman (MS) treated the species at some length, the essence of which is:

> Fairly common resident and apparently increasing slowly in numbers. I was told by Dr. Ch. H. Cook who had been a missionary among the Pima for more than forty years that the birds were more numerous when he first went there but that the Yaqui Indians had come in and trapped and killed them in large numbers for their red feathers. In my eight years' residence I could detect a positive increase in their numbers...the Pima are on good terms with their bird neighbors and live and let live in a way that is most admirable.
>
> The earliest nesting date was 25 April when eggs in advanced stage of incubation were found; 16 July was the latest date and the eggs were nearly fresh. Of 16 nests noted, three were in April, five in May, four in June, and four in July. Of the 16, nine were in *Baccharis,* four in mesquite, 2 in willows, and one in *Sarcobatus.* They seem to prefer building near water, hence the preponderance of the *Baccharis* for a nesting site. The nests found late in the season are of poor construction and much less material is used than in the earlier ones. The birds may be in a hurry as the season advances and also the nest should be more airy and slight to avoid the effects of the intense heat at that time of year. Four nests overhung running water. Three eggs form the usual set.

Modern: Common resident. The species occurs not only in the floodplain rancherias but also in lesser numbers in the bajadas (*Cercidium* and *Olneya*) and at the two marshes. In villages where the Pyrrhuloxia is common, few cardinals occur.

Change: Phillips et al. (1964:177) said the cardinal "Has considerably extended its range in past 75 years, both in Arizona and elsewhere." Phillips (1968:151) noted that "This gorgeous bird, apparently rare and local in the early 1870s and conceivably absent from Arizona a few years before, had spread by 1885 north to the Agua Fria River." In this light Gilman's insistence that the cardinal was increasing on the reservation (perhaps not for the reason given) is significant. The Riverine Pima and Papago morpheme appears to be a primary word and is the same among the lowland Pima Bajo of southern Sonora, Mexico, where presumably the cardinal has occurred for a longer time. I have identified a cardinal premaxilla from the floor of a pithouse dated ca. A.D. 1200 (Alder Wash Ruin, Hohokam, Upper San Pedro River). There is only one other archaeological report, also a premaxilla from the San Pedro (Big Ditch Site, Hohokam, A.D. 550–700, A. Ferg and Rea, MS).

Range expansion is not confined to the southwestern subspecies *C. c. superbus* Ridgway. The nominate eastern race of cardinal has extended northward steadily from the southern states to North and South Dakota and eastern Canada (A.O.U. Check-list, 1957; Bent et al., 1968).

Pyrrhuloxia, *Cardinalis sinuatus* Bonaparte

Ethnographic: While the Riverine Pima do not formally recognize this newly arrived species, both Papago and Pima Bajo of the lower Río Yaqui told me that they thought the Pyrrhuloxia would be included under *sipok,* 'Cardinal,' without any distinguishing specific marker. Pyrrhuloxia are quite common in lowland Pima Bajo fields.

Historic: Gilman's (MS) only record was "A pair seen near Sacaton in May 1910." Phillips et al. (1964:178) noted, "Then there are temporary extensions of breeding area; for instance, north to nine miles [14 km] east of Casa Grande [near the reservation], where a pair nested in one year, about 1924 (eggs collected by Dwight D. Stone)." Monson and Phillips (personal communication) did not see the species on the reservation in June of 1939 or 1947.

Modern: Common (even abundant) in villages where Milo maize (*Sorghum bicolor*) is regularly raised and fence rows provide protection. Rare and erratic elsewhere. I collected a male at Komatke (AMR 508, 15 April 1964) where it had been present two months, *fide* Apache Joseph Kanseah. But the species was not established there. At Komatke I found only one or two individuals in six of thirteen years, and none at all at Co-op or Maricopa colonies, where there were suitable fields. However, about Vahki, Sacate, Bapchule, Casa Blanca, Sacaton, and Santan, Pyrrhuloxias are common and far outnumber the cardinal. Curiously, I have found only cardinals at Sacaton Flats until 23 June 1974, when I located a male Pyrrhuloxia at the Olberg bridge. A winter roost (since destroyed) at Sacate in 1968 consisted of between thirty-five and fifty male Pyrrhuloxias, several females, and a few cardinals. Nesting time seems variable. At Sacate we located seven nests of territorial Pyrrhuloxias and one of the cardinal in the fence rows of two fields 9 May 1968. These were still empty 24 May. But at Sacaton on 21 June 1970 young Pyrrhuloxias had already fledged.

The species continues to expand: At Komatke in April 1978 I found at least three males and a female in fence rows. In October 1979 at least one male was at Barehand Lane.

Taxonomy: Though until recently placed in different genera (A.O.U. Check-list, 1957), the Pyrrhuloxia and the two cardinals are closely related species. A hybrid *C. cardinalis* x *C. sinuatus* (AMR 1577, immature female, 23 February 1967) was taken by Anthony de la Torre at Komatke. The specimen has the bill color of *cardinalis* and the plumage generally of *sinuatus* but is intermediate between the two species in crest shape, bill shape, wing chord, tail, and bill dimensions.

Change: The citation of Gilman's Sacaton record as the northern breeding limit of the species by the A.O.U. Check-list (1957) is misleading, as the species clearly was not established here, nor was there any evidence of breeding. Anderson (*in* Bent et al., 1968:27) wrote, "Why the Pyrrhuloxia has not established itself in the irrigated farmlands along the Gila River westward to the Colorado River is not known. Certainly it cannot be because of the higher temperatures of the lower elevation, because the species occurs in Sonora, Mexico, at probable equally high temperatures and also at sea level." The rather massive colonization of suitable parts of the reservation is a recent phenomenon.

Common Grosbeak, *Pheucticus ludovicianus* (Linnaeus)

Historic: Gilman (MS) found the western form, or Black-headed Grosbeak, a "Common migrant, appearing early in May and again through July, August, and September, though few in number. Seventeen September the latest date."

Modern: Common transient. Both sexes (apparently in equal numbers) migrate across the reservation, primarily in April and August.

Taxonomy: Two grosbeak taxa (*P. ludovicianus* and *P. melanocephalus* (Swainson) *sensu* A.O.U. Check-list 1957) meet and interbreed in a narrow zone in the Great Plains. West (1962) considered the two conspecific, Short (1969) preferred keeping them as superspecies, and Anderson and Daugherty (1974:10) concluded they "constitute a semispecies and together are a superspecies." I favor the first treatment, while admitting that the interbreeding is not as extensive and as random (*fide* Anderson and Daugherty [1974]) as in *Colaptes,* for instance. But equally abrupt zones and nonrandom mating are demonstrable between many other avian populations (and of course, with *Homo sapiens*).

I find two races of Black-headed Grosbeaks distinguishable in fully adult birds. Grosbeaks breeding along the Pacific Coast are small billed; maximum bill dimensions of adult males (n = 23) are: bill (from nares) 12.9, width of mandible 11.75, and height of bill at base 14.0. These birds are *P. l. maculatus* (Audubon), migrating commonly through western Arizona and the deserts of southern California (ARIZ, LACM, UCLA). One reservation specimen (AMR 1483, 29 April 1967, adult male, Bapchule) is *maculatus*. Birds breeding in the interior West are stout-billed *P. l. melanocephalus* (Swainson). Though I have made no attempt to determine the relative proportions of the two subspecies migrating through the reservation, the one adult female (AMR 1574, 19 April 1967, Komatke) is the interior race.

I saw a female Rose-breasted Grosbeak, *P. l. ludovicianus,* at the Salt-Gila confluence on 26 June 1977. This eastern form is casual in Arizona, with most records from May and June (Phillips et al., 1964).

Blue Grosbeak, *Passerina caerulea* (Linnaeus)

Historic: Breninger (1901:46) noted in September, "One seen; probably an immature bird." Gilman (MS) said, "I hardly know in what terms to express this bird as he was seen so seldom. Dates seen were as follows: 1 June [1914, female], 8 and 12 July; 17 September; 20 September [1910, female]."

Modern: Fairly common migrant, a few occurring in summer. Migrants appear, usually in weedy edges of fields with sunflowers, during May and again from 15 August (AMR 1660, adult male, Barehand Lane, where the species did not breed) to the end of September, with one specimen (ARP, St. John's Indian School, Komatke) caught by hand 22 October 1964, the latest fall migrant specimen for Arizona. In 1974 and 1975 there were indications that the species might breed on the reservation. At the Salt-Gila confluence 23 June 1974, Johnson, Simpson, and I observed adult males that appeared to be defending intraspecific territories. I collected an adult male and female the following day, but the female was not yet in breeding condition (largest ovum 2.2 mm). At Blackwater, Sally Pablo and I found an immature male 23 July 1974 that appeared to be holding a territory in sunflowers at the edge of a Sudan grass field, but we discovered no female or nest. A female at a small pond in mesquite at Co-op Colony on 6 June 1975 may have been a late migrant. An adult male at Barehand Lane on 22 June 1975 appeared not to be localized. Thus there is no conclusive evidence yet that the species breeds on the reservation.

Taxonomy: Reservation specimens are the large-billed *P. c. interfusa* (Dwight and Griscom). I agree with Phillips et al. (1964) and Mayr and Short (1970) in merging *Guiraca* Swainson (1827) in *Passerina* Vieillot (1816) on the basis of vocalizations, ethology, and osteology. I consider *caerulea* derived within this group. Some would include the Neotropical *Cyanocompsa* in this group. Although the birds appear superficially similar, the skeletons of *C. cyanea* and *C. cyanoides* can be distinguished readily by the configuration of the posterior iliac crest. Furthermore, the feather texture, rectrix shape, and overall coloration pattern of *Cyanocompsa* are quite different from those of *Passerina (sensu lato). Pheucticus, Pitylus,* and *Caryothraustes* are still more distinctive osteologically, the latter having a more primitive condition of the humeral head. I would keep *Cyanocompsa* and *Passerina* separate pending a more extensive comparison of all the "finches," particularly *Loxigilla, Melanospiza,* and *Rhodothraupis.* Raikow (1978:31) found that this subfamily "cannot be distinctly separated from either the emberizine finches or the tanagers on the basis of hindlimb myology."

Change: No doubt the Blue Grosbeak formerly bred along the Gila when the riparian community was intact, as it still does on the lower Verde River and on the

Gila above The Buttes. But in the disturbed and drying conditions of Gilman's day, as in the 1960s and 1970s, there appears to be only a scattering of apparently non-breeding birds during the summer.

Common Bunting, *Passerina cyanea* (Linnaeus)

Historic: Gilman (MS) found the western form, the Lazuli Bunting, a "Regular migrant, appearing about 20 April and seen up to 18 May. Appearing again about the middle of July and seen last 25 August."

Modern: Fairly common migrant. Extreme dates are 4 and 16 May and 24 July and 9 August to 29 October. All specimens and observations are for wet areas with weeds (usually sunflowers), mostly at Barehand Lane and the Salt-Gila confluence.

Taxonomy: The above dates apply to the common western form, *P. c. amoena* (Say). Johnson, Simpson, and I observed an adult and a first-year male Indigo Bunting, *P. c. cyanea,* at the Salt-Gila confluence 23–24 June 1974, and collected the immature (AMR 4610, testes 7 × 8 mm, cloaca enlarged). An adult male Lazuli was also present, but we saw no females of either race. The Indigo range has expanded greatly in Arizona in the past half-century (Phillips, 1968), and two intermediate males (AMR, Fort Apache; ARP, Flagstaff) have been taken where both forms occurred together. But the interrelationships between these taxa in Arizona remain to be investigated.

Green-tailed Towhee, *Pipilo chlorurus* (Audubon)

Historic: Breninger (1901:46) noted "One seen in peach orchard" in September. Gilman (MS) called this towhee a "Common migrant being seen as early in fall as 5 September and as late in spring as 28 May. A few may winter, as I have seen them 16 January on one occasion."

Modern: Common migrant, uncommon in winter. Peak migrations are throughout September and April. My latest spring specimen is 6 May 1975 atop the Sierra Estrella in a gooseberry patch. Here the birds were still numerous and singing, though they had already abandoned the lowlands. Only a few towhees winter, these keeping to damp and shady microhabitats.

Taxonomy: Osteologically *Chlorura* Sclater (1862) is simply a *Pipilo* Vieillot (1816), and I agree with the merger proposed by Sibley (1955), Phillips et al. (1964), and Mayr and Short (1970), and finally the A.O.U. Check-list Committee (1976).

Rufous-sided Towhee, *Pipilo erythrophthalmus* (Linnaeus)

Historic: Gilman (MS) found this towhee a

> Regular winter visitant in small numbers. Seen in the months from 8 October to 10 April, and two secured 12 November and 9 January. In his 'Birds of Arizona' Mr. Swarth states that the bird does not visit the lowlands, being apparently resident where found. The bird secured 9 January was submitted to him and pronounced to be this form [*P. e. montanus*], hence I have presumed that the others seen were the same.

Modern: Uncommon and irregular in winter. Specimens extend from 10 October to 12 April. Singles or pairs appeared in fall or winter of 1965, 1967, 1970–71, 1975–76, 1979. Only in the winter of 1972–73 were towhees actually numerous and widespread (Sierra Estrella, Co-op, Bapchule, Vahki, Sacaton, Casa Blanca). They were already common by 29 October (flock of seven at the marsh and another along the stream at Barehand Lane).

Taxonomy: Eight of eleven specimens are females. Of these, six are the commonly wintering race *P. e. montanus* Swarth (breeding in the southern Rocky Mountains and the Southwest), one (AMR

4206, 29 October 1972) is *P. e. curtatus* Grinnell (breeding in the Great Basin), and one is intermediate *montanus* x *curtatus*. The males are apparently *montanus* (but not safely distinguishable from *curtatus*). Identifications are by Phillips and Rea.

Canyon Towhee, *Pipilo fuscus* Swainson

Modern: Fairly common. Canyon Towhees are resident in the Sierra Estrella, sometimes found as low as the very uppermost part of the bajadas. Here they are at home among perennial grasses and shrubs (*Vauquelinia, Penstemon, Celtis, Crossosoma, Bernardia,* etc.).

Taxonomy: The problem of evaluating the taxonomic status of closely related species with disjunct distributions is well known. Following Oberholser (1919), the Canyon Towhees (*fuscus* group) of the interior have been considered conspecific with the allopatric Brown Towhees (*crissalis* group) of the Pacific status. Oberholser's reason (1919:211), that *albigularis* of southern Baja California "intergrades individually with *Pipilo fuscus mesoleucus* of northwestern Mexico and Arizona," is morphologically and geographically incorrect (see map in Davis, 1951:44). As Marshall (1960:49) pointed out, "In the field, the Canyon Towhee is not even recognizably the same species as the populations of California; it is rather the Abert Towhee which in form, posture, voice, and abundance seems the counterpart of the birds of coastal California." The Canyon Towhee is essentially a bird of steep desert slopes with grasses, not a bird of the floodplains and gardens as is the coastal bird. Even its call note is entirely different. Therefore, I consider *Pipilo crissalis* (Vigors) to be a distinct species. The two are morphologically distinguishable as well (if the specimen's wing formula has not been mutilated by the pernicious practice of stripping the ulna): the *crissalis* group has the outer primary much shorter than the longest secondaries while the *fuscus* group has the outer primary about equal to or longer than the longest secondaries; in relation to the wing, the tail is much longer in the *crissalis* group. The two groups are also distinguishable by osteology: mandible and at least some postcranial elements (humerus, femur, and probably scapula, coracoid, and ulna). The eggs of the interior group are much more heavily marked, with a very pale bluish white or light pearl gray ground, while the ground of the *crissalis* group is light greenish blue (Bendire 1890:23, 26). *Pipilo c. albigula* Baird and *P. f. mesoleucus* Baird I consider to be examples of parallel evolution. The cape bird has the voice, structure, and eggs of the other contiguous Pacific towhees rather than the interior ones it superficially resembles in color (Marshall, 1964).

The Sierra Estrella population is *P. f. mesoleucus* Baird, 1854, found in Arizona and most of New Mexico. This is one of the westernmost points on the Gila where the interior form has been found. I have examined the type and type series of *P. f. relictus* van Rossem (Harquahala Mountains, 1946) as well as comparable museum-aged *mesoleucus* (all UCLA) and find this race not distinguishable (though the type and possibly several other specimens are indeed darker).

Change: No change is implied. Gilman did not find Canyon Towhees on the reservation, but there is no evidence that he worked into the higher mountains where the species is restricted to slopes with admixtures of Upper Sonoran shrubs and grasses.

Abert's Towhee, *Pipilo aberti* Baird

Ethnographic: *bíchput;* [pl.] *bíbichput*

The *bíchput* is a common bird, living and nesting in Pima fence rows and in brush fences. It was regularly trapped for food or captured at night by torch hunting after rains (Rea, 1979).

Historic: Breninger (1901:46) found the Abert's Towhee "Common along the river and about mesquite growths." Gilman (MS) recorded it as a "Very abundant resident. Nests early and late and often." He made notes on 125 nests ("It was about as much sport finding them as going to gather eggs in a hen-house") from 28 February to 4 September. The major distribution of nests was: March, 15 percent; April, 22 percent; May, 18 percent; June, 30 percent; July, 19 percent. Primary nest locations (descending order) were *Prosopis, Baccharis,* and *Lycium.*

Modern: Abundant on floodplains, particularly where fence rows and other brush afford protection near fields.

Taxonomy: The Abert's Towhee is a polytypic species, with at least two recognizable subspecies. Controversy surrounds the question of which is the nominate population (van Rossem, 1946; Davis, 1951; Phillips, 1962; Hubbard, 1972). Phillips (1962) identified the middle Gila population with the more cinnamon, perhaps somewhat grayer birds of the Tucson area. I have examined twenty-five fresh fall to mid-winter skins taken between 1964 and 1976 from the Santa Cruz (Tucson), San Pedro, and Gila drainages. Aside from the nomenclatural problem involved, I find most of the towhees from the reservation and east to the mouth of the San Pedro (Winkelman) to be darker brown and less grayish on the crown, nape, and back than Tucson area birds. For this reason I apply no subspecific name to the middle Gila population extending from the mouth of the Salt to the mouth of the San Pedro.

Lark Bunting, *Calamospiza melanocorys* Stejneger

Ethnographic: *áwawtortupwa* [fixed plural]

The Lark Bunting was a common winter bird and an important food trapped in the days when the Pima raised wheat and other grain crops more abundantly. This species was referred to by several additional names, evidently an indication of its importance as a food item:

chíchu tópiwa (no gloss available)

s'bában mákam ('very,' plural 'coyote' [-like], 'eater')

kauk áwtam ('hard,' 'bones')

Historic: In February 1854, Heermann (1859) "first observed [the Lark Bunting] on approaching the Pimos villages, associated with large flocks of sparrows, gleaning grain and grass seed from the ground. When started it would fly but a short distance before again resuming its occupation."

Gilman (MS) noted the Lark Bunting a

> Common winter visitant it might be termed, though its status is most peculiar, and I cannot be certain of a diagnosis of the case. On 10 August I saw eight of them together at Blackwater and thought it might be a family group as two looked different from the others. A week later I saw many of them on the north side of the river and 30 August several were at Sacaton. . . . Some seasons they remain later in spring than in others and their numbers vary in different years. In 1913–1914 they were very abundant and stayed late in spring to 17 May.

Modern: Common to abundant in winter, but irregular. My outside dates (with specimens) are 22 September to 11 May. Lark Buntings appear to be more common and in larger flocks in the central and eastern Pima villages than in the northwest section. At Komatke I saw the species during only two winters and once in spring (solitary female, 10 April 1978, SD). I found them at Co-op Colony only in the winter of 1977–1978. Four Lark Buntings were on the Salt River flats near 107th Avenue on 11 May 1978 (SD, Rea and Paul Schneider), my only record there.

Savannah Sparrow, *Ammodramus sandwichensis* (Gmelin)

Historic: Breninger (1901:46) observed "A few perched on the wire fences," in September. Gilman (MS) called it "Fairly common in winter" and secured four specimens 12 January.

Modern: Fairly common in winter. Outside dates are 11 October to 28 March (specimens). Though it is most frequent about alfalfa fields, I took one in a mesquite bosque with White-crowned Sparrows.

Taxonomy: I follow Phillips et al. (1964), Murray (1968), Dickerman (1968), Mayr and Short (1970), and other authors in considering *Passerculus* Bonaparte (1838) best merged in *Ammodramus* Swainson (1827).

Phillips identified my several specimens as *A. s. nevadensis* (Grinnell) and *A. s. anthinus* (Bonaparte), both common winter residents in Arizona. Grinnell identified three of Gilman's specimens as *alaudinus* (=*anthinus* of A.O.U. Check-list, 1957) and one as *nevadensis*. *Anthinus* breeds in interior Alaska and northwest Canada, *nevadensis* in the Great Basin and Great Plains.

Vesper Sparrow, *Pooecetes gramineus* (Gmelin)

Historic: In September Breninger (1901:46) found Vesper Sparrows "Very common among the bushes of the desert." Gilman (MS) considered them an "Abundant winter visitant arriving in September and leaving in April." The latest specimen record is 15 April 1939 (female, CGR).

Modern: Fairly common. My outside dates are 26 September to 13 March. Vesper Sparrows are usually found in fields, particularly those planted in permanent pasture.

Taxonomy: Four of the reservation specimens are the Great Plains–Great Basin race *P. g. confinus* Baird, which winters commonly in Arizona. Two (AMR 1747, 7 October 1967; 2506, 14 November 1968) are *P. g. altus* Phillips (*in* Phillips, Marshall, and Monson) breeding in the Southwest. There are few other Arizona lowland specimens of *altus*. Identifications are by Phillips and Hubbard.

Lark Sparrow, *Chondestes grammacus* (Say)

Historic: Breninger (1901:46) recorded in September "Only a few seen; date rather early." Gilman (MS) found Lark Sparrows "Abundant during the fall and winter but rarely seen in late spring. Occasionally seen during late July and August but none observed breeding." His notes give several for 27–28 July and many for 3–25 August 1914 at Sacaton.

Modern: Fairly common winter visitant. Postbreeding birds arrive in July (12 July 1972, two collected from group of four; one 24 July 1967). Several birds may be found on the reservation almost any winter. My late dates are 12 March 1964 (specimen) and 17 March 1973 (small flocks at Vahki and Casa Blanca).

Chipping Sparrow, *Spizella passerina* (Bechstein)

Historic: Breninger (1901:46) found Chipping Sparrows "numerous" in September. Gilman (MS) called them "Common from September until March."

Modern: Common to abundant migrant, fairly common in winter. They are seldom as common as the Brewer's Sparrow. My outside dates (probably not representative) are 10 September to 30 April and once 15 May (loose flock of about a dozen in willows). During spring migration this sparrow forages atop the blooming mesquites, much like the associated parulids. And in September I took a female foraging high in a tamarisk with a Wilson's Warbler.

Taxonomy: Six of eight reservation specimens are *S. p. arizonae* Coues, breeding throughout the Southwest, Great Basin, and the Rocky Mountains (identifications by Phillips, Parkes, and Rea). One strangely colored specimen (AMR 4737, Salt-Gila confluence, 15 May 1975) has been identified by Dr. Kenneth Parkes as *S. p. boreophila* Oberholser, which has a breeding range (as restricted by Parkes) in the northern part of the continent, from eastern Alaska to interior Quebec, exclusive of the Rockies (*contra* Oberholser, 1974). The next latest specimen (AMR 1001, Komatke, 30 April 1966) Parkes considers nearest *boreophila,* possibly intermediate toward nominate *passerina.*

Brewer's Sparrow, *Spizella breweri* Cassin

Historic: Breninger (1901:46) found the Brewer's Sparrow in September, "The most common species; seen everywhere." Gilman (MS) considered it a "Common winter visitant." His early dates are 15 August, 10 and 18 September.

Modern: Common to abundant migrant, fairly common in winter. My outside dates are 12 September (AMR 1708, Barehand Lane) to 20 May 1972 (AMR 3897, male, Komatke).

Taxonomy: Two subspecies of Brewer's Sparrow are currently recognized: nominate *breweri* of the Great Basin desert and western Great Plains and *S. b. taverneri* Swarth and Brooks, breeding in the alpine zone of western Canada. *Taverneri* has broad blackish dorsal streaking on a gray ground color. Several heavily blackish-streaked gray May specimens (AMR 759, 3897, both Komatke; JSW 1035, Blue Point Cottonwoods) appear to be approaching *taverneri* but have long and somewhat to quite heavy bills.

The nominate form appears to be a composite of two distinct races. The more common migrant in the Southwest has a pale sandy dorsal ground color with narrow brownish streaks, pale auriculars, and buffy-cream wing bars and tertial edgings. There are four reservation specimens of this form (26 September to 14 November). The apparently undescribed form has the dorsal ground color reddish-washed except for a contrasting gray medial crown stripe, normal to heavy umber dorsal streaking, dark auriculars, and cream-cinnamon wing bars and tertial edgings. This form somewhat parallels *S. pallida* (Swainson) in coloration and pattern; partially or completely molted specimens (SD, WSU) presumably still on their breeding grounds, taken from late July to early September in eastern parts of Washington, Oregon, and California, suggest its western origin. There are several reservation specimens of this richly colored population (AMR 1708, 28 September 1964; AMR 632, 21 November 1965, and probably AMR 529, 25 April 1964). By spring the color differences are of little or no value.

Rufous-crowned Sparrow, *Aimophila ruficeps* (Cassin)

Modern: Fairly common in Sierra Estrella; probably common at upper elevations on north-facing slopes. I have taken no females or juvenals so do not know the breeding season, but on 22 July 1974 and 6 May 1975 males were not very responsive to tape recordings of their songs, though they were in breeding readiness.

Taxonomy: Two races are recognized in Arizona by the A.O.U. Check-list (1957) and tentatively by Phillips (*in* Phillips et al., 1964): *A. r. scottii* (Sennett 1888) of New Mexico and most of Arizona and *A. r. rupicola* van Rossem (1946) of the western mountains of Arizona. With a good series of Rufous-crowned Sparrows (AMR, GM, ARIZ) from western New Mexico, northern Sonora, and both eastern and western Arizona, I am unable to recognize any subspecific difference in either fall or spring specimens. In fact, the only really dark-backed bird of the series of 27 specimens is AMR 4812 (8 December 1975) from the bottom of Salt

River Cañon, Arizona. Hubbard (1975) reviewed most populations and also considers *rupicola* a synonym of *scottii*. A lowland specimen (AMR 885, immature male, 24 October 1965, Bapchule), collected in forbs at the edge of the dried Gila River, was identified by Phillips as *A. r. eremoeca* (Brown) and referred to the population found in the Arbuckle Mountains, Oklahoma. This record is accidental for Arizona and the subspecies new to the state. Phillips (personal communication) considers it possibly the only specimen of a true migrant/vagrant of the species (see also Hubbard, 1975).

Sage Sparrow, *Amphispiza belli* (Cassin)

Historic: Gilman (MS) called the Sage Sparrow a "Fairly common winter visitant, being seen most frequently on dry uplands away from the river." His early dates are 23 October 1910 and 8 October 1914.

Modern: Common in winter. This species is rather strongly restricted to *Atriplex-Ambrosia* or *Suaeda-Salicornia-Atriplex* communities in the floodplains and lower bajadas, and is not likely to be encountered elsewhere. My outside dates (specimens) are 19 October to 4 March.

Taxonomy: Two well-marked races occur. Nine specimens (7 November to 4 March) are the larger *A. b. nevadensis* (Ridgway), breeding in the Great Basin. Only three (19 October to 13 November) are the smaller *A. b. canescens* Grinnell of interior southern California, the easternmost Arizona records.

Black-throated Sparrow, *Amphispiza bilineata* (Cassin)

Historic: Breninger's (1901:46) September record for *"Amphispiza belli nevadensis"* must surely belong here: "A few seen on the desert." (He does not mention *bilineata*.) Gilman (MS) recorded the Black-throated Sparrow a "Fairly common resident. One nest found 26 April . . . Seen in small numbers in winter."

Modern: Common. This is a species most frequently found on the bajadas and into the mountains. But in the heat of the summer it may be seen about water holes or ditches. It appears that most of the breeding coincides with the summer rains, as most of my observations of birds still in juvenal plumage are from mid-August through the end of September. (Breninger's September birds were probably juvenals.)

Taxonomy: Reservation birds are the large *A. b. deserticola* (Ridgway), which is more strongly vinaceous washed on the dorsum and flanks. Grayer populations also called *deserticola* (A.O.U. Check-list, 1957; Phillips et al., 1964) occur from southern Nevada and adjacent Arizona and southwestern New Mexico.

Dark-eyed Junco, *Junco hyemalis* (Linnaeus)

Archaeological: The fragmentary *Junco* reported by McKusick from Snaketown is undoubtedly referable to this species.

Ethnographic: I have been unable to obtain a native lexeme for this species. It is a readily trapped winter species that must have been taken regularly by the Pima.

Historic: Gilman (MS) unfortunately said little of the junco. His early dates are 16 October 1910 and 25 October 1914. Grinnell called two of Gilman's December specimens *J. h. thurberi* Anthony, and Swarth thought the October specimens to be *J. h. shufeldti* Coale "but in the stage of plumage could not be certain." I have not relocated these skins.

Modern: Common to abundant, with numbers varying from year to year, and even from week to week. Outside dates (specimens) are 2 October to 2 April.

Taxonomy: In spite of extensive interbreeding of Dark-eyed Junco populations, there has been considerable inertia

(Miller, 1941; and others) about applying the biological species concept to this bird. Various authors (Mayr, 1942; Dickinson, 1953; Phillips, 1961; Phillips *in* Phillips et al., 1964; Short, 1969; Mayr and Short, 1970) have acknowledged the biological situation. Nevertheless, "*J. caniceps,* currently under study, is maintained for the present as a separate species" by the A.O.U. Check-list Committee (1973:418). However, the red-backed *caniceps* interbreeds with Sierra Nevadan *thurberi* in southern Nevada (called *J. h. "mutabilis"* van Rossem) as well as with *mearnsi* in northern Utah (called *J. h. "annectens"* Baird). In the treatment of reservation specimens, I have followed the most recent revision (Phillips, 1961; Phillips *in* Phillips et al., 1964), except that I do not recognize the population of secondary intergrades *"mutabilis"* as a valid subspecies. Evidently *caniceps* genes are now swamping out the traces of *thurberi* influence at the type locality (specimens, AMR, UNLV).

J. h. thurberi Anthony, breeding in the Sierra Nevada, is represented by seven specimens, from 15 October to 2 April.

J. h. shufeldti Coale, breeding in southern British Columbia and the western part of the northern Rocky Mountains of the United States, is represented by ten specimens from 15 October to 12 March. Three additional specimens of *J. h. "montanus"* Ridgway are included here, taken 5 November to 15 March.

J. h. simillimus Phillips, breeding along coastal Oregon and Washington, is represented by ten specimens taken from 3 November to 27 January. Regarding the questionable application of the name *shufeldti* to a coastal or interior race, see Miller (1941:394), Phillips (1961), and Browning (1974). Three additional specimens (15 October to 10 December) are *simillimus* intermediate toward a neighboring race.

J. h. henshawi Phillips (="*cismontanus*" Dwight of Miller [1941] and A.O.U. Check-list [1957]) breeds from southern Yukon south through interior northern and central British Columbia. A single specimen (AMR 930, male) was taken 11 March 1966.

J. h. mearnsi Ridgway breeds in the eastern part of the northern Rocky Mountains. It is represented by a single specimen (AMR 3812) taken 12 March 1972 and an apparent sight record 25 December 1973.

J. h. caniceps (Woodhouse) is represented by five specimens taken 2 October to 12 March. It breeds in the central Rocky Mountains (Colorado, Utah, northern New Mexico) reaching northern and northeastern Arizona as somewhat intermediate toward *J. h. dorsalis* Henry.

A sight record for an apparent *J. h. hyemalis* (Linnaeus), an eastern race, is 6 April 1971 (present with several *caniceps*).

White-crowned Sparrow, *Zonotrichia leucophrys* (Forster)

Ethnographic: *támtol;* [pl.] *táwtamtol*

Within the memory of my consultants the *táwtamtol* were regularly trapped and eaten by the Pima. They consider it to have been far more common when grain crops were grown more frequently.

Historic: Breninger (1901:46) in September noted "a few" of the dark-lored type and found the pale-lored birds "along with the above; later this is the commoner form." Gilman recorded the dark-lored race "Common in fall and spring. Appearing late in September and seen as late as 20 May in spring. No certain winter occurrence." Actual dates given are 29 and 30 September, 14 October, and in spring 14, 17, and 18 May. The status given for the pale-lored far northwestern race is "Common winter visitant, arriving in the fall about the same time as [the dark-lored bird] but not staying so late in the spring, 1 May being the latest date noted." Gilman saw the first of the pale-lored birds 23 September, several two days later, and many by 8 October.

Modern: Common to abundant in winter, but varying greatly from place to place, and season to season, with the greatest numbers being spring and fall transients.

Taxonomy: There is a lack of agreement concerning which of the two Canadian races should bear the oldest name *Z. l. leucophrys* Forster. It appears to me that the most convincing arguments are presented by Todd (1953), Phillips et al. (1964), and especially Godfrey (1965), for applying the name *leucophrys* to the more western race, of which *Z. l. gambelii* Nuttall falls a synonym. This leaves the black-lored northeastern race with the name *Z. l. nigrilora* Todd (=*leucophrys* Foster of A.O.U. Check-list [1957], of Banks [1964], *et auctorum*). The large-billed, dark-lored birds of the interior West are *Z. l. oriantha* Oberholser (included in *leucophrys* by Banks [1964]). Colors of the flanks, rump, and streaks of the interscapulars, and ground color, as well as general ventral color, are important in identifying White-crown subspecies, as noted by Godfrey (1965) and others. Cortopassi and Mewaldt (1965) mapped the breeding distributions of the races (*sensu* A.O.U., 1957).

Oriantha occurs on the reservation as a transient. Specimen dates (AMR, CGR; all males but one) extend from 29 April to 12 May. I have never found it in the fall. The common wintering bird is the pale-lored form with a small, yellowish bill, breeding in Alaska and western Canada. This is *leucophrys* (= *gambelii, auctorum*). Specimen dates extend from 22 September to 15 April (AMR) and exceptionally to 28 April (CGR). Identifications are by Rea, Banks, Phillips, and Hubbard.

Harris' Sparrow, *Zonotrichia querula* (Nuttall)

Historic: Gilman (MS; see also 1914b) reported "One secured 16 March 1913 from a small group of White-crowned Sparrows."

Modern: Rare. Simpson (personal communication) observed two immature Harris' Sparrows 16 November 1974 in Maricopa Colony west of 91st Avenue.

Lincoln's Sparrow, *Passerella lincolnii* (Audubon)

Historic: Gilman (MS) recorded this species simply "Spring and fall migrant. Secured in March and November."

Modern: Common winter resident in marshes and along streams and overgrown irrigation ditches. Fairly common transient even in drier fields. I have no satisfactory fall arrival dates. Wintering specimens extend from 9 December to 27 March, but I have seen individuals as late as 7 May at the Salt-Gila confluence in low emergent vegetation over sluggish side channels.

Taxonomy: The generic limits of a number of emberizid species remain to be determined. On the basis of external morphology, osteology, and behavior I have merged *Melospiza* Baird (1858) in *Passerella* Swainson (1837), as proposed by Linsdale (1928a, 1928b) and Mayr and Short (1970). Paynter (1964) submerged both these genera into *Zonotrichia* Swainson (1832), and Dickerman (1961) and Short and Simon (1965) have documented frequent hybridization of *Zonotrichia* with *Junco* Wagler (1831). While it is conceivable that *Zonotrichia, Junco,* and possibly *Emberiza* Linnaeus (1758), ultimately may be considered congeneric, I consider this premature until detailed comparative studies of postcranial osteology and behavior of both Old and New World taxa are completed.

Seven reservation specimens are *P. l. lincolnii,* the expected race.

Song Sparrow, *Passerella melodia* (Wilson)

Historic: Breninger (1901:46) noted "Several seen among the willows and others heard." Gilman (MS) recorded the following (edited):

> Abundant resident. Nesting in numbers along the "Little River." The birds frequent water so much that they seem partly aquatic, hopping into shallow water in search of food, even alighting and walking on water vegetation that barely reaches the surface of the water. Of 24 nests noted, 8 were in the month of April, 11 in May, 4 in June, and 1 in July. Most of the nests contained three eggs and three had sets of four. Nineteen of the 24 nests were placed in batamote shrubs, *Baccharis glutinosa*; two in *Pluchea sericea*; one in *Sarcobatus,* one in tree tobacco, *Nicotiana glauca*; and one in a pile of drifted brush and trash standing in the edge of the stream. The nests were from three to eight feet [1–2.5 m] from the ground with an average of four and one half feet [1.4 m]. The Dwarf Cowbirds patronize this Song Sparrow to some extent.

Modern: Locally common. Breeds in the effluent-supported vegetation along parts of the Salt River and formerly at Barehand Lane. Transients occur rarely even in weedy fields. The breeding season is prolonged, at least in some years. A female (AMR 1669) taken at Barehand Lane 20 August 1967 had two yolks and two recently ruptured follicles, and males were still territorial. Birds in full juvenal plumage were collected from 9 June to 25 July.

Taxonomy: The breeding race is *P. m. fallax* (Baird), as are also the great majority of wintering birds. The dark birds collected from 23 October to 15 March are *P. m. montana* (Henshaw) and *P. m. fisherella* Oberholser, of the Rocky Mountains and Great Basin, respectively.

Change: The Song Sparrow, now dependent on vegetation around permanent wastewater, is entirely absent from the Santa Cruz and virtually all the middle Gila rivers. It persists at or has recolonized only the lower Salt and a few runoff marshes, where it may occur in numbers.

[Swamp Sparrow, *Passerella georgiana* (Latham)]

Modern: Sight records believed to apply to this species are for 14 March 1971, 26 November 1972, and 31 December 1975, but no specimen of the suspect birds was secured for verification.

Chestnut-collared Longspur, *Calcarius ornatus* (Townsend)

Historic: Gilman (1910) took three specimens from a flock of about seventy-five at Sacaton 25 October 1909. They were in a field of dry grass and alfalfa.

Modern: Rare. Dr. Klaus Immelmann of West Berlin and I observed a small group of this longspur, accompanied by some Vesper Sparrows, walking in *Atriplex* near the Maricopa cemetery, 9 March 1975. This is my only observation of longspurs on the reservation.

Icteridae

Hooded Oriole, *Icterus cucullatus* Swainson

Archaeological: I have reexamined the reputed eight Hooded Oriole bones excavated from Snaketown. Seven of these are the notoriously nondiagnostic distal ends of tibiotarsi, and I would hesitate to assign them to family. The remaining specimen is a cranium. The configuration of its postorbital process, zygomatic process, temporal fossa, and the anterior edge of the occipital wing (overhanging the tympanic recess) are definitely not those of *Icterus.* I reassign the cranium to the Troglodytidae, cf. *Salpinctes* or *Catherpes.*

Historic: Gilman (MS) found this oriole a

> Common summer visitant nesting mostly in cottonwood trees unless there are Washington palm trees available which are always preferred. . . . Only host of Bronzed Cowbird and is also parasitized by Dwarf Cowbirds. They nest very late, July being the time of most of it. Six July is the earliest egg date. They arrive in the spring generally the first week in April.

Modern: Rare summer resident. I have found this species only at Komatke (palms in old convent gardens), though it is to be expected about more urbanized settlements, as at Sacaton. But my search of apparently suitable nesting locations has failed to reveal the bird or its conspicuous nest. An early date at Komatke was two males, evidently transients, 19 April 1964. The pair nesting there maintains a territory from at least 24 June to 4 August. In 1969 the "pair" consisted of an adult male, an immature male, and a female; each evening they roosted together in the same palm, but in separate nests. Simpson (personal communication) observed an adult male 16–23 November 1974 on the lower Salt.

Taxonomy: I find all clean spring specimens (AMR, SD) from Baja California (exclusive of the Cape Region), southern California (at least to coastal San Diego County), Arizona, and New Mexico to be indistinguishable. These are *I. c. nelsoni* Ridgway (see also Phillips, 1963; and Blake, 1968).

Change: The Hooded Oriole is another essentially Mexican species dynamically extending its range northward (summarized by Phillips et al., 1964; and Phillips, 1968). It was found in central Arizona by early ornithologists (Palmer, Scott, Mearns). The earliest authentic specimens of this oriole on the Gila drainage appear to be those of Edward Palmer from old Camp Grant on the lower San Pedro (confluence with Aravaipa Creek) in 1867 and in the Gila headwaters, New Mexico, by Frank Stephens (SD, 10 June 1876). Since Gilman's time this oriole has suffered an enormous reduction in population numbers, attributable to loss of riparian timber (and perhaps to excessive cowbird parasitism).

Orchard Oriole, *Icterus spurius* (Linnaeus)

Modern: Accidental. I collected a fully adult male (AMR 2444) in a weedy field at Komatke 22 October 1968. There are two previous specimens for the state (Phillips et al., 1964).

Taxonomy: The proposed race *I. s. affinis* (Lawrence) is unsatisfactory (see Ridgway, 1902; Dickerman and Warner, 1962). Allopatric *fuertesi* Chapman is undoubtedly conspecific with *I. spurius*. The females are indistinguishable and the biology of the two taxa is similar (Graber and Graber, 1954).

[Scott's Oriole, *Icterus parisorum* Bonaparte]

Historic: Gilman (MS) found the Scott's Oriole a "Spring migrant, once 31 March, usually seen from first or second week of April; also one male 19 July and one male immature eating watermelon, the black of throat increasing, 15 to 24 September 1914."

Change: I have never found this oriole on the reservation, even as a migrant. Although there are small islands of oak, juniper, agave, and *Canotia* atop the Sierra Estrella where one might expect this species, I have not found the bird or its nests there. But there are no tall yuccas there to attract it.

Northern Oriole, *Icterus galbula* (Linnaeus)

Ethnographic: *váchukukĭ*; [pl.] *vápchukukĭ*

This species, nesting abundantly about villages of the Pima, is well known to them, but the Pima appear to be totally unfamiliar with the Hooded Oriole. Sylvester Matthias once brought me a palm frond with a fiber nest of the Hooded

Oriole, asking me what bird had made it. He (and other consultants) were familiar with the dark horsehair nests of the *váchukukĭ*. The group of Pima singers from Sacaton Flats is known as the "Oriole Singers." All orioles are called *wákokam* at Quítovac Papago oasis and *vá:kuagam* by lowland Pima Bajo. Both names mean 'water gourd owner' in reference to nests.

Historic: Breninger (1901) did not note the Bullock's or western form of the Northern Oriole in September. Gilman (MS) found it a "Common summer visitant, nesting in great numbers. Much more abundant than Hooded. Usually appearing last week in March though once 17 March saw them and another time [as late as] 5 April. Usually leave the last of August, though one male was seen as late as 22 September. Most of the nesting is in May, though a few extend into June."

Modern: Common spring migrant and summer resident. My outside specimen dates are 25 April 1964 (fat migrant *parvus*) to 10 August 1968. Migrants pass through *Prosopis* bosques in numbers during April, continuing at least to 10 May (two adult males together in bajada). But by the last of April the breeding birds have already established territories and begun nesting. Nests are most frequently placed in tamarisk trees. Both species of cowbirds frequent the nesting areas.

Taxonomy: A complex of three orioles, *galbula, bullockii* (Swainson), and *abeillei* (Lesson), have distinctive color patterns in the adult males but the females are morphologically similar. These three interbreed freely and are considered conspecific (see Phillips, 1961; Mayr and Short, 1970).

Within the western *bullockii* complex van Rossem (1945) described a small subspecies *I. g. parvus* of the lower Colorado River Valley and the Pacific Coast. The A.O.U. Check-list (1957) recognized it, Sibley and Short (1964) ignored it, Rising (1970) considered it invalid, and Phillips et al. (1964:169) called it "rather tenuous." I have re-

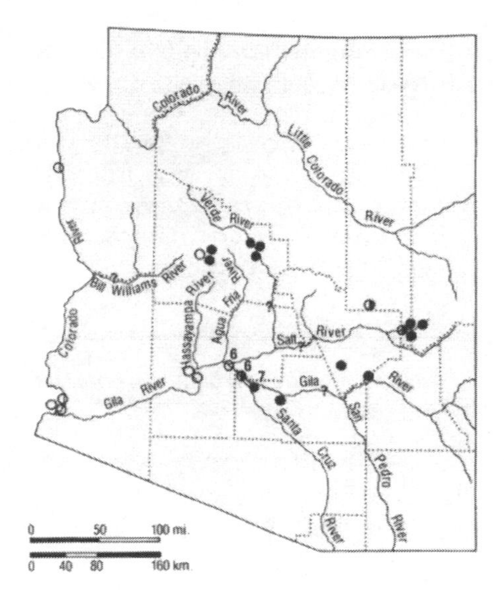

Northern Orioles (breeding)

○ = Icterus galbula parvus ● = I.g. bullockii
◐ = intermediate ? = breeding race unknown

measured breeding orioles from various parts of the West, including twenty breeding males (San Diego and Imperial counties, California) that van Rossem used in his description of *parvus*. I consider the race valid and useful; Blake (1968) is probably correct in restricting the range. I find about one millimeter overlap in wing lengths of adult males.

The racial status of orioles breeding in central Arizona was unknown (see Phillips et al., 1964, map p. 169). Therefore, I collected known breeding orioles (see map) after they had established territories in late April and May. Two adult males and a female from Arlington, Arizona, 28 miles (45 km) west of the Salt-Gila confluence are typical *parvus*. The Komatke population is intermediate (ten adult males, four immature males, three females). Of these, six are *parvus*, six are intermediate, and five are *bullockii*. An adult male from Bapchule and one from northeast of Casa Grande Ruins are *bullockii*. This eastern race breeds in pure form at San Carlos, Fort Apache, and Prescott, Arizona (AMR), but the subspecific status is unknown from the mouth of the San Pedro River to the eastern Pima villages.

Change: Pimas consider the species to have been much more abundant formerly, which no doubt was true when there were cottonwoods along the river and tamarisk trees about most of the houses. Bullock's Oriole pairs are heavily attended by Bronzed Cowbirds, *contra* Phillips et al. (1964).

Eastern Meadowlark, *Sturnella magna* (Linnaeus)

Modern: Fairly common (erratic?) winter resident. The precise status of this species remains to be determined by regular collecting. It may be present only during some winters. I first noticed about thirty Eastern Meadowlarks in stubble fields of Co-op Colony 9 December 1972. (An apparent Eastern from Vahki 4 November 1972 proved in the hand to be the Northwest race of the Western.) Eastern Meadowlarks could not be found in these fields later in December. In the bare fields at the Chandler-reservation boundary I collected an immature male (AMR 4691) from a mixed flock of the two species 19 January 1975.

Taxonomy: The one reservation specimen is the paler *S. m. lilianae* Oberholser. I have not seen material to judge the utility of the separation of this race from *S. m. hoopesi* Stone of Texas and south.

Change: I doubt that any real change in status is involved with this species. Rather, I suspect that it is present only certain winters and is easily overlooked.

Western Meadowlark, *Sturnella neglecta* Audubon

Archaeological: Two meadowlark specimens, both mandibles, were identified from Snaketown by McKusick. The species (*neglecta* or *magna*) is not determinable.

Ethnographic: *tósif;* [pl.] *tótosif*

The Pima relate that the *tósif* nested in their fields. That this species is intended, not *S. magna,* is evident from the Pima rendition of the bird's song: *vátto chí:pia sí:kul kúli.* (The first word of the song is 'ramada' and the last word 'old man' but the middle two words are perhaps forgotten as no one could tell me their meaning.) The name in Papago is *tósiw.*

Historic: In September Breninger (1901:46) recorded "A dozen or so seen in a field." Gilman (MS) thought "Possibly a few breed as I have seen them in June and August, though rarely." He considered them abundant from 1 October on. He recorded two seen 8 September 1914 and four near Mesa 20 August that year.

Modern: Fairly common. Migrants or winter birds, verified by specimens, span from 29 October to 7 April. At this time meadowlarks occur either in fallow fields or in open *Atriplex-Ambrosia-Lycium* flats, occasionally in larger fields, particularly of alfalfa. At Casa Blanca 3 June 1968 about fifteen pairs appeared to be nesting colonially in a grain field and feeding in the adjacent alfalfa. We secured an adult territorial male (AMR 2170). But it does not occur in numbers at the western end of the reservation, where territorial males in alfalfa fields are few (1975) or none (1976). A meadowlark singing at Komatke on 20 May 1979 may have been territorial. It was in the *Plantago* and *Schismus* flats where Horned Larks were courting.

Taxonomy: Most reservation specimens are the nominate race, breeding in Arizona and widely in the West. Two 1972 birds (AMR 4199, 29 October; 4213, 4 November) are the darker, richer *S. n. confluenta* Rathbun of the Pacific Northwest west of the Cascades. This race is apparently casual in Arizona (Phillips et al., 1964). Reservation specimens were compared with fall skins from the Puget Sound lowlands (UW). Pima fire drives (Rea, 1979) no doubt helped to maintain local grasslands (see p. 39), increasing the numbers of meadowlarks and grassland sparrows.

As with flickers, the paler interior meadowlark populations are highly migratory throughout the West. The reluctance to recognize the coastal race (e.g.,

Blake, 1968; Mayr and Short, 1970) is based no doubt on comparing composite samples *after* migration has begun. Because of extensive wear of the diagnostic dorsal feathers, only fall and winter specimens are taxonomically useful. The breeding range of *confluenta* is poorly known because molting or just molted meadowlarks taken before migration is well underway are exceedingly rare in collections. However, as was carefully noted in the original description (Rathbun, 1917), the breeding population east of the Cascades (WSU) is the nominate race, not *confluenta* as assigned by Jewett et al. (1953:585) and the A.O.U. Check-list (1957).

Red-winged Blackbird, *Agelaius phoeniceus* (Linnaeus)

Archaeological: Four specimens (long bones) of this species from Snaketown were identified by McKusick.

Ethnographic: *sháwshaiñ*

Though this term refers specifically to the red-wing, native consultants believed it might also be expanded without the addition of a qualifying term to include both the Brewer's Blackbird and the Yellow-headed Blackbird, common winter visitants which often associate with the "true" *sháwshaiñ* and which frequent similar habitats.

Historic: Breninger (1901) noted the species was "seen among rushes and in flocks in cornfields." Gilman (MS) found it a "Common resident and breeding abundantly." His egg dates are from 28 April to 15 July, presumably in 1910.

Modern: Common to abundant. Red-wings on the reservation nest wherever there is suitable vegetation for the nest placement (saltcedars, cattails, oat fields, thistles, mustard) over standing water to afford some protection. This blackbird has nested earlier on the reservation than is known for the rest of Arizona. According to Phillips et al. (1964:166), "The earliest eggs were found 28 April (at Sacaton-Gilman), and our earliest young out of the nest are 30 May (*1952* [specimen] at Imuris, Sonora—ARP) and 3 June (1933 at Binghampton Pond, Tucson—A. H. Anderson)." By 11 May 1966 some young were already out of the nest at Barehand Lane while other nests had nestlings. In 1968, in a nearby grain field, females were collecting nesting material at least by 8 April, and a female carrying plant strips was collected 10 April. Eggs were hatching, and females and a nestling were collected 25 April. Some young were on the wing by 12 May, when fifty to sixty females a minute were crossing the road to gather food in a pasture. I shot two flying young (AMR 2139, 2140). On 14 May students and I found about twenty nests with three to five eggs, others with nestlings, and many additional nests from which the young had already fledged. In 1974 young were flying by 22 May. Thus I took a nestling earlier than Gilman's early state egg date, and juvenals in three different years were out of the nest several weeks earlier than previously recorded for Arizona. I have no records of summer nesting.

After the nesting season same-sex and same-age flocks gather, usually retiring to less disturbed areas (as along the Salt River) late in summer while they molt. Migrants greatly augment the numbers of local birds, starting at least by mid-September. In winter the species becomes abundant in fields, and literally thousands formerly roosted at the Barehand Lane marsh.

Taxonomy: Recent revisions of this widespread, geographically plastic species include van Rossem (1926, California races), Phillips and Dickerman (*in* Phillips et al., 1964, races reaching Arizona), and Oberholser (1974, races reaching Texas). I am primarily responsible for the specimen identification of the two southwestern breeding races, Phillips and Dickerman for other races.

The breeding race of the reservation is *A. p. sonorensis* Ridgway. The bill is long and slender in both sexes, and the females are pale, particularly in feather edgings. Oberholser (1974) was completely in error in calling the Phoenix area birds *A. p. utahensis* Bishop, a thicker-billed race (perhaps not recognizable). Breeding red-wings from the reservation are indistinguishable in color and bill shape from the Tucson area *sonorensis* (ARIZ).

The other southwestern race is currently called *A. p. fortis* Ridgway. It has a deeper, shorter bill, the adult male's epaulets are deeper red, less orange, and the females are a bit darker. This is the breeding race of New Mexico and the northern half of Arizona. These have been separated from true *fortis* of the western Great Plains as *A. p. heterus* Oberholser (1974, type from Fort Wingate, McKinley Co., New Mexico, 23 June), which I suspect may prove valid. However, careful comparisons will have to be made with breeding red-wings of Utah, Colorado, and western Kansas and Nebraska before a conclusion can be reached. (I have long believed that bills of breeding birds from Acoma and Zuni, New Mexico, and northern Arizona west to Prescott, all AMR, are too thin for true *fortis*.) *Fortis (sensu lato)* arrives as early as 11 September (AMR 1703, immature female, not typical) and stays as late as 11 April. *Sonorensis* and *fortis* probably form the bulk of the red-wings wintering on the reservation. I have no reservation records of the large, robust-billed true *fortis* of the Great Plains.

A. p. nevadensis Grinnell of the western Great Basin is represented by two adult females: AMR 1610, 28 October 1966; 4337, 20 January 1973.

A. p. arctolegus Oberholser of the far north is represented by six females (AMR 4188, 4286, 4330, 4333, 4334, 4335) taken from 29 October 1972 to 20 January 1973, and an atypical female (AMR 1612, approaching *nevadensis*) taken 28 October 1966. This subspecies has not been reported previously from Arizona.

A. p. neutralis Ridgway breeds in southern coastal California. I identified one specimen (AMR 4264, adult female, 10 December 1972) as this race; Phillips verified this using US material. There is a previous Arizona record.

A. p. caurinus Ridgway breeds in the coastal Northwest and is the darkest, most richly colored western race. One specimen (AMR 1482, 29 April 1967) was identified as this race by Phillips, Dickerman, Laybourne, and Banks. It was in a colony of red-wings at Bapchule, where we shot it for a female cowbird! The unusually late date probably is explained by the loss of most right primaries, just growing back in. There is one previous record of *caurinus* in Arizona and another in nearby Sonora.

Yellow-headed Blackbird, *Xanthocephalus xanthocephalus* (Bonaparte)

Archaeological: McKusick identified one mandible from Snaketown. I identified a humerus (size of male) from the Las Colinas site.

Ethnographic: See previous species for generic extension of taxon. Frank Jim, a Papago, called the Yellow-head *ge shawshaiñ,* 'large blackbird.'

Historic: Breninger (1901) did not see the species in September. Gilman (MS) recorded the Yellow-head as "Decidedly erratic; believe...a few spend the entire winter." It was not nesting. Individual dates he gave are 26 June (earliest lowland record), 7 September, 1 and 25 October, 15 November, 9 February, 20 March, 1 and 15 April.

Modern: Common to abundant postbreeding flocks. Flocks of adult and immature (first-year) males arrive at marshes (Barehand Lane, Salt River) at least by mid-July, rapidly increasing in numbers. Females arrive by the end of the month. By 1 August 1972 thousands were roosting in the cattails at Barehand Lane marsh. Yellow-heads molting wing and tail feathers throughout August and bob-tailed immatures not yet in definitive plumage give the erroneous impression of local breeding. (The species bred at least once in the Phoenix region, at Youngstown, specimen in fully juvenal plumage, AMR, destroyed at University of Arizona.) The enormous flocks may continue through December. But I have no records of the species in January or February. My only spring records consist of an immature male and female 9 March, two probable immature males 12 May, and a single female at Barehand Lane 22 May 1974 (when the marsh was beginning to dry and there was no remaining surface water). I saw several silent adult males at a cattail marsh on the Salt in June 1976. Some nonbreeding individuals may summer where dairies and feed lots are near marshes, as west of 91st Avenue on the Salt. For example, Simpson and I saw about a hundred individuals (adult and first-year males and females) here on 25 June 1977. The birds were not singing and were not paired. These may already have left their breeding grounds at higher elevations.

Taxonomy: Osteologically, the genus *Xanthocephalus* is nearest *Agelaius* but has no close affinities with the superficially similar *Gymnomystax* of the tropics.

Change: Most of the reservation marsh habitat has been destroyed, and this blackbird now occurs primarily along cattail fringes of the Salt.

Brown-headed Cowbird, *Molothrus ater* (Boddaert)

Historic: Breninger (1901:45) noted them "Seen among Red-wings; probably the northern form" in September. Gilman (MS) found them

> Common summer visitant and a few remaining throughout the year. Flocks [of]...25 spent the entire winter of 1915 around the Santan school grounds and another flock at Sacaton. Numerous during late spring and early summer and congregating in large flocks of over 100 sometimes, after the young have left the nest and are able to shift for themselves. This assembling is very noticeable late in August and soon afterwards the numbers decrease and comparably few stay through the winter. The Abert's Towhee [is]...most often victimized. Next in frequency were Song Sparrows as hosts, closely followed by gnatcatchers....Then came Hooded Oriole with smaller percentage. Only two varieties of notes were heard: the ordinary shrill squeak, and a short gurgle, heard only in courtship. With few exceptions only one egg [is laid] in any bird's nest. A gnatcatcher's nest with two of the cowbirds', two in the Abert's Towhee, two in one Song Sparrow's nest, and four in a Hooded Oriole's nest were the only exceptions found.

Modern: Uncommon winter, abundant summer resident. Two different populations occur on the reservation. Breeding cowbirds arrive and begin displaying the first half of March. A lone adult male 10 March 1975 displaying in a mature mesquite at Komatke evoked aggressive responses from a Bendire's Thrasher, Phainopepla, and Cactus Wren (all early nesters). But a silent, solitary female (SD, local race) foraging in the lower Salt cottonwoods on 2 April 1978 was still in prebreeding condition, and another taken there 11 May had not yet ovulated (largest yolk 4 mm). These small breeding birds are virtually gone by August. Wintering cowbirds are considerably larger in size but decidedly uncommon. These are usually females, often accompanying flocks of other icterids. However, on 27 December 1980 near Barehand Lane marsh I found a flock of about three thousand large cowbirds feeding near sheep in alfalfa. Nearly all were males, with but a sprinkling of females and some female and immature (but no adult) male Yellow-headed Blackbirds.

Taxonomy: The small breeding race, *M. a. obscurus* (Gmelin), is present from at least 10 March to 5 August (specimens). Two winter females (31 October and 9

December) are the quite large *M. a. artemisiae* Grinnell, breeding in the Great Basin and northward. *M. a. ater,* intermediate in size but with a stubby bill, breeds in the eastern woodlands and the southern Great Plains. A shot into the edge of the 27 December Barehand Lane flock yielded thirty-two specimens (SD, one female, four immature males, twenty-seven adult males). Of these, twenty-one are *artemisiae,* three are *artemisiae* x *ater,* three are *artemisiae* x *obscurus,* three are nominate *ater,* and two appear to be *obscurus* at the upper size limit for that form.

Bronzed Cowbird, *Molothrus aeneus* (Wagler)

Historic: This Mexican species was discovered in Arizona simultaneously by Gilman (1909d, 1910) at Sacaton and by Visher (1910) in Tucson. Gilman's birds were first seen 9 May 1909, remained to at least mid-August, but were not there when Gilman returned by 21 September. At Santan and Sacaton he found no eggs of this cowbird in twenty-five Northern Oriole nests examined in 1913 nor in twenty-eight examined in 1914. But both Bronzed and Brown-headed eggs (even together) were deposited in Hooded Oriole nests.

Gilman (MS) recorded the early status of the species as: "Summer visitant in limited numbers and breeding. Seen each summer from 1909....Eggs found only in nests of Arizona Hooded Orioles. ...Appears in May, 5th being the earliest and 25th the latest date of arrival, and they stay until summer is over, 30 September being the latest date seen."

Modern: Fairly common in summer. My extreme dates and specimens are 18 May to 24 July only. This cowbird is not nearly so "rural" as the Brown-headed, and it stays primarily about Pima rancherias with livestock or urbanized communities with lawns. It does not winter on the reservation and only rarely about Phoenix stockyards (Johnson and Roer, 1968).

Taxonomy: I agree with the merger of *Tangavius* (1839) into *Molothrus* (1832), following Parkes and Blake (1965). The dimorphic Arizona race *T. a. milleri* van Rossem becomes *M. a. loyei* Parkes and Blake.

Change: After its arrival early this century the Bronzed Cowbird spread rapidly for two decades (Phillips, 1946, 1968), then stabilized. The major change on the reservation since Gilman's time is its switching from the now almost extirpated Hooded Oriole to the Bullock's (Northern) Oriole as principal host species.

Brewer's Blackbird, *Euphagus cyanocephalus* (Wagler)

Ethnographic: See Red-winged Blackbird account.

Historic: Gilman (MS) called it an "Abundant winter visitant, arriving about the middle of September [15 September 1914] and remaining until about the middle of April, though sometimes remaining later, as 30 April 1914, for instance."

Modern: Common to abundant winter visitant. The Brewer's Blackbird occurs in greatest numbers about cattle feedlots on the Laveen and west Phoenix area and at roosts in cattails and saltcedars. My outside dates for the rest of the reservation are 30 September to 30 March.

Taxonomy: The taxonomy of Brewer's Blackbirds is complicated by a lack of satisfactory color variation in males and by radical seasonal color changes in females, with spring birds becoming much blacker with wear, at the same time losing the brown or gray tones of their anterior parts. Were this not enough, there is protracted migration and few specimens have any label annotations indicating fat or gonad condition. This revision therefore is tentative, based on 177 females taken at various seasons.

Females of the Pacific slope are small (usually wing under 114 and bill depth under 8) and moderately pale. These are *E. c. minusculus* Oberholser, 1920. It is not known how far east and north this race extends, but some birds from the lowlands of Oregon and Washington are also small, but blackish. Breeding birds from the southern Rocky Mountain area (at least Colorado, Arizona, and New Mexico) are large (usually wing over 118 and bill depth over 8) and pale (abdomen slaty and anterior parts tipped with ashy brown or gray). These are *E. c. cyanocephalus.* The type unfortunately is a male that was wintering in Mexico, but as first reviser Oberholser has restricted the name to this large interior population, which is migratory. Most of the wintering Arizona lowland specimens, including those from the reservation, are this form. There is a tendency for birds from the Great Basin to be blacker, especially in spring, and more strongly brownish washed across the upper parts (throat, chest, head, nape). These have been named *E. c. aliastus* Oberholser, 1932. This trend appears to be too weak to segregate freshly molted fall females, even on their breeding grounds. If "*aliastus*" were recognized, then several southwestern wintering females must be referred to it (6 December, Pinetop, Navajo County, Arizona; 15 February, Palo Verde, on the middle Gila, Arizona; 21 April, Apache, Grant County, New Mexico) and even 6 April 1978, Michoacán, Mexico. The blackest, most metallic spring females I have seen are from the northeast (eastern Alberta, Saskatchewan, Minnesota, and Wisconsin). These are *E. c. brewsteri* (Audubon, 1843), *fide* Oberholser. A satisfactory revision must await the collection of females early in the breeding season and also immediately after basic molt throughout the species' range. Meanwhile a conservative course might be to recognize all larger, darker birds, from the Pacific Northwest to the Great Lakes, as *brewsteri,* all large, pale birds of the Rocky Mountains as nominate *cyanocephalus* and all small, thin-billed, rather pale birds of the California lowlands as *minusculus.*

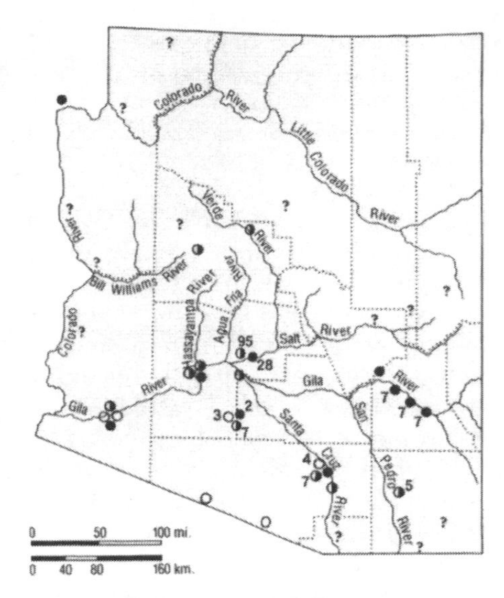

Great-tailed Grackles (1960s and 1970s)

● = Quiscalus mexicanus monsoni
○ = Q.m. nelsoni
◑ = Q.m. monsoni x nelsoni **?** = race unknown

Great-tailed Grackle, *Quiscalus mexicanus* (Gmelin)

Ethnographic: No name for these newly arrived birds in the Pima lexicon. But at Quitovac, the ancient oasis village in northern Sonora, Papago Luciano Noriego told me that grackles were "*shawshaiñ tambien pero no hay antes.*" Apparently grackles, both species of cowbirds, and the three species of blackbirds are included in the ethnotaxon *shawshaiñ* by Papago and lowland Pima Bajo.

Modern: Uncommon, breeding at Sacaton. A first-year male (AMR 2036) appeared at Komatke 13 March 1968. Later in the spring a female, accompanied by a large male, examined thick tamarisks about St. John's Indian School but departed shortly after. One or two females were apparently feeding young west of Sacaton 21 June 1970. Grackles bred about the Sacaton agency at least by 1973, fledging young 21 June (*fide* Harley and Charlie Jim; adults of both sexes seen by me later in the day).

Taxonomy: The two races involved in the Southwest invasion are *Q. m. monsoni* (Phillips) and *Q. m. nelsoni* (Ridgway). They have met and interbred extensively in southern Arizona (Rea, 1969). The Phoenix population is mixed, but *monsoni* influence is predominant. And the one reservation specimen is typically *monsoni* somewhat intermediate toward *nelsoni* (wing chord, 174; tail, 175; bill, 31.0). I have studied the behavior, vocalizations, and osteology of *Q. lugubris* (Swainson), *Q. quiscula* (Linnaeus), and *"Cassidix" mexicanus* and agree they are congeneric, as proposed by Yang and Selander (1968), Blake (1968), and others.

Change: The colonization locally on the reservation is part of a widespread invasion of the Southwest. Modifications by man, particularly agriculture, cattle feedlots, and the planting of exotic trees, promote grackle colonization (see chapter 5).

Carduelidae

[Evening Grosbeak, *Coccothraustes vespertinus* (W. Cooper)]

Modern: Erratic winter visitant. I took no specimen. Sight records (all at St. John's Indian School, Komatke) are for 22 October and 11 November 1963 and 18 September 1964.

Taxonomy: I find nothing in the cranial or postcranial skeletal anatomy to suggest that *Hesperiphona* Bonaparte, 1851, is generically distinct from *Cocothraustes* Brisson, 1760. Its only merit seems to be euphony over cacophony, which, alas, has no bearing in nomenclature.

Purple Finch, *Carpodacus purpureus* (Gmelin)

Modern: Rare. I saw this finch but once (AMR 4266, immature male, Barehand Lane, 10 December 1972).

Taxonomy: *C. p. rubidus* Duvall of the Northwest is not recognized either by the A.O.U. Check-list (1957) or by Howell (1968). I find this race to be satisfactory. The one reservation specimen is *C. p. californicus* Baird, the paler, more widespread race.

Cassin's Finch, *Carpodacus cassinii* Baird

Historic: In September Breninger (1901:46) recorded "A few seen perched in a mesquite tree." Though he also noted the more common House Finch, the Cassin's Finch observation requires specimen substantiation.

Modern: Rare in winter. Two specimens (AMR 4247, 4248, immature males) were taken from a flock of three at Komatke 26 November 1972. I saw another purple finch (adult male, either *C. cassinii* or *C. purpureus)* at Vahki 24 December 1972.

Taxonomy: Two races appear to be identifiable in this species, nominate *cassinii* of the more southern Rocky Mountains and *C. c. vinifer* Duval of the interior Northwest. One reservation specimen is *cassinii* (paler, wing chord 91.3+), and the other is *vinifer* (darker, wing chord 86.5). Identifications are by Rea and Phillips.

House Finch, *Carpodacus mexicanus* (Muller)

Historic: Breninger (1901:46) recorded "Several seen." Gilman (MS) called it "Resident in small numbers and breeding. Does not nest so close to buildings as in southern California, though it has one nesting place that appears to be a favorite in both states: *Opuntia bigelovii.*"

Modern: Common year round. This finch occurs not only on the floodplains and bajadas but well up into the mountains. In agricultural areas it gathers into enormous flocks in mid-winter.

Pine Siskin, *Carduelis pinus* (Wilson)

Historic: Gilman (MS) recorded the siskin as "Seen but twice: 12 October several seen and 16 May a small flock noted." The year is not given.

Modern: Erratically common to abundant in winter. I found this species only in 1969 and 1972, but it probably occurred in 1975 (when I did not visit suitable fields on the reservation but saw large flocks in western Phoenix). My outside dates (and specimens) are from 29 October to 24 December. The flocks, often accompanied by *C. psaltria* and *C. tristis,* feed in weedy fields left fallow, especially those overgrown with *Helianthus.* Apparently the flocks wander after depleting the local seed supply, as I have been unable to discover siskins in January.

Taxonomy: The breeding race of the United States in general is *C. p. pinus. C. p. macropterus* (Du Bus) breeds in Mexico north to the mountains of extreme southern Arizona (Sutton, 1943) and supposedly floods northward in winter (Phillips et al., 1964). Of the fifteen extant reservation specimens (others destroyed, University of Arizona), six are small *pinus,* seven are intermediate, and two are long-winged like *macropterus.* Phillips (personal communication) now considers the long-winged southwestern specimens not satisfactorily separable from nominate *pinus*; true *macropterus* are dark headed.

Lesser Goldfinch, *Carduelis psaltria* (Say)

Historic: In September Breninger (1901:46) stated that this goldfinch was "Seen feeding in a field of sunflowers." Gilman (MS) noted "A few seen at various seasons as: October, November, February, March, April, and June. No nests found and none seen between June and October.

Modern: Fairly common. My records for this species away from the lower Salt include each month from 30 September to 14 March and again from 19 May to 21 June (postbreeding?), with no records in April or July through most of September. I found no evidence of breeding until 1978. At the Salt-Gila confluence on 2 April a pair was copulating atop a tall willow and the next day the female was seen incubating. The pair copulated at about 20-minute intervals. She was apparently still incubating on 11 April. Farther upstream on 1 September 1980 Lesser Goldfinches were abundant but erratic in young willow thickets, and Takashi Ijichi took a bright immature female (SD).

Change: The maturing of willow and cottonwood trees on the lower Salt appeared to be the necessary condition for this species to breed. I searched for pairs breeding here in earlier years but found none. It is doubtful that it breeds elsewhere on the reservation.

American Goldfinch, *Carduelis tristis* (Linnaeus)

Modern: Uncommon and erratic. I saw this species in but two winters: 14 February 1971 (Barehand Lane) and 26 November to 24 December 1972 (widespread in Pima villages). It was always much less numerous in mixed flocks than *C. psaltria.*

Taxonomy: The usual race in Arizona is *C. t. pallidus* Mearns, breeding in the Great Basin and Rocky Mountains. The three February specimens and one of the 24 December 1972 specimens are this large, pale race. But three others selectively collected from the December flock (AMR 4287–4289, immature female, adult female, adult male, respectively) are the relatively darker and smaller *C. t. tristis* of the eastern half of North America. This race is casual in Arizona with only one "perfectly typical" previous specimen *fide* Phillips et al. (1964:187).

Lawrence's Goldfinch, *Carduelis lawrencei* (Cassin)

Historic: Gilman (MS) recorded this California breeding finch as "Rarely seen: three in October, four in February, 15 in one flock during March and twice in April." He took a female (SBC) at Blackwater on 1 December 1907.

Modern: Uncommon and erratic. I observed the species in but four winters (maximum of three individuals) as follows: 3 November 1963, 11 January 1967, 24 February 1969, and 29 October to 25 December 1972.

Passeridae (=Ploceidae)

House Sparrow, *Passer domesticus* Linnaeus

Historic: This Eurasian species was not found by Breninger (1901). It arrived during Gilman's (MS) time: "Not noted at Sacaton in the fall of 1907 nor that winter, but late in spring three or four of them were seen. At Casa Grande on the Southern Pacific Railroad, I saw three of the birds 16 October 1907. This was only 15 miles [24 km] from Sacaton, so possibly those at Sacaton came from Casa Grande. The spring of 1910 saw a small colony at Sacaton and they increased and spread to Santan in a short time."

Modern: Common. The species is resident about the Pima villages and many ranches, where there is farming activity, often horses, and tamarisk trees for nesting. But it is still absent from habitats not constantly modified by man (mountains, bajadas, marshes).

Change: This non-endemic species was introduced into North America during the last century. Its colonization of the reservation, documented by Gilman, is in accord with data for the rest of Arizona (see Phillips et al., 1964:162).

Reference Material

APPENDIX A

Piman Orthography

Comparative Piman Orthographies

Symbol Used Here*	Approx. English Equivalent	I.P.A. (Pike, 1947)	Russell (Dolores) 1908	Saxton & Saxton (1969)†
'	[glottal stop]	ʔ	'	'
a	p*o*t	a	a	a
e	ros*e*s [with lips spread]	ï [approx.]	ʊ [or] ʋ	e
i	*e*vent	i	i	i
o	b*oa*t	o	o	o
aw	*aw*l	)	â	o
u	b*oo*t	ü	u	u
:	[lengthened vowel]	.	[none]	[vowel + h]
ĭ [etc.]	[nonvoiced terminal vowel]	I [etc.]	[none]	[not distinguished]
j	*j*ump	ǰ	[none]	j
ch	*ch*ile	č	tc	ch
sh	*sh*ip [with tip of tongue higher than in English]	š	c	sh
s	*s*ip	s	s	s
l	[between r and l]	ř	r [or] l	l
v	*v*inegar	v	v	[none]
w	*w*inter	w	w	w
ḍ	[retroflexed *d*]	ḍ	td	D
ñ	ca*ny*on	ñ	ny	[positional]
t	*t*one	ţ	t	t
g	gun	g	g	g

*Basically the Alvarez and Hale (1970) system, except where Riverine Pima distinguishes additional phonemes. Only sounds likely to cause problems for English speakers are included here.

†Subsequently modified in various works.

Diphthongs in Riverine Pima with Approximate English Equivalents

Symbol Used Here	Approximate English Equivalent
ai	b*i*ke
au	*ou*ch!
ei	no English equivalent
oi	b*oy*
ui	g*ooey*

Most authors (e.g., Alvarez and Hale, 1970; Bahr et al., 1974; Densmore, 1929; Saxton and Saxton, 1969, 1973) transcribing northern Piman have worked with one or more of the eight dialects (Saxton and Saxton, 1969) of desert Piman, *táwbano áwawtam,* now politically designated "Papago." This study deals with still another dialect, of the Riverine Pimans, *ákimul áwawtam.* These are the "Pima" of current political designation.

The Pima distinguish several significant sounds apparently not so distinguished in any of the Papago dialects (except perhaps the northernmost). In Pima [v] is a phoneme distinct from [w], whereas only the phoneme [w] occurs in Papago (except for certain northern villages settled by Kohadk). [f] is an allophone of [v] occurring syllable terminally. Most authors have chosen to write the intermediate sound [ř] as [l]; there is no English equivalent. In Pima it is near a single flap [ř] of Pike (1947) and is termed "initial uvular r." [aw] may be an allophone of [o] in Papago, but since it appears to me to be an important dialectical difference, I have preserved it in my Pima transcription. [aw] is used here as a single vowel, as in English *aw*l or *aw*ful, not the glide of the Alvarez and Hale system, as in English w*ow*. The latter sound, which occurs also in Riverine Pima, is here written [au] as in *haupal* or *baui* ('coral bean'). [û] and [ʌ] may be allophones of [a] in certain environments, but these three are distinct from [u]. Any Piman vowel may be lengthened, indicated by [:] without an intervening glottal stop, as in English 'far' or 'moon' pronounced slowly. In certain environments, [k] and [g] may be allophones in Pima, written as [ɣ] in Russell; in these cases I have depended on native consultants to tell which fits better. A number of morpheme-terminal vowels in Riverine Pima are nonvoiced and aspirated, even when united with another morpheme. [ŋ] is a significant (though rare) sound in Pima but does not appear in the avian ethnotaxonomy. [d], the retroflexed [ḍ], and [t] are significant sounds in the northern Piman languages. Riverine Pima uses [t] in a number of bird names where Papago uses [ḍ].

Since ethnotaxonomies are of such great importance to comparative linguistics, and since the prehistoric movements and interrelationships of Piman-speaking people in western Mexico is so poorly understood, I think it best to err on the cautious side in transcribing apparent dialectical peculiarities of Riverine Pima. All quoted material appears here in its original orthography.

APPENDIX B

Riverine Pima Ethnotaxonomy of Birds

Pima Name	English Name	Scientific Name
u'uhik	Bird [Life Form]	Aves
chuáwgiakam vákoiñ	Brown Pelican	*Pelecanus occidentalis*
ñúi	Turkey Vulture	*Cathartes aura*
s'chuk ñúi	Black Vulture	*Coragyps atratus*
háupal	Red-tailed Hawk	*Buteo jamaicensis*
vá:kaf	Harris' Hawk	*Parabuteo unicinctus*
ba:k	Golden Eagle	*Aquila chrysaetos*
se:p wéhadam	Marsh Hawk	*Circus cyaneus*
vakoiñ ba:k	Osprey	*Pandion haliaetus*
oam ñúi	Crested Caracara	*Caracara cheriway*
ví:sukĭ	Prairie Falcon	*Falco mexicanus*
chúduk vi:sukĭ	(?Peregrine) Falcon	*Falco* (*peregrinus*?)
sísik	American Kestrel	*Falco sparverius*
kákachu	Gambel's Quail	*Callipepla gambelii*
tóva	Common Turkey	*Meleagris gallopavo*
kó:makĭ vakoiñ	Great Blue Heron	*Ardea herodias*
tóa vákoiñ	Common Egret	*Ardea alba*
(same?)	Snowy Egret	*Ardea thula*
tash cúnam vákoiñ	Black-crowned Night Heron	*Nycticorax nycticorax*
haiñ júlshap	Sandhill Crane	*Grus canadensis*
váchpik	American Coot	*Fulica americana*
chívichuich, sívichich	Killdeer	*Charadrius vociferus*
shú:dakĭmat	sandpiper (all species)	Scolopacidae (part?)
chúvul ú'uhik	Black-necked Stilt	*Himantopus mexicanus*
vápkukĭ	ducks, swans, geese	Anatidae
hiá:lak	Snow Goose	*Anser caerulescens*
awkakoi	White-winged Dove	*Zenaida asiatica*
háwhi	Mourning Dove	*Zenaida macroura*
chúhupĭ, [also] vúhigam	Common Ground Dove	*Columbina passerina*
gúgu	Inca Dove	*Scardafella inca*
s'náwkadam u'uhik	parrot	(all species)
á:ḍo	macaw	*Ara* spp.
kádgam	Yellow-billed Cuckoo	*Coccyzus americanus*
táwḍai	Greater Roadrunner	*Geococcyx californianus*
é:'et váhudam	Barn Owl	*Tyto alba*
chúkut	Great Horned Owl	*Bubo virginianus*
kú:kul, [also] kú:kvul	Common Screech Owl	*Otus asio*
kókoho	Burrowing Owl	*Athene cunicularia*

APPENDIX B (Continued)

Pima Name	English Name	Scientific Name
kó:logam	Nuttall's Poorwill	*Phalaenoptilus nuttallii*
ñúput	Lesser Nighthawk	*Chordeiles acutipennis*
vípismal	hummingbird (all species)	Trochilidae
báifchul	Belted Kingfisher	*Megaceryle alcyon*
kúdat	Northern Flicker	*Colaptes auratus*
híkvik	Gila Woodpecker	*Melanerpes uropygialis*
chúhugam	Ladder-backed Woodpecker	*Dendrocopos scalaris*
chíkukmal	Western Kingbird	*Tyrannus verticalis*
hé:vel maws	Say's Phoebe	*Sayornis saya*
hévachut [also] hévaichut	Barn Swallow	*Hirundo rustica*
gídval	swallow or Purple Martin (?)	Hirundinidae
unmú'ukik	? Steller's Jay	?*Cyanocitta stelleri*
schuk ú'uhik	Common Raven	*Corvus corax*
hávaiñ	Common Crow	*Corvus brachyrhynchos*
gáw 'uguf	Loggerhead Shrike	*Lanius ludovicianus*
gí:sop	Verdin	*Auriparus flaviceps*
háwkut	Cactus Wren	*Campylorhynchus brunneicapillus*
vávas	Rock Wren	*Salpinctes obsoletus*
	[? also Cañon Wren]	*Catherpes mexicanus*
shu:g	Northern Mockingbird	*Mimus polyglottos*
bet kúishnam	Bendire's Thrasher	*Toxostoma bendirei*
kud vik	(? Curve-billed) Thrasher	*Toxostoma (curivirostre?)*
kúli huch	(? Crissal) Thrasher	*Toxostoma (crissale?)*
chuf chíñgam	(? Crissal) Thrasher	*Toxostoma (crissale?)*
kómakĭ ú'uhik	(? Le Conte's) Thrasher	*Toxostoma lecontei*
básho sve:k	Northern Robin	*Turdus migratorius*
schuk máw awkum gí:sop	Black-tailed Gnatcatcher	*Polioptila melanura*
kwígam	Phainopepla	*Phainopepla nitens*
hé'et kávult	Summer Tanager	*Piranga rubra*
sípok	Northern Cardinal	*Cardinalis cardinalis*
bíchput	Abert's Towhee	*Pipilo aberti*
áw'awtórtupwa, [also] s'bában mákam	Lark Bunting	*Calamospiza melanocorys*
támtol	White-crowned Sparrow	*Zonotrichia leucophrys*
tósif	Western Meadowlark	*Sturnella neglecta*
sháwshaiñ	Red-winged Blackbird	*Agelaius phoeniceus*
váchukukĭ	Northern (Bullock's) Oriole	*Icterus galbula*

Bibliography

Adams, C. T. 1955. Comparative osteology of the night herons. *Condor* 57:55–60.

Aldrich, J. W. 1968. Population characteristics and nomenclature of the Hermit Thrush. *Proc. U. S. Nat. Mus.* 124:1–33.

———, and A. J. Duvall. 1958. The races of the Mourning Dove: their distribution and migration. *Condor* 60:108–128.

Alvarez, A., and K. Hale. 1970. Toward a manual of Papago grammar: some phonological terms. *Int. J. Amer. Linguist.* 36:83–97.

Amadon, D. 1944. The genera of Corvidae and their relationships. *Amer. Mus. Novit.* 1251:1–21.

American Ornithologists' Union. 1931. *Check-list of North American birds.* 4th ed. Lancaster Press, Lancaster, Pa.

———. 1957. *Check-list of North American birds.* 5th ed. Baltimore Press, Baltimore, Md.

American Ornithologists' Union Check-list Committee. 1973. Thirty-second supplement to the American Ornithologists' Union Check-list of North American birds. *Auk* 90:411–419.

———. 1976. Thirty-third supplement to the American Ornithologists' Union Check-list of North American birds. *Auk* 93:875–879.

Anderson, B. W., and R. J. Daugherty. 1974. Characteristics and reproductive biology of grosbeaks (*Pheucticus*) in the hybrid zone in South Dakota. *Wilson Bull.* 86:1–11.

Anza, J. B. 1930. *Anza's California expeditions: diaries of Anza, etc.* Vol. 2. [tr. and ed.] H. E. Bolton. Univ. of California Press, Berkeley.

Arizona Water Commission. 1975. *Summary: Phase 1—Arizona State Water Plan: Inventory of Resource and Uses.* Ariz. Water Comm., Phoenix.

Arvey, D. M. 1951. Phylogeny of the waxwings and allied birds. *U. Kansas Publ. Mus. Nat. Hist.* 3:473–530.

Audubon, J. J. 1841. *Birds of America.* Vol. 3. J. B. Chevalier Co., Philadelphia, Pa.

Audubon, J. W. 1906. *Audubon's Western Journal: 1849–1850.* [ed.] Maria R. Audubon. Arthur H. Clark Co., Cleveland, Ohio.

Austin, G. T. 1968. Additional bird records for southern Nevada. *Auk* 85:692.

———. 1971. On the occurrence of eastern wood warblers in western North America. *Condor* 73:455–462.

———, and A. M. Rea. 1976. Recent southern Nevada bird records. *Condor* 78:405–408.

———, E. L. Smith, and S. Speich. 1972. New Arizona bird records. *Calif. Birds* 3:43–44.

Bahr, D. M., J. Gregorio, D. Lopez, and A. Alvarez. 1974. *Piman shamanism and staying sickness* (ka:cim mukidag). Univ. of Arizona Press, Tucson.

Bailey, A. M. 1928. A study of the Snowy Herons of the United States. *Auk* 45:430–440.

Baird, W. M., and S. F. Baird. 1843. Description of two species, supposed to be new, of the genus *Tyrannula* Swains, found in Cumberland County, Pennsylvania. *Proc. Acad. Nat. Sci.* Philadelphia 1:283–285.

Baird, S. F. (with cooperation of J. Cassin and G. N. Lawrence). 1858. Birds. In *Pacific Railroad reports, explorations and surveys for a railroad route from the Mississippi River to the Pacific Ocean.* U. S. War Dept., Washington, D. C.

Bakus, G. J. 1962. Early nesting of the Costa Hummingbird in southern California. *Condor* 64:438–439.

Balda, R. P., B. C. McKnight, and C. D. Johnson. 1975. Flammulated Owl migration in the southwestern United States. *Wilson Bull.* 87:520–533.

Bangs, O., and T. E. Penard. 1921. Descriptions of six new subspecies of American birds. *Proc. Biol. Soc. Wash.* 34:89–92.

Banks, R. C. 1964. Geographic variation in the White-crowned Sparrow, *Zonotrichia leucophrys. Univ. Calif. Publ. Zool.* 70:1–123.

———. 1970. Molt and taxonomy of Red-breasted Nuthatches. *Wilson Bull.* 82:201–205.

Barger, R. L., and P. F. Ffolliott. 1971. Prospects for cottonwood utilization in Arizona. *Progr. Agric. in Arizona* 23:14–16.

Bartlett, J. R. 1854. *Personal narrative of explorations and incidents in Texas, New Mexico, California, Sonora, and Chihuahua.* Vol. 2. D. Appleton and Co., New York and London.

Behle, W. H. 1942. Distribution and variation of the Horned Larks (*Otocoris alpestris*) of western North America. *Univ. Calif. Publ. Zool.* 46:205–316.

———. 1960. The birds of southeastern Utah. *Univ. Utah Biol. Ser.* 12:1–56.

———. 1968. A new race of the Purple Martin from Utah. *Condor* 70:166–169.

———. 1973. Clinal variation in White-throated Swifts from Utah and the Rocky Mountain region. *Auk* 90:299–306.

———. 1976. Systematic review, intergradation, and clinal variation in Cliff Swallows. *Auk* 93:66–77.

Bendire, C. E. 1890. Notes on *Pipilo fuscus mesoleucus* and *Pipilo aberti,* their habits, nests, and eggs. *Auk* 7:22–29.

Bennett, C. F. 1968. Human influences on the zoogeography of Panama. *Ibero-Americana* no. 51.

Bent, A. C. 1926. Life histories of North American marsh birds. *Bull. U. S. Nat. Mus.* 135.

———. 1940. Life histories of North American cuckoos, goatsuckers, hummingbirds and their allies. *Bull. U. S. Nat. Mus.* 176.

———, and collaborators. 1968. Life histories of North American cardinals, grosbeaks, buntings, towhees, finches, sparrows, and allies. Oliver L. Austin, Jr. [ed.]. *Bull. U. S. Nat. Mus.* 237.

Bishop, L. B. 1900. Descriptions of three new birds from Alaska. *Auk* 17:113–120.

Blake, E. R. 1968. Family Icteridae, pp. 138–201. In R. A. Paynter, Jr. [ed.] *Check-list of birds of the world.* Vol. 14. Mus. Comparative Zoology, Cambridge, Mass.

Bock, W. J. 1958. A generic review of the plovers (Charadriinae, Aves). *Bull. Mus. Comp. Zool.* 118:27–97.

Bond, R. M. 1943. Variation in western Sparrow Hawks. *Condor* 45:168–185.

Bowden, C. 1977. *Killing the hidden waters.* Univ. Texas Press, Austin.

Breninger, G. F. 1898. The Ferruginous Pygmy Owl. *Osprey* 2:128.

———. 1901. A list of birds observed on the Pima Indian Reservation, Arizona. *Condor* 4[=3]:44–46.

Bringas, D. M. 1977. *Friar Bringas reports to the King: methods of indoctrination on the frontier of New Spain 1796–97.* [transl. and ed.] D. S. Matson and B. L. Fontana. Univ. of Arizona Press, Tucson.

Brodkorb, P. 1942. Notes on some races of Rough-winged Swallow. *Condor* 44:214–217.

———. 1949. Variation in the North American forms of Western Flycatcher. *Condor* 51:35–39.

———. 1957. New passerine birds from the Pleistocene of Reddick, Florida. *J. Paleontol.* 31:129–138.

———. 1968. Birds, pp. 270–451. *In* W. F. Blair [ed.] *Vertebrates of the United States.* 2nd ed. McGraw-Hill, New York.

———. 1971. Catalogue of fossil birds: Part 4 (Columbiformes through Piciformes). *Bull. Florida State Mus. Biol. Ser.* 15:163–266.

———. 1978. Catalogue of fossil birds, Part 5 (Passeriformes). *Bull. Florida State Mus. Biol. Sci.* 23:139–228.

Browder, J. A. 1973. Long-distance movements of Cattle Egrets. *Bird-banding* 44:158–170.

Brown, H. 1900. The conditions governing bird life in Arizona. *Auk* 17:31–34.

Browning, M. R. 1974. Taxonomic remarks on recently described subspecies of birds that occur in the northwestern U. S. *Murrelet* 55:32–38.

———. 1976. The status of *Sayornis saya yukonensis* Bishop. *Auk* 93:843–846.

———. 1977. Geographic variation in *Contopus sordidulus* and *C. virens* north of Mexico. *Great Basin Naturalist* 37:453–456.

———. 1979. A review of geographic variation in continental populations of the Ruby-crowned Kinglet (*Regulus calendula*). *Nemouria* 21:1–9.

Bryan, K. 1922. Erosion and sedimentation in the Papago country, Arizona, with a sketch of the geology. *USGS Bull.* 730-B:19–90.

———. 1925. Date of channel trenching (arroyo cutting) in the arid southwest. *Science* 62 (1607):338–344.

Budowski, G. 1956. Tropical savannas: a sequence of forest felling and repeated burning. *Turrialba: Revista Interamericana de Ciencias Agricolas* 6:23–33.

Burkham, D. E. 1972. Channel changes of the Gila River in Safford Valley, Arizona, 1846–1970. Gila River Phreatophyte Project. *Geol. Surv. Prof. Pap.* 655-G, U. S. Govt. Printing Office, Washington, D.C.

Burleigh, T. D. 1960a. Three new subspecies of birds from western North America. *Auk* 77:210–215.

———. 1960b. Geographic variation in the Western Wood Peewee (*Contopus sordidulus*). *Proc. Biol. Soc. Wash.* 73:141–146.

Campbell, K. E., Jr. 1979. The non-passerine Pleistocene avifauna of the Talara Tar Seeps, northwestern Peru. *Life Sciences Contrib. Royal Ontario Mus.* no. 118.

Carothers, S. W., and J. R. Haldeman. 1967. New records of northern Arizona birds. *Plateau* 40:41–43.

———, and R. P. Balda. 1973. Breeding birds of the San Francisco Mountain area and the White Mountains, Arizona. *Mus. Northern Ariz. Tech. Ser.* 12:1–54.

———, R. R. Johnson, and S. W. Aitchison. 1974. Population structure and social organization in southwestern riparian birds. *Amer. Zool.* 14:97–108.

Cooch, G. 1961. Ecological aspects of the Blue–Snow Goose complex. *Auk* 78:72–89.

Cook, S. F. 1949. The historical demography and ecology of the Teotlalpan. *Ibero-Americana* no. 33.

Cooke, F., and F. G. Cooch. 1968. The genetics of polymorphism in the goose, *Anser caerulescens. Evolution* 22:289–300.

Cornwallis, R. K. 1964. Irruption, pp. 403–406. *In* A. L. Thompson [ed.] *A new dictionary of birds.* McGraw-Hill Book Co., New York.

Cortopassi, A. J., and L. R. Mewaldt. 1965. The circumannual distribution of White-crowned Sparrows. *Bird-banding* 36:141–169.

Crosby, A. W., Jr. 1972. The Columbian exchange: biological and cultural consequences of 1492. *Contributions in American Studies* no. 2. Greenwood Press, Westport, Conn.

Crosby, G. T. 1972. Spread of the Cattle Egret in the Western Hemisphere. *Bird-banding* 43:205–212.

Davis, G. P., 1982. *Man and wildlife in Arizona: the American exploration period, 1824–1865.* [Eds.] N. B. Carmony and D. E. Brown. Arizona Game and Fish Department, Phoenix.

Davis, J. 1951. Distribution and variation of the brown towhees. *Univ. Calif. Publ. Zool.* 52:1–120.

———, and L. Williams. 1957. Irruptions of the Clark Nutcracker in California. *Condor* 59:297–307.

de Boer, L. E. M. 1976. The somatic chromosome complements of 16 species of Falconiformes (Aves) and the karyological relationships of the order. *Genetica* 46:77–113.

Deignan, H. G., R. A. Paynter, Jr., and S. D. Ripley. 1964. Prunellidae, Turdinae, Orthonychinae, Timaliinae, Panurinae, Picathartinae, Polioptilinae, pp. 1–502. *In* E. Mayr and R. A. Paynter, Jr. [eds.] *Check-list of birds of the world.* Vol. 10. Mus. Comp. Zool., Cambridge, Mass.

Delacour, J. 1941. On the species of *Otus scops. Zoologica* 26:133–142.

———. 1951. The significance of the number of toes in some woodpeckers and kingfishers. *Auk* 68:49–51.

———, and E. Mayr. 1945. The family Anatidae. *Wilson Bull.* 57:3–55.

Densmore, F. 1929. Papago music. *Smithsonian Inst. Bureau Amer. Ethnol. Bull.* no. 90.

Dickerman, R. W. 1961. Hybrids among the fringillid genera *Junco–Zonotrichia* and *Melospiza*. *Auk* 78:627–632.

———. 1963. The grebe *Aechmophorus occidentalis clarkii* as a nesting bird of the Mexican Plateau. *Condor* 65:66–67.

———. 1968. A hybrid Grasshopper Sparrow x Savannah Sparrow. *Auk* 85:312–315.

———, and D. W. Warner. 1962. A new Orchard Oriole from Mexico. *Condor* 64:311–314.

Dickinson, J. C., Jr. 1953. Report on the McCabe Collection of British Columbian birds. *Bull. Mus. Comp. Zool.* 109:123–205.

Dilger, W. C. 1956a. Hostile behavior and reproductive isolating mechanisms in the avian genera *Catharus* and *Hylocichla*. *Auk* 73:313–353.

———. 1956b. Relationships of the thrush genera *Catharus* and *Hylocichla*. *Syst. Zool.* 5:174–182.

DiPeso, C. D. 1976. Medio Period commerce, pp. 141–192. *In* C. D. DiPeso, J. B. Rinaldo, and G. J. Fenner [eds.] *Casas Grandes: a fallen trading center of the Gran Chichimeca.* Vol. 8. Amerind Foundation, Inc., Dragoon, Arizona.

Dobyns, H. F. 1978. Who killed the Gila? *J. Ariz. Hist.* 19:17–30.

———. 1981. From fire to flood: historic human destruction of Sonoran Desert riverine oases. *Ballena Press Anthro. Pap.* no. 20.

Doelle, W. H. n.d. *The adaptation of wheat by the Gila Pima: a study in agricultural change.* Unpublished typescript deposited Arizona State Museum Library, Tucson.

Dorst, J. 1962. *The migration of birds.* C. D. Sherman [tr.]. Riverside Press, Cambridge, Mass.

Drewien, R. C., C. D. Littlefield, L. H. Walkinshaw, and C. E. Braun. 1975. Conservation committee report on status of Sandhill Crane. *Wilson Bull.* 87:297–302.

Duce, J. T. 1918. The effect of cattle on the erosion of canyon bottoms. *Science* 47(1219):450–452.

Durivage, J. E. 1937. Letters and journal of John E. Durivage, pp. 159–255. *In* R. P. Bieber [ed.] *Southern trails to California in 1849.* Arthur H. Clark Co., Glendale, Calif.

Eccleston, R. 1950. *Overland to California on the southwestern trail, 1849. Diary of Robert Eccleston.* G. P. Hammond and E. H. Howes [eds.] Bancroft Library Publ. no. 2, Univ. of California at Berkeley.

Eckman, E. C., M. Baldwin, and E. J. Carpenter. 1923. Soil survey of the middle Gila Valley area, Arizona, pp. 2087–2119. *In* M. Whitney [ed.] *Field operations of the Bureau of Soils, 1917.* 19th Report, U. S. Dept. Agriculture, Bureau of Soils.

Eisenmann, E. 1962. On the genus "*Chamaethlypis*" and its supposed relationship to *Icteria*. *Auk* 79:265–267.

Emory, W. H. 1848. Notes of a military reconnaissance from Ft. Leavenworth, in Missouri, to San Diego, in California. *House Executive Documents* no. 41:5–126. 30th Congress, 1st session.

Evans, G. W. B. 1945. *Mexican gold trail: the journal of a Forty-niner.* G. S. Dumke [ed.]. Huntington Library, San Marino.

Evans, R. 1967. Nest site movements of a Poor-will. *Wilson Bull.* 79:453.

Ezell, P. H. 1963. The Maricopas: an identification from documentary sources. *Anthropological Papers, Univ. of Arizona,* no. 6.

Ferguson, D. E. 1967. A possible case of egg transport by a Chuck-will's Widow. *Wilson Bull.* 79:452–453.

Fewkes, J. W. 1912. Casa Grande, Arizona. *28th Report of the Bureau of American Ethnology,* pp. 25–180.

Font, P. 1930. Font's complete diary of the Second Anza Expedition. H. E. Bolton [tr. and ed.]. *Anza's California expeditions,* Vol. '4. Univ. of California Press, Berkeley.

———. 1931. *Font's complete diary: a chronical of the founding of San Francisco:* H. E. Bolton [tr. and ed.]. Univ. of California Press, Berkeley.

Formosov, A. N. 1933. The crop of cedar nuts, invasions into Europe of the Siberian Nutcracker (*Nucifraga caryocatactes macrorhynchus*) and fluctuations in numbers of the squirrel (*Sciurus vulgaris* L.). *J. Anim. Ecol.* 2:70–81.

Friedmann, H. 1950. Birds of North and Middle America. *Bull. U. S. Nat. Mus.* no. 50, pt. 11.

———, L. Griscom, and R. T. Moore. 1950. Distributional check-list of the birds of Mexico. Part 1. *Pacific Coast Avifauna* no. 29.

Fritts, H. C. 1965. Tree-ring evidence for climatic changes in western North America. *Monthly Weather Review* 7:421–443.

Gadow, H. 1883. *Catalogue of the Birds in the British Museum.* Vol. 8. Brit. Mus., London.

Garcés, F. T. H. 1900. *On the trail of a Spanish pioneer.* 2 vols. E. Coues [tr.]. F. P. Harper, New York.

———. 1968. *Diario de exploraciones en Arizona y California en los años de 1775 y 1776. Introducción y notas de John Galvin.* Univ. Nac. Auton. Méx., D. F., Mex.

Gabrielson, I. N., and F. C. Lincoln. 1959. *The birds of Alaska.* Stackpole Co., Harrisburg, Pa.

Gaines, D. 1974. Review of the status of the Yellow-billed Cuckoo in California: Sacramento Valley populations. *Condor* 76:205–209.

Giff, J. 1980. Pima Blue Swallow songs of gratitude, pp. 127–139. *In* F. Barkin and E. Brandt [eds.], *Speaking, singing and teaching: a multidisciplinary approach to language variation.* Proc. 8th Ann. Southwestern Areal Language and Linguistics Workshop. Ariz. State Univ. Anthrop. Res. Pap. No. 20.

Gilman, M. F. 1909a. Among the thrashers in Arizona. *Condor* 11:49–54.

———. 1909b. Some owls along the Gila River in Arizona. *Condor* 11:145–150.

———. 1909c. Nesting notes on the Lucy Warbler. *Condor* 11:166–168.

———. 1909d. Red-eyed Cowbird at Sacaton, Arizona. *Condor* 11:173.

———. 1910. Notes from Sacaton, Arizona. *Condor* 12.45–46.

———. 1911a. Notes from Sacaton, Arizona. *Condor* 13:35.

———. 1911b. Doves on the Pima Reservation. *Condor* 13:51–56.

———. 1914a. Breeding of the Bronzed Cowbird in Arizona. *Condor* 16:255–259.

———. 1914b. Notes from Sacaton, Arizona. *Condor* 16:260–261.

———. 1915a. A forty acre bird census at Sacaton, Arizona. *Condor* 17:86–90.

———. 1915b. Woodpeckers of the Arizona lowlands. *Condor* 17:151–163.

———. *Birds of the Pima Indian Reservation in Arizona.* (Unpublished MS; deposited with Dr. A. R. Phillips, Nuevo León, Mex.)

Gladwin, H. S., E. W. Haury, E. B. Sayles, and N. Gladwin. 1937. Excavations at Snaketown, Material Culture. *Medallion Papers,* no. 25. Gila Pueblo, Globe, Arizona.

Godfrey, W. E. 1965. [Review of] Geographic variation in the White-crowned Sparrow *Zonotrichia leucophrys.* [by] R. C. Banks. *Auk* 82:510–511.

Goodwin, D. 1967. *Pigeons and doves of the world.* Trustees Brit. Mus. (Nat. Hist.), London.

Gordon, B. L. 1957. Human geography and ecology in the Sinu Country of Colombia. *Ibero-Americana* no. 39.

Graber, R. R., and J. W. Graber. 1954. Comparative notes on Fuertes and Orchard Orioles. *Condor* 56:274–282.

Granger, B. H. 1960. *Arizona Place Names.* Univ. of Arizona Press, Tucson.

Greenway, J. C., Jr. 1967. Sittidae, pp. 124–149. *In* R. A. Paynter, Jr., and E. Mayr [eds.] *Check-list of birds of the world.* Vol. 12. Mus. Comp. Zool., Cambridge, Mass.

Griffin, J. S. 1943. *A doctor comes to California.* California Historical Society, San Francisco.

Grinnell, J. 1904. Midwinter birds at Palm Springs, California. *Condor* 6:40–45.

———, and A. H. Miller. 1944. The distribution of the birds of California. *Pacific Coast Avifauna* no. 27.

Hanson, H. C., and E. D. Churchill. 1961. *The plant community.* Reinhold, New York.

Hardy, J. W. 1969. A taxonomic revision of the New World jays. *Condor* 71:306–375.

Hargrave, L. L. 1939. Bird bones from abandoned Indian dwellings in Arizona and Utah. *Condor* 41:206.

———. 1970. Mexican macaws: comparative osteology and survey of remains from the Southwest. *Anthropological Papers, Univ. of Arizona* no. 20.

———, and S. D. Emslie. 1979. Osteological identification of Sandhill Crane vs. turkey. *Amer. Antiquity* 44:295–299.

Harris, B. B. 1960. *The Gila Trail: the Texas Argonauts and the California Gold Rush.* R. H. Dillon [ed.]. Univ. of Oklahoma Press, Norman.

Harris, D. R. 1966. Recent plant invasions in the arid and semi-arid Southwest of the United States. *Ann. Assoc. Amer. Geogr.* 56:408–423.

Hastings, J. R., and R. M. Turner. 1965. *The changing mile: an ecological study of vegetation change with time in the lower mile of an arid and semiarid region.* Univ. of Arizona Press, Tucson.

Haury, E. W. 1950. *The stratigraphy and archaeology of Ventana Cave, Arizona.* Univ. of Arizona Press, Tucson.

———. 1976. *The Hohokam: desert farmers and craftsmen. Excavations at Snaketown, 1964–1965.* Univ. of Arizona Press, Tucson.

Hayden, C. 1965. A history of the Pima Indians and the San Carlos irrigation project. [Compiled in 1924.] *Senate Documents,* no. 11, 89th Congress, 1st session. U. S. Govt. Printing Office, Washington, D. C.

Heermann, A. L. 1859. Report upon the birds collected on the survey. *In Zoological report Explorations and surveys for a railroad route from the Mississippi River to the Pacific Ocean.* 10 (part 4), no. 1:9–20; no. 2:29–80. U.S. War Dept., Washington, D.C.

Hellmayr, C. E. 1934. Catalogue of birds of the Americas. *Field Mus. Nat. Hist. Publ., Zool. Ser.* 13, pt. 11:1–662.

Hine, R. V. 1968. *Bartlett's West. Drawing the Mexican Boundary.* Yale Univ. Press, New Haven.

Hoffmann, R. 1927. *Birds of the Pacific States.* Houghton Mifflin Co., Boston, Mass.

Howell, T. R. 1968. Subfamily Carduelinae [New World forms], pp. 207–299. *In* R. A. Paynter [ed.]. *Check-list of birds of the world.* Vol. 14. Mus. Comp. Zool., Cambridge, Mass.

Hubbard, J. P. 1972. The nomenclature of *Pipilo aberti* Baird (Aves: Fringillidae). *Proc. Biol. Soc. Wash.* 85:131–138.

———. 1975. Geographic variation in non-California populations of the Rufous-crowned Sparrow. *Nemouria* 15:1–28.

———. 1976. The nomenclatural history of the Crissal Thrasher (Aves : Mimidae). *Nemouria* 20:1–7.

———, and R. S. Crossin. 1974. Notes on northern Mexican birds: an expedition report. *Nemouria* 14:1–41.

Huey, L. M. 1939. Birds of the Mount Trumbull Region, Arizona. *Auk* 56:320–325.

———.1942. A vertebrate faunal survey of the Organ Pipe Cactus National Monument, Arizona. *Trans. San Diego Soc. Nat. Hist.* 9:353–376.

Humphrey, R. R. 1958. The desert grassland: a history of vegetational change and an analysis of causes. *Bot. Rev.* 24:193–252.

Hurley, R. J., and E. C. Franks. 1976. Changes in the breeding ranges of two grassland birds. *Auk* 93:108–115.

International Commission of Zoological Nomenclature. 1963. *International Code of Zoological Nomenclature adapted by the 15th International Congress of Zoology.* International Trust for Zoological Nomenclature, London.

Jaeger, E. C. 1948. Does the Poor-will "hibernate"? *Condor* 50:45–46.

———. 1949. Further observations on the hibernation of the Poor-will. *Condor* 51:105–109.

Jehl, J. R., Jr. 1968. Relationships in the Charadrii (Shorebirds): a taxonomic study based on color patterns of the downy young. *San Diego Soc. Nat. Hist. Mem.* 3:1–54.

Jewett, S. G., W. P. Taylor, W. T. Shaw, and J. W. Aldrich. 1953. *Birds of Washington state.* Univ. of Wash. Press, Seattle.

Johannessen, C. L. 1963. Savannas of interior Honduras. *Ibero-Americana* no. 46.

Johnsgard, P. A. 1965. *Handbook of waterfowl behavior.* Cornell Univ. Press, Ithaca, New York.

Johnson, N. K. 1976. Breeding distributions of Nashville and Virginia's Warblers. *Auk* 93:219–230.

Johnson, R. R. 1977. Foreword [unpaginated.] *In* R. R. Johnson and D. A. Jones. Importance, preservation and management of riparian habitat. *USDA Forest Service Gen. Tech. Rept.* RM-43.

———, and B. Roer. 1968. Changing status of the Bronzed Cowbird in Arizona. *Condor* 70:183.

———, and J. M. Simpson. 1971. Important birds from Blue Point Cottonwoods, Maricopa Co., Arizona. *Condor* 73:379–380.

———. *Annotated check-list of the birds of Blue Point Cottonwoods.* (Unpublished MS; deposited with J. M. Simpson, Phoenix, Ariz.)

———, J. M. Simpson, and J. R. Werner. *Birds of the Salt River Valley.* (Unpubl. MS; dep. J. M. Simpson, Phoenix, Ariz.)

Johnston, A. R. 1848. Journal of Captain A. R. Johnston, first dragoons. *House Executive Documents,* no. 41:567–614, 30th Congress 1st session. U.S. Govt. Printing Ofc., Washington, D.C.

Johnston, R. F. 1961. The genera of American ground doves. *Auk* 78:372–378.

Jollie, M. 1976. A contribution of the morphology and phylogeny of the Falconiformes. Part 1. *Evolutionary Theory* 1:1–14.

———. 1977. A contribution of the morphology and phylogeny of the Falconiformes. Parts 2, 3, 4. *Evolutionary Theory* 2:15–342.

Keith, A. R. 1968. A summary of extralimital records of the Varied Thrush, 1848 to 1966. *Bird-banding* 39:245–276.

Kessel, B. 1953. Distribution and migration of the European Starling in North America. *Condor* 55:49–67.

Kino, E. F. 1919. *Kino's historical memoir of Pimeria Alta.* Vol. 1. H. E. Bolton [tr. and ed.]. A. H. Clark, Cleveland, Ohio.

Kroeber, A. L. 1934. Uto-aztecan languages of Mexico. *Ibero-Americana* 8:1–28.

Lack, D. 1970. *The natural regulation of animal numbers.* University Press, Oxford.

Ligon, J. D. 1974. Comments on the systematic relationships of the Piñon Jay (*Gymnorhinus cyanocephalus*). *Condor* 76:468–470.

Linsdale, J. M. 1928a. Variations in the Fox Sparrow (*Passerella iliaca*) with reference to natural history and osteology. *Univ. Calif. Publ. Zool.* 12:251–392.

———. 1928b. The species and subspecies of the fringillid genus *Passerella* Swainson. *Condor* 30:349–351.

Lowe, C. H. [ed.]. 1964. *The vertebrates of Arizona. Annotated check lists of the vertebrates of the state: the species and where they live.* Univ. of Arizona Press, Tucson.

Manje, J. M. 1954. *Luz de tierra incógnita.* [tr.] H. J. Karns and Associates, Tucson, Arizona Silhouettes.

Marshall, J. T., Jr. 1960. Interrelations of Abert and Brown Towhees. *Condor* 62:49–64.

———. 1964. Voice in communication and relationships among Brown Towhees. *Condor* 66:345–356.

———. 1967. Parallel variation in North and Middle American screech-owls. *Monograph Western Found. Vert. Zool.* no. 1.

Mason, J. A. 1917. Tepecano, a Piman language of western Mexico. *Ann. New York Acad. Sci.* 25:309–416.

Mathiot, M. n.d.[1979–1980?] A dictionary of Papago usage, Ku-ʔu. *Language Science Monographs* vol. 8/2. Indiana Univ. Publs.

Mayr, E. 1942. [Review of] Speciation in the avian genus *Junco.* [by] A. H. Miller. *Ecology* 23:378–379.

———, and J. C. Greenway, Jr. [eds.] 1962. *Check-list of birds of the world,* Vol. 15. Mus. Comp. Zool. Cambridge, Mass.

———, and L. L. Short. 1970. Species taxa of North American birds: a contribution to comparative systematics. *Publ. Nuttall Ornith. Club* no. 9.

McKusick, C. R. 1980. Three groups of turkeys from southwestern archaeological sites. Pp. 225–235. *In* K. E. Campbell, Jr. [ed.]. Papers in avian paleontology honoring Hildegarde Howard. *Contrib. Sci. Natur. Hist. Mus. Los Angeles County* no. 330.

Meinertzhagen, R. 1926. Introduction to a review of the genus *Corvus. Novitates Zoologicae* 33:57–121.

Miller, A. H. 1932. Bird remains from Indian dwellings in Arizona. *Condor* 34:138–139.

———. 1941. Speciation in the avian genus *Junco. Univ. Calif. Publ. Zool.* 44:173–434.

———, H. Friedmann, L. Griscom, and R. T. Moore. 1957. Distributional check-list of the birds of Mexico. Part 2. *Pacific Coast Avifauna* no. 33.

———, and L. Miller. 1951. Geographic variation of the Screech Owls of the deserts of western North America. *Condor* 53:161–177.

Miller, R. R. 1961. Man and the changing fish fauna of the American Southwest. *Pap. Michigan Acad. Sci., Arts, and Letters* 46:365–404.

Monson, G. 1968. The Arizona state bird-list, 1964–1967. *J. Ariz. Acad. Sci.* 5:34–35.

———, and A. R. Phillips. 1941. Bird records from southern and western Arizona. *Condor* 43:108–113.

Moore, R. T. 1951. Records of two North American corvids in Mexico. *Condor* 53:101.

Morony, J. J., Jr., W. J. Bock, and J. Ferrand, Jr. 1975. *Reference list of the birds of the world.* American Museum Natural History, New York.

Murray, B. G., Jr. 1968. The relationships of sparrows in the genera *Ammodramus, Passerherbulus,* and *Ammospiza* with a description of a hybrid Le Conte's x Sharp-tailed Sparrow. *Auk* 85:586–593.

Neff, J. A. 1940a. Range, population and game status of the western White-winged Dove in Arizona. *J. Wildlife Management* 4:117–127.

———. 1940b. Notes on nesting and other habits of the White-winged Dove in Arizona. *J. Wildlife Management* 4:279–290.

Nentvig, J. 1894. Rudo ensayo. [tr.] E. Guteras. *Records of the American Catholic Historical Society of Philadelphia* 5:109–264.

Nowak, E. 1975. *The range expansion of animals and its causes (as demonstrated by 28 presently spreading species from Europe).* Foreign Scientific Publ. Dept. Nat. Center for Sci. Technol. & Econ. Info., Warsaw, Poland.

Oberholser, H. C. 1904. A revision of the American Great Horned Owls. *Proc. U. S. Nat. Mus.* 27:177–192.

———. 1918. New light on the status of *Empidonax traillii* (Audubon). *Ohio J. Sci.* 18:85–98.

———. 1919. Description of a new subspecies of *Pipilo fuscus. Condor* 21:210–211.

———. 1920. *Toxostoma crissalis* versus *Toxostoma dorsalis. Auk* 37:303.

———. 1930. Notes on a collection of birds from Arizona and New Mexico. *Sci. Publ. Cleveland Mus. Nat. Hist.* 1:83–124.

———. 1934. A revision of the North American House Wrens. *Ohio J. Sci.* 34:86–96.

———. 1937. A revision of the Clapper Rails (*Rallus longirostris* Boddaert). *Proc. U.S. Nat. Mus.* 84:313–354.

———. 1974. *The bird life of Texas.* [Ed.] B. Kinkaid, Jr., Univ. of Texas Press, Austin.

Olson, S. L., and A. Feduccia. 1980. *Presbyornis* and the origin of the Anseriformes (Aves:Charadriomorphae). *Smithsonian Contrib. Zool.* no. 323.

Osgood, W. H. 1901. New subspecies of North American birds. *Auk* 18:179–185.

Ouellet, H. 1977. Relationships of woodpecker genera *Dendrocopos* and *Picoides. Ardea* 65:165–183.

Palmer, R. S. 1962. *Handbook of North American birds.* Vol. 1. Yale Univ. Press, New Haven, Conn.

Parkes, K. C. 1953. Some bird records of importance from New York. *Wilson Bull.* 65:46–47.

———. 1959. Systematic notes on North American birds. Part 3: The northeastern races of the Long-billed Marsh Wren (*Telmatodytes palustris*). *Ann. Carnegie Mus.* 35:275–281.

———, and E. R. Blake. 1965. Taxonomy and nomenclature of the Bronzed Cowbird. *Fieldiana-Zoology* 44:207–216.

Payne, R. B. 1974. Species limits and variation of the New World Green Herons *Butorides virescens* and Striated Herons *B. striatus*. *Bull. Brit. Ornith. Club* 94:81–88.

———. 1979. Family Ardeidae, pp. 193–244. *In* E. Mayr and G. W. Cottrell [eds.]. *Check-list of birds of the world.* Vol. 1, second edition. Mus. Comp. Zool., Cambridge, Mass.

———, and C. J. Risley. 1976. Systematics and evolutionary relationships among the herons (Ardeidae). *Misc. Publ. Mus. Zool.,* Univ. Mich., no. 150.

Paynter, R. A., Jr. 1960. Family Troglodytidae, pp. 379–440. *In* E. Mayr and J. C. Greenway, Jr. [eds.] *Check-list of birds of the world.* Vol. 9. Mus. Comparative Zoology, Cambridge, Mass.

———. 1964. Generic limits of *Zonotrichia. Condor* 66:277–281.

———, and R. W. Storer. 1970. *Check-list of birds of the world.* Vol. 13. Mus. Comparative Zoology, Cambridge, Mass.

Pennington, C. W. 1969. *The Tepehuan of Chihuahua, their material culture.* Univ. Utah Press, Salt Lake City.

———. 1979. *The Pima Bajo of central Sonora, Mexico.* Vol. 2: *Vocabulario en la lengua Névome.* Univ. of Utah Press, Salt Lake City.

Pfefferkorn, I. 1949. Sonora, a description of the province. [tr. and ann.] T. E. Treutlein. *Coronado Cuarto Centennial Publications 1540–1940.* [Ed.] G. P. Hammond. Vol. 12. Univ. of New Mexico Press, Albuquerque.

Phillips, A. R. 1942. Notes on the migrations of the Elf and Flammulated Screech Owls. *Wilson Bull.* 54:132–137.

———. 1946. *The birds of Arizona.* Ph.D. Dissertation, Cornell Univ., Ithaca, New York.

———. 1947. The races of MacGillivray's Warbler. *Auk* 64:296–300.

———. 1947. Status of the Anna Hummingbird in southern Arizona. *Wilson Bull.* 59:111–113.

———. 1948. Geographic variation in *Empidonax traillii. Auk* 65:507–514.

———. 1950a. The Great-tailed Grackles of the Southwest. *Condor* 52:78–81.

———. 1950b. The pale races of the Steller Jay. *Condor* 52:252–254.

———. 1959. The nature of avian species. *J. Arizona Acad. Sci.* 1:22–30.

———. 1961. Notas sistemáticas sobre aves mexicanas. 1. *Anal. Inst. Biol. Mex.* 32:333–381. [Publ. 30 March 1962]

———. 1962. Notas sistemáticas sobre aves mexicanas. 2. *Anal. Inst. Biol. Mex.* 33:331–372.

———. 1964. Notas sistemáticas sobre aves mexicanas. 3. *Rev. Soc. Mex. Hist. Nat.* 25:217–242.

———. 1966. Further systematic notes on Mexican birds. *Bull. Brit. Ornith. Club* 86:86–94, 103–112, 125–131, 148–159.

———. 1968. The instability of the distribution of land birds in the Southwest, pp. 129–162. *In* A. H. Schroeder [ed.] *Collected papers in honor of Lyndon Lane Hargrave.* Pap. Archaeol. Soc. New Mexico no. 1.

———. 1973. On the supposed genus *Petrochelidon. Bull. Brit. Ornith. Club* 93:20.

Phillips, A. R. *(Continued)*

———. 1974. The first prebasic molt of the Yellow-breasted Chat. *Wilson Bull.* 86:12–15.

———, J. Marshall, and G. Monson. 1964. *The birds of Arizona.* Univ. of Arizona Press, Tucson.

———, S. Speich, and W. Harrison. 1973. Black-capped Gnatcatcher, a new breeding bird for the United States; with a key to the North American species of *Polioptila. Auk* 90:257–262.

Pike, K. L. 1947. *Phonemics, a technique for reducing languages to writing.* Univ. of Michigan Press, Ann Arbor.

Raikow, R. J. 1978. Appendicular myology and relationships of the New World nine-primaried oscines (Aves : Passeriformes). *Bull. Carnegie Mus. Nat. Hist.* No. 7.

Raitt, R. J. 1967. Relationships between black-eared and plain-eared forms of Bushtits (*Psaltriparus*). *Auk* 84:503–528.

———, and R. D. Ohmart. 1966. Annual cycle of reproduction and molt in Gambel Quail of the Rio Grande Valley, southern New Mexico. *Condor* 68:541–561.

Rand, A. L. 1944. A northern record of the flicker and a note on the cline *Colaptes auratus* cl. *auratus–luteus. Canadian Field-Naturalist* 58:183–184. [Published in 1945]

Rathbun, S. F. 1917. Description of a new subspecies of the Western Meadowlark. *Auk* 34:68–70.

Rea, A. M. 1969. *The interbreeding of two subspecies of Boat-tailed Grackle,* Cassidix mexicanus nelsoni and Cassidix mexicanus monsoni, in secondary contact in central Arizona. M.S. Thesis, Arizona State University, Tempe.

———. 1970. The status of the Summer Tanager on the Pacific slope. *Condor* 72:230–233.

———. 1972. Notes on the Summer Tanager. *Western Bird Bander* 47:52–53.

———. 1973. Turkey Vultures casting pellets. *Auk* 90:209–210.

———. 1979. Hunting lexemic categories of the Pima Indians. *Kiva* 44:113–119.

———. 1981a. Resource utilization and food taboos of Sonoran Desert peoples. *J. Ethnobiology* 1:69–83.

———. 1981b. Avian remains from Las Colinas, Appendix E, pp. 297–302. *In* L. C. Hammack and A. P. Sullivan [eds.] The 1968 Excavations at Mount 8, Las Colinas Ruins Groupe, Phoenix, Arizona. *Arizona State Museum Archaeological Ser.* No. 154.

———. 1983. Cathartid affinities: a brief overview. *In* S. R. Wilbur and J. A. Jackson [eds.] *Vulture biology and management.* Univ. Calif. Press.

———. in press. [miscellaneous taxonomic revisions]. *In* A. R. Phillips, *Known Birds of North and Middle America.*

———. *Animal foods of the Pima Indians.* [Unpublished MS; copy deposited Antevs Library, Dept. of Geosciences, Univ. of Arizona]

Reichhardt, K. L., B. Schladweiler, and J. L. Stelling. 1978. *An inventory of riparian habitats along The San Pedro River.* Univ. Ariz. Office Arid Land Studies, Tucson.

Reid, J. C. 1858. *Reid's tramp or a journal of the incidents of ten months travel through Texas, New Mexico, Arizona, Sonora, and California.* John Hardy and Co., Selma, Alabama (reprinted 1935, Steck Co., Austin Texas).

Ricklefs, R. E. 1973. *Ecology.* Chiron Press, Newton, Mass.

Ridgway, R. 1894. On geographical variation in *Sialia mexicana* Swainson. *Auk* 11:145–160.

———. 1902. The birds of North and Middle America. *Bull. U.S. Nat. Mus.* no. 50, pt. 2.

———. 1904. The birds of North and Middle America. *Bull. U.S. Nat. Mus.* no. 50, pt. 3.

———. 1907. The birds of North and Middle America. *Bull. U.S. Nat. Mus.* no. 50, pt. 4.

———. 1914. The birds of North and Middle America. *Bull. U.S. Nat. Mus.* no. 50, pt. 6.

Ripley, S. D. 1952. The thrushes. *Postilla: Yale Univ.* 13:1–48.

Rising, J. D. 1970. Morphological variation and evolution in some North American orioles. *Syst. Zool.* 19:315–351.

Root, C. T. 1955. Comparative osteology of the night herons. *Condor* 57:55–60.

Rubink, D. M., and K. Podborny. 1976. *The southern Bald Eagle in Arizona* (a status report). Endangered Species Report 1. U.S. Fish and Wildlife Service, Albuquerque, New Mexico.

Russell, F. 1908. The Pima Indians. *Annual Report of the Bureau of American Ethnology* 26:3–389. [Reprinted by Univ. of Arizona Press, Tucson, 1975.]

Sauer, C. O. 1966. *The early Spanish Main.* Univ. of California Press, Berkeley.

Saunders, G. B. 1968. Seven new White-winged Doves from Mexico, Central America, and the southwestern United States. *North American Fauna* no. 65.

Saxton, D., and L. Saxton. 1969. *Dictionary: Papago and Pima to English, English to Papago and Pima.* Univ. of Arizona Press, Tucson.

———. 1973. O'othham hoho'ok a'agitha. *Legends and lore of the Papago and Pima Indians.* Univ. of Arizona Press, Tucson.

Scott, W. E. D. 1886. On the avi-fauna of Pinal County, with remarks on some birds of Pima and Gila Counties, Arizona. With annotations by J. A. Allen. *Auk* 3:249–258; 383–389; 421–432.

Sedelmayr, J. 1955. *Jacobo Sedelmayr, 1744–1751. Missionary, frontiersman, explorer in Arizona and Sonora.* Four original manuscript narratives [including 1744 draft of the *relación*]. [Tr. and ed.] P. M. Donne. Arizona Pioneers' Historical Society, Tucson.

Selander, R. K. 1964. Speciation in wrens of the genus *Campylorhynchus. Univ. Calif. Publ. Zool.* 74:1–259.

Sheridan, D. 1981. *Desertification of the United States.* Council on Environmental Quality, U. S. Government Printing Office, Washington, D.C.

Short, L. L., Jr. 1965. Hybridization in the flickers (*Colaptes*) of North America. *Bull. Amer. Mus. Nat. Hist.* 129:307–428.

———. 1968. Variation of Ladder-backed Woodpeckers in southwestern North America. *Proc. Biol. Soc. Wash.* 81:1–10.

———. 1969. Taxonomic aspects of avian hybridization. *Auk* 86:84–105.

———, and S. W. Simon. 1965. Additional hybrids of the Slate-colored Junco and the White-throated Sparrow. *Condor* 67:438–442.

Sibley, C. G. 1955. The generic allocation of the Green-tailed Towhee. *Auk* 72:420–423.

———, and L. L. Short, Jr. 1964. Hybridization in the orioles of the Great Plains. *Condor* 66:130–150.

Simpson, J. M., and J. R. Werner. 1958. Some recent bird records from the Salt River Valley, central Arizona. *Condor* 60: 68–70.

Snow, D. W. 1967a. Family Aegithalidae, pp. 52–61. *In* R. A. Paynter, Jr., and E. Mayr [eds.] *Check-list of birds of the world.* Vol. 12. Mus. Comp. Zool., Cambridge, Mass.

———. 1967b. Family Remizidae, pp. 62–70. *In* R. A. Paynter, Jr., and E. Mayr [eds.] *Check-list of birds of the world.* Vol. 12. Mus. Comp. Zool., Cambridge, Mass.

Spicer, E. H. 1962. *Cycles of conquest. The impact of Spain, Mexico, and the United States on the Indians of the Southwest.* Univ. of Arizona Press, Tucson.

Spier, L. 1933. *Yuman tribes of the Gila River.* Univ. of Chicago Press, Chicago.

Stein, R. C. 1958. The behavioral, ecological and morphological characteristics of two populations of Alder Flycatcher, *Empidonax traillii* (Audubon). *N. Y. State Mus. Sci. Serv. Bull.* no. 37.

———. 1963. Isolating mechanisms between populations of Traill's Flycatchers. *Proc. Amer. Phil. Soc.* 107:21–50.

Sutton, G. M. 1943. Records from the Tucson region of Arizona. *Auk* 60:345–350.

———. 1944. The kites of the genus *Ictinia. Wilson Bull.* 56:3–8.

———. 1967. *Oklahoma birds: their ecology and distribution with comments on the avifauna of the southern Great Plains.* Univ. of Oklahoma Press, Norman.

———, and A. R. Phillips. 1942. June bird life of the Papago Indian Reservation, Arizona. *Condor* 44:57–65.

Swainson, W. 1831. [Taxonomic description]. Pp. 144–146. *In* W. Swainson and J. Richardson. *Fauna Boreali-Americana,* Part 2., The Birds.

Taylor, W. K., and H. Hanson. 1970. Observations on the breeding biology of the Vermilion Flycatcher in Arizona. *Wilson Bull.* 82:315–319.

Todd, W. E. C. 1953. Further taxonomic notes on the White-crowned Sparrow. *Auk* 70:370–372.

———. 1963. *Birds of the Labrador Peninsula and adjacent areas.* Univ. of Toronto Press, Toronto.

Turner, H. S. 1966. *Original journals of Henry Smith Turner, with Stephen Watts Kearny to New Mexico and California.* D. L. Clarke [ed.], Univ. of Oklahoma Press, Norman.

Underhill, R. M. 1946. *Papago Indian religion.* Contrib. to Anthro. no. 33. Columbia Univ. Press, New York.

United States Department of Commerce, Weather Bureau. 1964–1973. *Climatological data: Arizona, annual summary.* Vol. 68–77.

Van Devender, T. R. 1973. *Late Pleistocene plants and animals of the Sonoran Desert: a survey of ancient packrat middens in southwestern Arizona.* Ph.D. Dissertation, Univ. of Arizona, Tucson.

———. 1977. Holocene woodlands in the southwestern deserts. *Science* 198 (4313):189–192.

van Rossem, A. J. 1926. The California forms of *Agelaius phoeniceus* (Linnaeus). *Condor* 28:215–230.

———. 1934. Notes on some types of North American birds. *Trans. San Diego Soc. Nat. Hist.* 7:347–362.

———. 1936. Notes on birds in relation to the faunal areas of south-central Arizona. *Trans. San Diego Soc. Nat. Hist.* 8:121–148.

———. 1942. Notes on some Mexican and Californian birds, with descriptions of six undescribed races. *Trans. San Diego Soc. Nat. His.* 9:377–384.

———. 1945. A distributional survey of the birds of Sonora, México. *Occ. Pap. Mus. Zool., Louisiana State Univ.* no. 21.

———. 1946. Two new races of birds from the lower Colorado River Valley. *Condor* 48:80–82.

———. 1947. The distribution of the Yuma Horned Lark in Arizona. *Condor* 49:38–40.

Van Tyne, J., and G. M. Sutton. 1937. The birds of Brewster County, Texas. *Misc. Publ. Mus. Zool. Univ. Mich.* 37:1–119.

Velarde, L. 1931. Padre Luís Velarde's relación of Pimería Alta. [Ed.] R. K. Wyllys. *New Mexico Historical Review* 6:111–157.

Visher, S. S. 1910. Notes on the birds of Pima County, Arizona. *Auk* 27:279–288.

Warner, R. E. 1968. The role of introduced diseases in the extinction of the endemic Hawaiian avifauna. *Condor* 70:101–120.

Wauer, R. H., and R. C. Russell. 1967. New and additional records of birds in the Virgin River Valley. *Condor* 69:420–423.

Webb, G. 1959. *A Pima remembers.* Univ. of Ariz. Press, Tucson.

West, D. A. 1962. Hybridization in grosbeaks (*Pheucticus*) of the Great Plains *Auk* 79:399–424.

Wetmore, A. 1926. *The migrations of birds.* Harvard Univ. Press, Cambridge, Mass.

———. 1960. A classification for the birds of the world. *Smithsonian Misc. Coll.* 139:1–37.

———. 1962. Systematic notes concerned with the avifauna of Panamá. *Smithsonian Misc. Coll.* 145:1–14.

———. 1964. A revision of the American vultures of the genus *Cathartes. Smithsonian Misc. Coll.* 146:1–18.

———. 1965. The birds of the Republic of Panamá. Part 1. Tinamidae (Tinamous) to Rynchopidae (Skimmers). *Smithsonian Misc. Coll.* Vol. 150:1–483.

Winter, J. C. 1973. Cultural modifications of the Gila Pima: A.D. 1697–A.D. 1846. *Ethnohistory* 20:67–77.

Yang, S. Y., and R. K. Selander. 1968. Hybridization in the grackle *Quiscalus quiscula* in Louisiana. *Syst. Zool.* 17:107–143.

Zimmerman, D. A. 1973. Range expansion of Anna's Hummingbird. *American Birds* 27:827–835.

General Index

Primary entries in the species accounts are shown in boldface type. The Index of Piman Words and Expressions begins on page 285.

Abbreviations citing collections, 119
Acacia greggii, 37, 43, 45, 73, 202
Accipiter cooperii, 62, 70, 71, 83, 100, **132**; *striatus,* 100, **131**
Accipitridae, 131–34
Accipitriformes, 131–38
Actitis macularia, 103, 107, **151–52**
Adobe (sandwich) houses, 48, 50
Aechmophorus occidentalis, **125–26**
Aegithalidae, 202–03
Aegithalos, 202; *caudatus,* 202; *minimus,* 65, 66, 67, 72, **202–03**, 310
Aeronautes saxatalis, 70, 72, 106, **174**
Agave, 42, 43, 242
Age categories, 120
Agelaius, 99, 247; *phoeniceus,* 73, 84, 102, 103, 104, 105, 107, **245–46**
Agriculture. *See* Mechanized farming; Native agriculture
Aimophila quinquestriata, 88; *ruficeps* 106, **237–38**
Ajaia, 131
Alaudidae, 191–94
Alcedinidae, 177–78
Alfalfa, 47, 50, 59, 60, 146, 151, 164, 171, 192, 215, 224, 236, 241, 244, 247
Allenrolfea occidentalis, 36, 46
Amaranthus spp., 47
Amazilia violiceps, 88
Ambrosia, 73, 108, 170, 201, 211, 238, 244; *ambrosioides,* 45 (*see also* Canyon bursage); *deltoidea,* 43, 46, 47; *dumosa,* 42, 43 (*see also* White bursage)
American Ornithological Union Check-list, 120
Ammodramus sandwichensis, 103, **236**
Amphispiza belli, 100, **238**; *bilineata,* 73, 100, **238**
Anas, 158; *acuta,* 79, 80, **156**; *americana, 158; clypeata,* 106, **157**; *crecca,* 104, 106, **157**; *cyanoptera,* 75, 103, 104, 106, **157**; *discors,* 103, 104, **157**; *platyrhynchos,* **156**
Anatidae, 155–58
Ani, Groove-billed, **164**
Animal husbandry: Anglo, 25–29; Apache, 25; Pima, 31
Anser albifrons, 79, 80, **155–56**; *caerulescens,* 79, 80, **156**; *hyperborea,* **156**
Anseriformes, 155–58
Anthoscopus, 203
Anthus spinoletta, 103, 107, **217**
Anza, Captain Juan Bautista de, 20, 50
Anza Expedition, 19
Aphelocoma, 199; *coerulescens,* 66, 67, 83, 93, 95, **198**
Apodidae, 174
Apodiformes, 174–76
Aquila chrysaetos, 78, 80, 81, **133–34**
Ara macao, 163
Archaeological sites, Hohokam, 79, 88, 139; Alder Wash Ruin, 230; Big Ditch, 230; Casa Grande Monument, 3; Casa Grande Ruins, 17, 18, 22, 23, 39, 50, 222; Las Colinas, 106, 139, 140, 199, 246; Pueblo Grande, 116; Snaketown, 10, 36, 79, 116, 139, 140, 142, 147, 153, 155, 156, 178, 198, 199, 208, 238, 241, 244, 245, 246; South Santan Salvage, 116, 139, 147
Archilochus alexandri, 62, 63, 70, 86, 104, **175**; *anna,* 88, **176**; *costae,* 106, **175–76**
Ardea, 141; *alba,* **142–43**; *herodias,* 62, 77, 80, 81, **142**; *ibis,* **146**; *thula,* 83, **143–45**
Ardeidae, 141–47
Ardeiformes, 141–47
Ardeola, 141, 146; *striata,* 142; *virescens,* 62, 67, 75, 80, 81, 85, 103, **141–42**
Arrowweed, 2, 16, 32, 37, 39, 46, 47, 57, 62, 68. *See also Pluchea sericea*
Arroyo vegetation, 45
Arundo donax, 16, 57, 221
Ash, Arizona, 1, 37, 38, 181, 183

Asio flammeus, **171–72**; *otus,* **171**
Aster, spiny, 47, 49, 54, 68
Aster spinosus, 47, 54, 69
Asyndesmus lewis, 94, 95, **180**
Athene, 171; *cunicularia,* 100, 102, **170–71**
Atriplex, 32, 34, 47, 100, 108, 170, 176, 187, 191, 201, 211, 215, 238, 241, 244; *canescens,* 36, 46, 50, 57, 219; *lentiformis,* 36, 46, 47, 54, 57; *linearis,* 36, 57; *polycarpa,* 36, 46, 47, 53 (*see also* Saltbush)
Audubon, John W., 23
Auriparus, 203; *flaviceps,* 69, 84, 100, 123, **203–04**
Avocets, American, 53, 79, 103, 107, **153**
Aythya affinis, **158**; *collaris,* **158**; *valisineria,* **158**

Baccharis glutinosa, 23, 36, 37, 69, 230, 235, 241. *See also* Seepwillow
Bajada, 38, 43, 44, 105
Barehand Lane, 53–56, 80, 84–85, 104
Bartlett, John R., Boundary Commissioner, 24, 50; sketch by, 26
Batamote, 39, 241. *See also* Seepwillow
Beaver, and beaver trapping, 3, 21, 26, 29, 38
Becard, Rose-throated, 88, 89
Beloperone, 45
Bernardia, 234
Bernardia, 43, 234
Bird names (ethnotaxonomies): Névome (17th century Pima Bajo), 136, 137, 139, 162, 166; Northern Tepehuan, 127, 134, 137, 139, 140, 142, 166, 175, 177, 178, 199; Papago (*tówhano áw'awtam*), 117, 127, 130, 132, 133, 134, 135, 136, 137, 139, 140, 142, 149, 159, 160, 161, 162, 164, 166, 168, 170, 172, 173, 175, 178, 180, 181, 184, 197, 199, 201, 203, 206, 207, 208, 210, 216, 218, 230, 244, 246; Pima Bajo (lowland), 127, 130, 132, 134, 136, 137, 139, 140, 142, 149, 159, 160, 161, 164, 165, 166 173,175, 178, 180, 181, 199, 203, 218, 230; Riverine Pima, 257–58; Tepecano, 127, 136, 142, 166, 181
Bittern, 142; American, **146**; Least, 62, 75, 77, 80, 86, 103, **107**, 147
Blackbird, 74, 249; Brewer's, 93, 102, 103, 105, 245, **248–49**; Red-winged, 33, 37, 52, 59, 73, 74, 84, 85, 102, 103, 104, 105, 107, 109, **245–46**, 247, 248; Yellow-headed, 79, 81, 103, 104, 105, 109, 245, **246–47**
Blackwater (Pima settlement), 31, 36, 50, 61, 168
Bluebird, Mountain, 67, 95, **215**; Western, 67, 95, **215**
Blue Point Cottonwoods, 14, 61, 62, 67
Bombycilla cedrorum, 95, **217**
Bombycillidae, 217–19
Bosque. *See* Mesquite
Botaurinae, 142
Botaurus lentiginosus, **146**
Brant, 155
Branta canadensis, 79, 80, 155
Breeding, determination of, 120
Breninger, George F., observations of, 34, 35, 117
Bringas, Fr. Diego, O.F.M., 20, 50
Brittlebush, 42, 43, 44
Brown, Herbert, observations of, 25–28, 87
Bubo virginianus, 70, 71, 100, 101, 102, 104, 105, 106, **166–68**
Bubulcus, 141, 146
Bucephala albeola, **158**
Buckwheat brush, 43
Bufflehead, **158**
bullockii, Icterus g., 73, 243
Bulrush, 16, 54, 82. *See also* Emergent vegetation
Bunting, Common, **233**; Indigo, 86, **233**; Lark, 103, **235**; Lazuli, 71, 72, 74, **233**
Burrobush, 45, 46, 57
Bursage, 44, 64, 108; canyon, 45; triangle-leaf, 43, 46, 47; white, 42, 43
Bushtit, 65, 66, 67, 72, 95, **202–03**
Buteo albonotatus, **133**; *jamaicensis,* 101, 105, 106, **132**; *regalis,* **133**; *swainsoni,* **132–33**
Buteogallus anthracinus, 62, 66, 79, 88
Butorides, 141
Buttcs-Cochran Area, 14, 61, 65, 67–75

Cactus, barrel, 42; fishhook, 42; giant, 169, 178; hedgehog, 42, 53; prickly pear, 42, 43. *See also* Cholla; *Opuntia;* Saguaro
cafer, Colaptes a., 179
Calamospiza melanocorys, 103, **235**
Calcarius ornatus, **241**
Calendar stick, 30–31
Calidris, 152
Callipepla, 139; *gambelii,* 73, 83, 100, 101, 103, 104, 105, **139–40**; *squamata,* 139
Campylorhynchus brunneicapillus, 73, 100, 206
Canals. *See* Hohokam, canal system; Irrigation
Cane, 19, 39. *See also* Phragmites
caniceps, Junco h., 239
Canotia holacantha, 42, 242. *See also* Crucifixion thorn
Canvasback, **158**
Capella gallinago, 84, 103, 104, 151
Caprimulgidae, 172–73
Caprimulgiformes, 172–73
Caracara, Crested, 78, 81, 83, **135**
Caracara cheriway, 78, 83, **135**
Cardellina rubrifrons, 88
Cardinal, Northern, 49, 71, 73, 75, 84, 88, 101, 103, 105, 108, **229–30**, 231
Cardinalis cardinalis, 71, 73, 84, 88, 101, 103, 105, **229–30**, 231; *sinuatus,* 73, 83, 87, 103, **230–31**
Carduelidae, 250–52
Carduelis lawrencei, 93, 95, **252**; *pinus,* 94, 95, **251**; *psaltria,* 63, 64, 69, 71, 100, **251**; *tristis,* 95, **251**
Carex sp., 54, 74
carolinensis, Anas c., 157
Carpodacus cassinii, 95, **250**; *mexicanus,* 73, 100, 103, **250–51**; *purpureus,* 94, 95, **250**
Caryothraustes, 232
Casa Blanca (Pima settlement), 30, 36
Casmerodias, 141
Cassidix mexicanus, 250. *See also Quiscalus*
Castela emoryi, 37
Cathartes aura, 72, 83, 100, 106, **127–30**, 194
Cathartidae, 127–30
Catharus, 213; *guttatus,* 105, 213–14; *ustulatus,* 95, 214, 215
Catherpes mexicanus, 73, 106, 207–08, 241
Catsclaw acacia, 37, 42, 45, 72, 203
Cattail, 1, 16, 33, 53, 54, 55, 56, 57, 58, 62, 64, 78, 82, 86, 94, 99, 147, 148, 149, 158, 220, 226, 227, 245, 246, 247, 248

Cattle, wild, 25. *See also* Animal Husbandry
Celtis pallida, 42, 69, 73, 202, 234
Centurus, 180. *See also Melanerpes*
Cercidium, 73, 230; *floridum,* 37, 45; *microphyllum,* 37, 42, 43, 45. *See also* Paloverde
Cereus giganteus, 37, 43, 73, 87, 210. *See also* Saguaro
Certhia familiaris, 67, **204–05**
Certhiidae, 204–05
Ceryle, 177; *rudis,* 177
Chaetura vauxi, **174**
Chamise (chamizo), 19, 20. *See also Atriplex;* Saltbush
Charadriidae, 150–51
Charadriiformes, 150–54
Charadrius alexandrinus, **150**; *montanus,* **151**; *vociferus,* 57, 73, 83, 102, 103, 107, **150**
Chat, Yellow-breasted, 63, 65, 69, 71, 72, 74, 75, 77, 80, 83, 84, 85, 86, 104, 105, 107, **228**
Chelidonias niger, **154**
Chen, 156. *See also Anser*
Chickadee, Mountain, 67
Chloroceryle aenea, 177; *americana,* 177, 178; *inda,* 177
Chlorura, 233
Cholla, 64; buckhorn, 43, 44; pencil, 43; teddy bear, 42. *See also Opuntia*
Chondestes grammacus, 73, **236**
Chordeiles acutipennis, 57, 73, 78, 100, 101, 104, 106, **173**
chuparosa, 45, 175
Cicada, 89
Ciconiiformes, 127–31
ciénaga (*ciénega*), 20, 27, 38, 39
Circus cyaneus, **134**
Cistothorus palustris, 73, 77, 79, 80, 104, 105, 207
Clematis cf. *drummondii,* 42
Clines, 123, 179
Coccothraustes vespertinus, 94, 95, **250**
Coccyzus americanus, 63, 73, 80, 85, 104, **163–64**
Cochlearius, 141, 142
Cochran (ghost town). *See* Buttes-Cochran
Cocklebur, 29, 47, 49, 54
Cocomaricopa, 18, 24, 25. *See also* Indians, Maricopa
Colaptes auratus, 73, 93, 101, 106, **178–79**, 231; *cafer* group, 93
Colinus, 139
Columbidae, 159–61
Columbiformes, 159–61
Columbina 161; *passerina,* 73, 85, 86, 101, 102, 103, 104, **160–61**
Condalia lycioides, 47, 50. *See also* Graythorn; *Ziziphus obtusifolia*
Condor, California, 127
Confluence: Salt-Gila, 18, 21–22, 30, 31, 38, 54, 56–60; Salt-Verde, 14, 61; Santa Cruz–Gila, 16, 34, 38; Vekol–Santa Rosa–Santa Cruz, 39
Contopus borealis, **190**; *mesoleucus,* 190; *sordidulus,* 100, **189–90**
Cooke, Colonel Philip St. George, 24
Coot, American, 52, 63, 74, 75, 77, 80, 81, 85, 103, **149–50**
Coraciiformes, 177–78
Coragyps atratus, 87, 88, **130**
Cormorant, Double-crested, 77, **127**
Corn (maize), 47, 48, 180, 201, 207, 245. *See also* Maize, Milo
Corvid, sp.?, 80
Corvidae, 197–201, 204
Corvus, 198; *brachyrhynchos,* 66, 67, 78, 80, **200–01**; *corax,* 63, 64, 83, 93, **199–200**; *cryptoleucus,* 200–01
Cottonwood, 1, 2, 16, 17, 18, 19, 20, 23, 24, 27, 28, 29, 32, 33, 34, 36, 37, 38, 47, 48, 56, 57, 58, 59, 60, 62, 64, 68, 79, 83, 86, 89, 101, 107, 108, 132, 135, 136, 138, 141, 142, 143, 146, 160, 163, 164, 166, 168, 169, 170, 178, 179, 180, 181, 182, 183, 188, 191, 198, 201, 202, 205, 209, 221, 223, 224, 229, 242, 244, 247, 251
Coturnicops noveboracensis, **149**
Cowbird, 84, 242, 243, 249; Bronzed, 66, 71, 72, 84, 87, 88, 102, 242, **248**; Brown-headed, 66, 71, 72, 73, 74, 84, 101, 102, 104, **247–48**; Dwarf (*See* Brown-headed); parasitism, 66, 72, 89, 216, 241, 242, 244, 247, 248; Red-eyed, 88 (*See* Bronzed)
Coyote, 21, 201, 235
Crane, Sandhill, 78, 79, 80, **147–48**
Creeper, Brown, 51, 67, 95, **204–05**
Creosote bush (*Hediondilla*), 37, 42, 43, 44
Crossbill, Red, 94
Crossosoma bigelovii, 42, 234
Crotophaga sulcirostris, **164**
Crow, Common, 66, 67, 78, 199, **200–01**
Crucifixion thorn, 37, 42
Cuckoo, Yellow-billed, 58, 63, 73, 74, 80, 83, 85, 86, 87, 104, 107, **163–64**
Cuculidae, 163–65
Cuculiformes, 163–65
Curlew, Long-billed, **151**
Cyanocitta, 199; *stelleri,* 93, 95, **197–98**
Cyanocompsa, 232; *cyanea* 232; *cyanoides,* 232

Dam, Ashurst-Hayden Diversion, 14, 67; Hoover, 2; Salt River, 57; San Carlos (Coolidge), 33
Date grove, 216
DDE, DDT, 83
Deer, 140; Mule, 42; White-tailed, 42
Dendrocopos, 181; *scalaris,* 70, 71, 87, 100, **181–82**
Dendroica coronata, 101, 102, 104, 105, **224**, 226; *graciae,* **225**; *nigrescens,* 101, 104, **224–25**, 226; *occidentalis,* **225**; *pensylvanica,* **225**; *petechia,* 63, 64, 71, 77, 79, 80, 87, **223**; *townsendi,* 101, 102, 104, 105, **225**
Desertification of Gila floodplains, causes of, 3
Desert lavender, 45, 175. *See also* Tree sage; *Hyptis emoryi*
Dichromanassa, 141
Distichlis, 24, 33. *See also* Grasses; Grasslands
Disturbance vegetation, 29
Ditches. *See* Irrigation
Dock, 33, 54, 55. *See also Rumex*
Dove, 84; Common Ground, 73, 85, 86, 101, 102, 103, 104, **160–61**; Inca, 73, 74, 84, 88, 102, **161**; Mourning, 69, 70, 73, 86, 100, 101, 159, **160**; White-winged, 33, 34, 70, 71, 73, 86, 88, 99, 101, 102, 103, 104, 105, 106, **159–60**, 164
Dowitcher, Long-billed, **153**
Drought, 25–29, 37, 40, 201
Duck, 52, 79, 155; Ring-necked, **158**; Ruddy, 85, 86, 103, 106, 107, **158**
Durivage, John, 23

Eagle, Bald, 63, 65, 81, 134; Golden, 78, 80, 81, **133–34**
Eccleston, Robert, 23

Echinocereus engelmannii, 42, 53
Ecotone, 2
Egret, 74, 120; Cattle, **146**; Common, **142–43**; Snowy, 83, **143–45**
Egretta, 141, 143
Elderberry, Mexican, 47, 53, 55, 56, 84, 218
Emberiza, 240
Emberizidae, 229–41
Emergent vegetation, 52, 54, 218, 223, 226, 228. *See also* Cattail; Rush; *Scirpus; Typha*
Emory, Lieutenant Colonel William H., 22, 50, 140
Empidonaces, 229
Empidonax difficilis, 106, **188–89**, 226; *griseus,* 187; *hammondii,* 186, **188**; *minimus* (*Tyrannula minima*), 188; *oberholseri,* **187**; *obscurus,* 187; *pusillus,* **188**; *traillii,* 72, 79, **189**, 226; *wrightii,* **187–88**
Encelia farinosa, 37, 42, 43. *See also* Brittlebush
Eremophila alpestris, 50, 84, 90, 103, **191–94**
Ereunetes, 152; *mauri,* 103, 107, **153**; *pusilla,* **152**
Erolia melanotos, **152**; *minutilla,* 103, 107, **152**
Ethics, 3, 107
Ethnoscience, 117. *See also* individual species accounts
Ethnotaxa. *See* Bird names
Euphagus, 99; *cyanocephalus,* 93, 102, 103, 105, 248–49
Eupoda, **151**
Evans, George, 23
Extirpations, 77–81

Falco columbarius, **136**; *mexicanus,* 81, 106, **135–36**; *peregrinus,* 136; *sparverius,* 83, 101, 102, 105, **136–38**
Falcon, Peregrine, 136; Prairie, 81, 106, **135–36**
Falconidae, 135–38
Falconiformes, 131–38
Feathers, use of, 127, 132, 133, 136, 140, 162, 230; tabooed, 113, 127, 139
Fence rows and living fences, 37, 47, 48, 49, 52, 75, 108, 170, 187, 198, 209, 216, 218, 220, 229–30, 231, 234
Finch, Cassin's, 94, 95, **250**; House, 73, 100, 103, **250–51**; Purple, 94, 95, **250**
Fire drives, 245. *See also* Grasslands
Flicker, 244; Gilded, 73, 87, 106, 137, 138, 168, **178–79**; Northern, 101, **178–79**; Red-shafted, 93, 138, **178–79**; Yellow-shafted, **179**
Flight years, 93–97
Floodplain habitats, 46–60
Floods and floodwater, 2, 7, 20, 28, 31, 37, 38, 56, 57, 60, 63, 68–69, 126
Florence (Anglo community), 10, 30, 81, 170, 177
Florida, 141
Flycatcher, Acadian, 188; Ash-throated, 70, 73, 75, 86, 101, 104, 105, 106, **183**, 191, 208; Brown-crested (*see* Wied's Crested Flycatcher); Dusky, **187**; Gray, **187–88**; Hammond's, **188**; Least, **188**; Olive-sided, **190**; Traill's, 72, 188, **189**; Vermilion, 62, 63, 70, 72, 80, **191**; Western, 106, 164, **188–89**; Wied's Crested, 62, 63, 70, 73, 74, 75 80, **183**; Willow, 51, 72, 79, 188, **189**
Font, Fr. Pedro, O.F.M., 19–20, 40, 46
Fouquieria splendens, 37, 42
Fraxinus velutina, 37. *See also* Arizona ash
fuertesi, Icterus s., 242
Fulica americana, 63, 75, 77, 80, 81, 85, 103, **149–50**

Gadsden Purchase of 1853, 25
Gallery forest, 7, 26. *See also* Riparian vegetation
Galliformes, 139–40
Gallinula chloropus, 63, 75, 77, 103, 104, **149**
Gallinule, Common, 63, 74, 75, 77, 78, 84, 85, 86, 103, 104, 109, **149**
gambelii, Zonotrichia l., 240
Garcés, Fr. Francisco, O.F.M., 19, 50
Genus: criteria for recognition, 122
Geococcyx californianus, 73, 100, 101, 102, 104, 105, **164**
Geokichla, 213
Geothlypia trichas, 73, 77, 80, 84 104, 105, **226–27**
Gila Crossing (Pima village), 9, 30, 31, 33, 46
Gila River Tribal Farm, 10
Gileños, 9, 18, 20, 30, 201. *See also* Pima Indians
Gilman, M. [Marshall] French, 28, 34, 36, 118. *See also* individual species accounts
Glaucidium brasilianum, 63, 65, 78, 80, 169–70
Glosses of native names, 116. *See also* individual species accounts
Gnatcatcher, Black-capped, 88; Black-tailed, 70, 73, 100, **216**, 247; Blue-gray, 66, 67, 216; California, 216
Goldfinch, American, 94, 95, **251**; Lawrence's, 93, 95, **252**; Lesser, 58, 63, 64, 69, 71, 73, 74 75, 87, 100, **251**
Goose, 79, 140, 155; Blue, **156**; Canada, 79, 80, **155**; Snow, 79, 80, 155, **156**; White-fronted, 70, 80, **155–56**
Gooseberry, 42, 92, 168, 213, 233
Grackle, Great-tailed, 73, 74, 84, 87, 88, 90, 164, **249–50**
Grasses, 24, 33, 34, 35, 42, 47, 48, 50, 157, 193, 232, 235. *See also* Grasslands
Grasslands, 1, 7, 17–25, 27, 31, 34, 39–40, 50, 76, 245
Graythorn, 33, 35, 39, 47, 49, 50, 68, 203, 211, 230. *See also Ziziphus*
Grebe, Eared, **125**; Pied-billed, 77, 80, 86, 107, **126**; Western, **125–26**
Grosbeak, Black-headed, 73, 100, **231**; Blue, 63, 64, 71, 74, 75, 79, 86, 103, 105, 120, **232–33**; Common, **231–32**; Evening, 94, 95, **250**; Rose-breasted, **232**
Gruidae, 147–48
Gruiformes, 147–50
Grus canadensis, 78, 80, 140, **147–48**
Guiraca, 232
Gull, **154**
Gymnogyps californianus, 127
Gymnomystax, 247
Gymnorhinus cyanocephalus, 93, 198–99

Habitat delimitation, 41; deterioration and population size changes,

81–84; protection, 107–10; utilization by birds, 98–107
Habitats, aquatic, 50–60, 103–05, 107; mountain, 42, 43, 106, 108
Hackberry, desert, 42, 43, 68, 72
Haliaeetus leucocephalus, 63, 65, 81, 134
Haplopappus heterophyllus, 47, 50, 54, 68
Harrier, Northern, **134**
Hawk, Bay-winged (*see* Harris' Hawk); Common Black, 62, 66, 75, 79, 81, 88, 89; Cooper's, 62, 70, 71, 73, 74, 75, 83, 92, 100, **132**; Ferruginous, **133**; Harris', 63, 65, 66, 73, 74, 80, 81, **133**; Marsh (*see* Harrier, Northern); Mexican Black (*see* Common Black Hawk); Pigeon (*see* Merlin); Red-tailed, 101, 105, 106, **132**, 166; Sharp-shinned, 92, 100, **131**; Sparrow (*see* Kestrel, American); Swainson's, **132–33**; Zone-tailed, **133**
Helianthus annuus, 47, 54, 57, 86, 251. *See also* Sunflower
Heron, 74, 78, 79, 120; Black-crowned Night, 62, 65, 75, 77, 78, 79, 80, 81, 103, 107, 141, **146**; Great Blue, 62, 77, 78, 79, 80, 81, **142**, 143, 146; Green, 58, 62, 67, 75, 79, 80, 81, 85, 86, 87, 103, 107,**141–42**
Hesperiphona, 250
Hesperocichla naevia, 213
Heterocnus, 141
Hibernation, Pima concept of, 172
Himantopus mexicanus, 57, 103, 107, **153–54**
Hirundinidae, 194–97
Hirundo albifrons, 196; *lunifrons* **196–97**; *pyrrhonota,* 196; *rustica,* 63, 78, 80, **195–96**
Hohokam, 9–10, 140; canal system, 9, 31, 34, 36; ceramics, 10 (*see also* Archaeological sites); Pima continuum, 10
Hummingbird, Anna's, 88, 89, **176**; Black-chinned, 62 63, 70, 74, 75, 86, 87, 104, **175**; Costa's, 43, 45, 99, 106, **175–76**; Rufous, **176**; Violet-crowned, 88, 89
Hummingbirds, in Pima culture, 174–75
Husbandry: Anglo, 25–29; Apache, 25; Pima, 31
Hybridization, 123, 231, 241
Hydranassa, 141
Hylocichla mustelina, 213
Hymenoclea monogyra, 57; *salsola pentalepis,* 45, 46. *See also* Burrobush
Hyptis emoryi, 37, 45, 175. *See also* Desert lavender; Tree sage

Ibis, White-faced, 120, **130–31**
Icteria virens, 63, 69, 71, 77, 80, 84, 104, 105, **228**
Icteridae, 79, 84, 241–50
Icterus albeillei, 243; *cucullatus,* 63, 66, 71, 88, **241–42**; *fuertesi,* 242; *galbula,* 66, 83, 84, 101, 102, 104, 105, **242–44**; *parisorum,* **242**; *spurius,* **242**
Ictinia, 89; *mississippiensis,* 64, 88
Immature (defined), 120
Indian Reservations: Ak Chin (Maricopa), 8, 11; Fort McDowell, 66, 67, 83; Gila River, 3, 7, 8, 10, 12–13, 62, 67, 70; Salt River, 11, 30
Indians: Apache, 9, 25, 30, 133, 200; Halchidomas, 30 (*see also* Indians, Maricopa); Hualapai, 9; Kohadt (Kwahatk), 9, 175; Maricopa, 9, 10, 31, 39, 139; Mountain Pima, 117; Papago, 9, 117, 256; Pima Bajo (lowland), 117; Sobaipuri, 9; Tepecano, 117; Tepehuan, 117; Yaqui, 230; Yavapai, 9. *See also* Bird names; Hohokam; Pima Indians
Intergradation, primary, 124; secondary, 90, 123, 138, 179, 231, 250
Intermediate, subspecific, defined, 124
Iridoprocne, 194
Ironwood, desert, 37, 44, 45, 132, 168, 181, 183, 188, 189
Irrigation, Pima, 9, 17–23, 25, 28, 32, 33, 36, 37, 38, 49, 74–75, 107, 108
Irruptions, 93
Ixobrychus exilis, 62, 77, 80, 86, 103, **147**
Ixoreus, 213
Jay, Piñon, 93, 94, **198–99**; Scrub, 66, 67, 83, 93, 95, **198**; Steller's, 93, 94, 95, **197–98**
Jimmy weed, 47, 54. *See also Haplopappus*
Johnston, Captain Abraham R., 22, 50
Jornada, defined, 21
Junco, 240; *caniceps* group, 239; *hyemalis,* 102, 103, 106, **238–39**
Junco, Dark-eyed, 102, 103, 106, **238–39**; Red-backed, 93
Juniper, one-seeded, 42, 218, 242
Justicia (*Beloperone*) *californica,* 45, 175
Juvenal (defined), 120

Kearny Expedition, 22
Kestrel, American, 73, 83, 101, 102, 105, **136–38**
Killdeer, 57, 73, 74, 83, 102, 103, 107, **150**, 154
Kingbird, Cassin's, **182–83**; Thick-billed, 88, 89; Tropical, 88, 89; Western, 72, 73, 83, 101, 102, 108, **182**
Kingfisher, Belted, 74, 77, 78, 79, 80, 81, 165, **177–78**; Green, 177
Kinglet, Ruby-crowned, 93, 100, **217**, 224, 226
Kino, Fr. Eusebio, S.J., 7, 9, 17, 22, 50
Kite, Mississippi, 64, 88, 89
Komatke, relocation of, 31, 44
Krameria parvifolia, 42, 43

Labels, specimen (data on), 124, 173, 248
Languages, Uto-Aztecan, 8–9, 117
Laniidae, 201–02
Lanius ludovicianus, 63, 65, 73, 100, 101, 102, 106, **201–02**
Laridae, 154
Larks, Horned, 50, 59, 84, 90, 91, 103, 120, **191–94**, 244
Larrea tridentata, 37, 42, 43
Larus sp., **154**
Laterallus jamaicensis, 148
Leucophoyx, 141
Life Zone, Lower Sonoran, 10; Upper Sonoran, 42
"Little Gila," 19, 35, 37
Limnodromus scolopaceus, **153**
Livestock accompanying expeditions, 17, 19, 22, 24. *See also* Animal husbandry; Cattle
Living fences. *See* Fence rows

Lobipes lobatus, **153**
Longspur, Chestnut-collared, **241**
Lophortyx, 139
Lower Sonoran Desert vegetation, 138
Loxia curvirostra, 94
Loxigilla, 232
Lycium, 32, 47, 100, 176, 209, 211, 215, 218, 235, 244; *andersonii,* 46; *berlandieri,* 37; *californicum,* 53; *cooperi,* 37; *exsertum,* 37; *fremontii,* 37, 42, 46, 47; *parishii,* 53; *torreyi,* 37

Macaw, Scarlet, 163; use in trade, 162
Maguey, 35. *See also Agave*
Maize, Milo, 47, 231. *See also* Corn
Mallard, **156**
Manje, Captain Juan Mateo, 17
Maps, 124
Mareca, 158
Maricopa Wells, 21, 23, 24, 27, 39. *See also* Santa Teresa Ciénaga
Marsh, Barehand Lane, 53–56
Martin, Purple, 73, 74, 195, **197**
Meadowlark, Eastern, **244**; Western, 50, 73, 100, 102, 103, 106, **244–45**
Measurements of bird specimens, 124
Mechanized farming, 46, 47, 50, 83, 210
Megaceryle, 177; *alcyon,* 77, 80, 81, **177–78**; *torquata,* 177
Melanerpes, 180; *formicivorus,* 95, **180**; *uropygialis,* 70, 71, 87, 88, 102, 104, 106, **180**
melanocephalus, Pheucticus l., 232
Melanospiza, 232
Meleagris gallopavo, **140**
Melospiza, 240
Merganser, Common, 80, **158**
Mergus merganser, 80, **158**
Merlin, **136**
Mesquite, 2, 3, 19, 20, 22, 24, 27, 30, 32–35, 37, 39, 45–56, 62, 64, 65, 68, 71, 72, 82, 83, 99, 101, 102, 140, 165, 181, 183, 187, 188, 191, 198, 201, 203, 205, 206, 209, 211, 213, 215–25, 228, 230, 232, 235, 236, 247, 250
Micrathene, 171; *whitneyi,* 62, 63, 79, 80, **170**
Migration, 118; differential (age/sex), 188, 189, 226, 246–47; racial patterns and, 189, 214, 215, 220, 221, 224, 226, 227, 229, 239–40
Mimidae, 208–12
Mimus gilvus, 209; *polyglottos,* 63, 65, 83, 85, 100, 101, 102, 105, 107, **208–09**
Mistletoe, 215, 218
Mniotilta varia, 96, **221**
Mockingbird, Northern, 63, 65, 73, 74, 83, 84, 85, 100, 101, 102, 105, 107, 108, **208–09**, 211
Molothrus, 99; *aenus,* 66, 71, 72, 84, 87, 88, 102, **248**; *ater,* 66, 71, 72, 84, 101, 102, 104, 105, **247–48**
Monoculture, 50
Motacillidae, 217
Mountains, 38, 106; Sacaton, 43; Santan, 43; South (Salt River), 26, 43; Superstition, 26. *See also* Habitats; Sierra Estrella
Myadestes townsendi, 94, 95, **216**
Mycteria americana, 147
Myiarchus cinerascens, 70, 86, 101, 104, 105, 106, **183**; *tyrannulus,* 62, 63, 70, 80, **183**
Myiophoneus, 213

Native agriculture, 9, 17, 19, 21, 25, 26, 46, 48, 74, 108. *See also* Irrigation
Native consultants, 116, 117, 216; Brown, David, 117, 168, 181, 184, 185, 196; Brown, Rosita, 117, 161, 168, 184, 196; Córdova, Maria, 117, 161, 165; Estrella, Pedro, 117, 181; Giff, Joseph, 33, 117, 136, 142, 143, 163, 165, 181, 191, 196; Giff, Ruth, 117; Jim, Frank, 117, 246; Lewis, Delores, 117, 135, 159, 161, 208, 210; Matthias, Sylvester, 31, 33, 117, 126, 134, 135, 139, 142, 143, 146, 147, 159, 161, 163, 165, 172, 173, 177, 196, 199, 201, 208, 243; Noriego, Luciano, 117, 135, 181, 184, 249; Pablo, Sally, 232
Neff, Johnson A., 34
New York Thicket, 3, 24, 33, 34, 35, 50
Nicotiana glauca, 57, 176, 241
Nighthawk, Lesser, 56, 57, 73, 78, 100, 101, 104, 106, 172, **173**
Nucifraga, 93; *columbiana,* 94, **201**
Numenius americanus, **151**
Nutcracker, Clark's, 94, **201**
Nuthatch, Red-breasted, 94, 95, **204**; White-breasted, 66, 67
Nuttallornis, 190
Nyctanassa, 141; *leuconotus,* 142; *magnificus,* 142; *violacea,* 141
Nycticorax nycticorax, 62, 65, 77, 80, 81, 103, 141, 146

Oak, 168, 205, 242; scrub live, 42, 92, 168
Oasis, 10, 49. *See also* Quitovac; Riparian community
Oasis effect, 49. *See also* Fence rows; Irrigation, Pima
Ocotillo, 20, 37, 42, 45, 176
Odocoileus virginianus, 42; *hemionus,* 42
Olneya tesota, 37, 45, 230. *See also* Ironwood
Olor columbianus, 80, **155**
Oporornis philadelphia, 226; *tolmiei,* 100, 101, **226**
Opuntia, 37, 181; *acanthocarpa,* 43; *arbuscula,* 37, 43; *bigelovii,* 37, 42, 250; *chlorotica,* 43; *fulgida,* 37; *leptocaulis,* 37; *versicolor,* 37; *violacea,* 42. *See also* Cactus, prickly pear; Cholla
Oreocichla, 213
Oreoscoptes montanus, **212**
Oriole, 83; Bullock's (*see* Northern Oriole); Hooded, 63, 66, 71, 72, 74, 88, 89, **241–42**, 243, 247, 248; Northern, 49, 66, 73, 84, 86, 101, 102, 104, 105, 108, **242–44**, 248; Orchard, **242**; Scott's, **242**
Osprey, 77, 79, **134–35**
Osteology: in higher taxonomy, 121–22, 139, 142, 143, 146, 152, 161, 171, 177, 180, 181–82, 190, 194, 195, 196, 199, 202, 203, 206, 208, 212, 213, 228, 232, 233, 234, 240, 247, 250; in subspecific taxonomy, 130, 138, 140, 148, 156, 178, 199–200
Otus asio, 70, 71, 101, 123, **168**, 173; *flammeolus,* **168–69**; *trichopsis,* 169
Ovenbird, 96, **225**
Overgrazing, 11, 28, 30, 31, 40
Ovis canadensis, 42
Owl, 83; Barn, 71, 101, 102, **165–66**; Burrowing, 100, 102, **170–71**; Common Screech, 70, 71, 73, 101, 106, **168**, 169; Elf, 62, 63, 73, 74, 79, 80, 168, 169, **170**;

Ferruginous Pygmy, 63, 65, 79, 80, 168, **169–70**; Flammulated Screech, **168–69**; Great Horned, 70, 71, 72, 73, 100, 101, 102, 103, 105, 106, 165, **166–68**; Long-eared, **171**; Pygmy (*see* Ferruginous Pygmy); Short-eared, **171–72**
Oxyura jamaicensis, 85, 103, 106, 158

Pachyramphus aglaiae, 88
Palm, 74, 242, 243
Paloverde, 37, 42–45, 64, 181, 189, 204, 218
Pandion haliaetus, 77, 79, **134–35**
Pandionidae, 134–35
Parabuteo unicinctus, 63, 66, 73, 80, 81, 133
Paridae, 202, 203
Parrots, **162–63**; Thick-billed, 94, 163. *See also* Macaw
Parula americana, **223**
Parulidae, 91, 94, 221–28
Parus gambeli, 67; *wollweberi,* 66, 67, 80, **202**
Passer domesticus, 73, 84, 102, 103, 236, **252**
Passerella, 240; *georgiana,* **241**; *iliaca,* 94; *lincolnii,* 77, 104, 105, **240**; *melodia,* 71, 77, 86, 94, 104, 105, **240–41**
Passeridae, 252
Passeriformes, 182
Passerina, 232; *caerulea,* 63, 64, 71, 79, 86, 103, 105, **232–33**; *cyanea,* **233**
Pelecanidae, 126–27
Pelecaniformes, 126–27
Pelecanus occidentalis, **126–27**; *erythrorhynchus,* **126**
Pelican, Brown, **126–27**; White, 126
Penstemon, bush, 43, 234
Petrochelidon, 196
Peucedramus taeniatus, 88
Pewee, Western Wood, 100, **189–90**
Pfefferkern, Fr. Ignaz, S.J., 18, 22
Phainopepla, 70, 72, 73, 84, 85, 101, 102, 106, 108, 201, **217–19**, 247
Phainopepla nitens, 70, 73, 84, 101, 102, 105, 106, **217–19**
Phalacrocoracidae, 127
Phalacrocorax auritus, 77, **127**
Phalaenoptilus nuttallii, 78, 106, **172–73**
Phalarope, Northern, **153**, 154; Wilson's, **153**
Phalaropus, 153; *tricolor,* **153**
Phasianidae, 139–40
Pheucticus, 232; *ludovicianus,* 73, 100, **231–32**; *melanocephalus,* **231**
Phoebe, Black, 72, 104, **184**; Eastern, 51, 95, **183**; Say's, 63, 65, 73, 74, 102, 106, **184–87**, 201
Phoenix (Anglo settlement), 8, 9, 10, 31; region, 66, 87, 90–91
Phragmites (*communis*) *australis,* 16, 33, 35, 37. *See also* Common reed
Picidae, 178–82
Piciformes, 178–82
Pickleweed, 46
Picoides, 181
Pima Butte, 9, 26, 27, 29, 31, 39
Pima Indians, Riverine (summarized), 7, 8–9, 10, 39, 88; creation myth, 175, 180; fishing, 17, 78–79, 127, 134; orthography, 255–56; song and legend, 129, 136, 140, 151, 162, 164, 177, 180, 184, 197, 203, 207, 208, 216, 243, 244. *See also* Bird names; Irrigation, Pima; Native agriculture; Native consultants; Rancherias and ranches
Pima Villages (historic Anglo place name), 7, 21, 22, 23, 38
Pimería Alta, 7, 9, 10, 20
Pimo. *See* Pima Indians
Pine, 140, 163
Pintail, 79, 80, **156**
Pipilo, 233; *aberti,* 69, 71, 83, 86, 101, 103, 104, 105, 123, **234–35**; *chlorurus,* 93, 100, **233**; *crissalis,* 234; *erythrophthalmus,* **233–34**; *fuscus,* 106, **234**
Pipit, Water, 103, 107, **217**
Piranga ludoviciana, 100, **228–29**; *rubra* 63, 71, 77, 80, 86, 104, **229**
Pitylus, 232
Plantago, 244
Platalea ajaja, **131**; *leucorodia,* 131
Plataleidae, 130–31
Platyrhynchus pusillus, 188
Plegadis chihi, **130–31**
Ploceidae (Passeridae), 252
Plover, Mountain, **151**; Snowy, **150**
Pluchea odorata, 55, 56; *sericea,* 16, 31, 32, 36, 37, 46, 47, 57, 62, 69, 85, 241. *See also* Arroweed
Podiceps nigricollis, **125**
Podicipedidae, 125–26
Podicipediformes, 125–26
Podilymbus podiceps, 77, 80, 86, **126**
Polioptila caerulea 66, 67, 216; *californica,* 216; *melanura,* 70, 73 100, **216**; *nigriceps,* 88; *plumbea,* 216
Polygonum sp., 54, 57, 74, 86. *See also* Smartweed; Emergent vegetation
Polypogon monspeliensis, 55
Pond, Base Meridian, 57, 85; Riggs Road, 53; Run-off Ponds, 50, 106
Pooecetes gramineus, 93, **236**
Poorwill, 78, 106, **172–73**, 208
Populus fremontii, 16, 47, 57, 62, 164, 175, 189, 203, 225, 228, 229. *See also* Cottonwood
Porzana carolina, 66, 77, 104, **148–49**
Progne subis, 73, **197**
Prosopis, 69, 85, 176, 187, 202, 209, 210, 211, 218, 219, 220, 227, 228, 235, 243; *juliflora,* 62; *odorata,* 36, 37 (*see also* Screwbean); *pubescens,* 33, 36, 46 (*see also* Screwbean); *velutina,* 36, 37, 45, 46, 47, 50 (*see also* Mesquite)
Protonotaria citrea, **221**
Psaltria, 202
Psaltriparus, 202
Psittacidae, 162–63
Psittaciformes, 162–63
Pyrocephalus rubinus, 62, 63, 70, 72, 80, **191**
Pyrrhuloxia, 52, 73, 74, 84, 87, 89, 103, 108, 230, **230–31**

Quail, Gambel's, 73, 83, 100–05, 108, **139–40**; Scaled, 139
Quail-brush, 46, 47, 57. *See also Atriplex*
Quercus turbinella, 42. *See also* Oak
Quiscalus lugubris, 250; *mexicanus,* 73, 84, 87, 88, **249–50**; *quiscula,* 250
Quitovac Oasis (Papago), 117. *See also* Native consultants, Luciano Noriego

Racial instability, 90
Rail, 74, 82; Black, 148; Clapper, 103, 109, **148**; Virginia, 63, 75, 77, 78, 85, 104, 107, 109, **148**; Yellow, **149**
Rallidae, 148–50
Rallus limicola, 63, 75, 77, 85, 104, **148**; *longirostris,* 103, **148**

Rancherias and ranches, 46, 48, 108. *See also* Native agriculture
Range extensions, 87–91
Ranunculus, 84. *See also* Emergent vegetation
ratany, 42, 43. *See also Krameria*
Raven, Common, 63, 64, 72–74, 83, 93, **199–200**; White-necked, 201
Recurvirostra americana, 103, 107, **153**
Recurvirostridae, 153–54
Redstart, American, **288**
Reed, common, 16, 18, 35; giant, 16, 33, 57
Regulus calendula, 93, 100, **217**
Relictual population or community, 42, 81, 108, 110, 140
Remizidae, 203–04
Rhodothraupis, 232
Rhynchopsitta pachyrhyncha, 94, 163
Ribes quercetorum, 42, 213. *See also* Gooseberry
Ridgwayia, 213; *pinicola,* 213
Río de la Asunción, 18, 61. *See also* Salt River
Riparian community, defined, 2
Riparian vegetation (major references), 38–39, 56–60, 61–75
Riparia riparia, **194**, 195
Rivers in desert, 1; Agua Fria, 2, 7, 22; Colorado, 2, 9, 21, 31, 159; Gila, 2, 7, 16–40, 46, 61, 67–75, 76–81, 82, 89, 105, 107, 109; Hassayampa, 2, 22; New, 22; Salt, 2, 7, 9, 10, 18, 25, 30, 34, 56–60, 61, 81, 85–87, 103–04, 107; San Pedro, 2, 7, 10, 17, 25, 30; Santa Cruz, 2, 7, 10, 23, 26, 27, 33, 34, 35, 38, 39, 46, 82, 107; Verde, 2, 18, 81, 83
Roadrunner, Greater, 73, 100, 101, 102, 104, 105, **164**
Robin, American, 115; Northern, 67, 93, 95, 115, **212–13**
Roosting, 99, 231, 242, 245, 248
Rosewood, Arizona, 42, 43
Rudo Ensayo, 18
Rumex sp., 33, 54, 55, 57, 84, 148
Rush, 1, 245. *See also* Emergent vegetation; *Scirpus*
Russel, Frank, 34–35, 78, 159, 162, 201
Russian thistle, 29, 35, 54. *See also* Tumbleweed

Sacate (Pima settlement), 9, 23, 29, 31, 39, 50, 52, 171, 231
Sacaton (Pima settlement, Tribal Agency), 11, 26, 28, 36, 37, 38, 50, 52, 53, 78, 88,164
Saguaro, 20, 26, 34, 37, 43, 44, 60, 64, 74, 79, 83, 99, 132, 136–38, 166, 168, 170, 179, 180, 182, 183, 197, 219
Salicornia, 238
Salinization and salinity, 2, 3, 19, 23, 24, 36, 39–40, 54, 55
Salix, 16, 31, 37, 57, 86, 87, 164, 175, 189, 203, 221–25, 228, 229. *See also* Willow
Salpinctes, 208, 241; *fasciatus,* 208; *guttatus,* 208; *maculatus,* 208; *obsoletus,* 106, **208**
Salsola iberica, 34, 35, 47, 53, 57, 86, 205. *See also* Russian thistle; Tumbleweed; Disturbance vegetation
Saltbush, 19, 32, 108, 173; desert, 46, 47; fourwing, 20, 29, 46, 50, 57. *See also* Chamise; Quail-brush
Saltcedar, 2, 3, 16, 27, 29, 31, 46, 48, 53, 55–57, 60, 62, 65, 68, 74, 82, 84–86, 89, 91, 99, 159, 166, 197, 198, 202, 203, 217, 220, 221, 223–25, 245, 248
Salt grass, 24, 33, 34. *See also Distichlis;* Grasslands
Salt River Project, 57
Sambucus mexicana, 47, 53, 84, 85, 218, 219, 223, 228
Sandpipers, 53, **151**; Least, 103, 107, **152**; Pectoral, **152**; Semipalmated, **152**; Solitary, **152**; Spotted, 103, 107, **151–52**; Western, 103, 107, **153**
Santa Cruz Village (Pima), 33, 37, 42, 78, 117
Santan (Pima settlement), 36, 46
Santa Teresa Ciénaga, 18, 21, 27, 39. *See also* Maricopa Wells
San Xavier del Bac (Papago mission), 3, 19, 38
Sapsucker, Red-breasted, **181**; Red-naped, **180**; Williamson's, 51, 95, **181**; Yellow-bellied, 93, 101, 102, 104, **180–81**
Sarcobatus, 230, 241
Sayornis nigricans, 72, 104, **184**; *phoebe,* 95, **183**; *saya,* 63, 65, 102, 106, **184–87**
Scardafella, 161; *inca,* 73, 84, 88, 102, **161**; *squamata,* 161
Scaup, Lesser, **158**
Schismus barbatus, 34, 35, 39, 244
Scirpus, 31, 74, 78, 81, 84, 86, 207, 226, 227; *acutus,* 54; *olneyi* 16. *See also* Bulrush; Emergent vegetation
Scolopacidae, 151–53
Scopus, 168
Screwbean, 33, 34, 36, 37, 46, 181, 203
Sedelmayr, Fr. Jacobo, S.J., 17, 22, 50
Sedge, 54. *See also* Emergent vegetation
Seepweed, 34, 35, 39, 46, 47, 49. *See also Suaeda*
Seepwillow, 1, 23, 29, 47, 68. *See also Baccharis glutinosa*
Seiurus aurocapillus, 96, **225**
Selasphorus rufus, 176
Setophaga ruticilla, **228**
Sewage effluent, 18, 52, 53, 57, 58, 75, 85–87, 90, 107, 219, 225
Shooting, as cause of species decline, 83, 133, 135, 140, 148
Shoveler, 106, **157**
Shrike, Loggerhead, 63, 65, 73, 74, 75, 100, 101, 102, 106, **201–02**, 221
Sialia currucoides, 67, 94, 95, **215**; *mexicana,* 67, 95, **215**
Sierra Estrella, 21, 24, 26, 27, 31, 34, 47, 55; habitat, 42–43, 50, 108
Siskin, Pine, 94, 95, **251**
Sisymbrium sp., 57
Sitta canadensis, 94, 95, **204**; *carolinensis,* 66, 67
Sittidae, 203, 204
Smartweed, 54. *See also* Emergent vegetation; *Polygonum*
Snaketown. *See* Archaeological sites
Snipe, Common, 84, 103, 104, **151**
Solanum elaeagnifolium, 54; *nodiflorum,* 55
Solitaire, Townsend's, 94, 95, **216**
Sora, 66, 77, 84, 85, 104, 107, **148–49**
Sorghum bicolor, 47, 89, 159, 231; *halapense,* 47
Sparrow, 235, 245; Black-throated, **238**; Brewer's, 100, **237**; Chipping, 92, 100 **236–37**; Desert, 73, 100 (*see also* Black-throated); English (*see* House Sparrow); Five-striped, 88; Fox, 94; Harris', **240**; House, 59, 73, 74, 84, 102, 103, **252**; Lark, 73, **236**; Lincoln's, 77, 99, 104, 105,

240; Rufous-crowned, 43, 106, 120, **237–38**; Sage, 100, **238**; Savannah, 103, **236**; Song, 7, 71, 72, 74, 77, 84, 85, 86, 94, 104, 105, 109, **240–41**, 247; Swamp, **241**; Vesper, 93, **236**, 241; White-crowned, 93, 236, **239–40**
Spatula, 158
Species, criteria for recognition, 122
Species diversity, effect of native agriculture on, 108
Specimen preparation, 124; piercing nares, 205; stripping ulna, 234. *See also* Labels, specimen
Speotyto, 171; *megalopeza,* 171
Sphyrapicus thyroideus, 95, **181**; *varius,* 93, 101, 104, **180–81**
Spizella breweri, 100, **237**; *pallida,* 237; *passerina,* 92, 100, **236–37**
Spoonbill, Roseate, **131**
Sporobolus wrightii, 35. *See also* Grasslands
Squawberry. *See Lycium*
Starling, Common, 59, 73, 74, 84, 102, 105, 106, 164, **219**
Status of species: descriptive phrases, 118
Steganopus, 153
Stelgidopteryx ruficollis, 63, 64, 72, 79, 107, **194–95**
Sterna forsteri, **154**
Stilts, Black-necked, 53, 57, 103, 107, **153–54**
Stork, Wood, 147
Strigidae, 166–72
Strigiformes, 165–72
Sturnella magna, 244; *neglecta,* 50, 73, 100, 102, 103, 106, **244–45**
Sturnidae, 219
Sturnus vulgaris, 73, 84, 102, 105, 106, **219**
Suaeda moquinii, 34, 35, 37, 46, 47, 50, 238. *See also* Disturbance vegetation; Seepweed
Subspecies: changes in distribution, 90–91; criteria for recognition, 122; function of, 122
Succession, avifaunal, at Barehand Lane, 84; on the lower Salt River, 85
Sunflower, wild, 2, 47, 48, 54, 57, 85, 232, 233, 251
Swallow, 52, 56, 174; Bank, **194**; Barn, 60, 63, 78, 80, 194, **195–96**; Cliff, **196–97**; Rough-winged, 63, 64, 72, 79, 107, **194–95**; Tree, **194**; Violet-green, 72, **194**
Swan, Whistling, 80, **155**
Swift, Vaux's, **174**; White-throated, 70, 72, 106, **174**
Sylviidae, 216–17

Tachycineta bicolor, **194**; *thalassina,* 72, **194**
Tamarisk, Athel, 32, 47, 48, 50, 51, 68, 71, 83, 108, 135, 168, 171, 181, 187, 197, 205, 220, 223–25, 236, 243, 244, 249, 252
Tamarix, 32, 62, 69, 86; *aphylla,* 50, 164, 181, 189; *ramosissima,* 16, 31, 46, 57. *See also* Saltcedar
Tanager, Summer, 51, 58, 63, 65, 71, 74, 75, 77, 78, 80, 86, 87, 104, 107, **229**; Western, 100, **228–29**
Tangavius, 248
Taxonomic sequences, changes in, 121–22
Taxonomy, 120–24
Teal, 56; Blue-winged, 103, 104, **157**; Cinnamon, 75, 103, 104, 106, 107, **157**; Green-winged, 104, 106, **157**
Telmatodytes, 207
Tern, Black, **154**
The Buttes. *See* Buttes
Thrasher, 84, 209; Bendire's, 52, 83, 100, 102, 108, **209–10**, 211, 247; Crissal, 72, 101, 104, 105, 209, **211–12**; Curve-billed, 73, 106, 209, **210**; Le Conte's, 47, 100, 108, 209, **210–11**; Sage, **212**
Thraupidae, 228–29
Threskiornithidae, 130–31
Thrush, 92; Aztec, 213; Hermit, 92, 95, 105, **213–14**; Swainson's Thrush, 95, 213, **214–15**; Varied, 51, 95, **213**
Thryomanes, 205
Tigrisomatinae, 142
Titmouse, Bridled, 66, 67, 80, **202**
Towhee, 83; Abert's, 52, 69, 71, 83, 86, 101, 103, 104, 105, 209, **234–35**, 247; Brown (*See* Canyon Towhee); Canyon, 43, 106, 120, **234**; Green-tailed, 92, 93, 99, 100, **233**; Rufous-sided, 95, **233–34**
Toxostoma, 209, 211; *bendirei,* 83, 100, 102, **209–10**, 212; *crissale,* 72, 101, 104, 105, **211–12**; *curvirostre,* 73, 106, **210**; *dorsale,* 211; *lecontei* 100, **210–11**
Tree sage, 37, 43, 45
Tree tobacco, 57, 89, 241
Tringa melanoleuca, **152**; *solitaria,* **152**
Tripsurus, 180
Trochilidae, 174–76
Troglodytes aedon, 205 (*see domesticus*); *bewickii,* 65, 67, 69, 70, **205–06**; *domesticus,* 100, **205**, 206; *tanneri,* 206; *sissoni,* 206; *troglodytes,* **205**
Troglodytidae, 205–08, 241
Tule marsh, 37, 78, 207. *See also* Emergent vegetation
Tumbleweed, 34, 47, 53, 57, 205. *See also* Disturbance vegetation; Russian thistle
Turdidae, 212–16
Turdus migratorius, 67, 93, 94, 95, **212–13**
Turkey, Common, **140**
Turner, Henry Smith, 23
Typha domingensis, 16, 31, 37, 53, 56, 57, 62, 74, 78, 84, 86, 148, 207, 226, 227. *See also* Cattail
Tyrannidae, 182–91
Tyrannus crassirostris, 88; *melancholicus,* 88; *verticalis,* 72, 83, 101, 102, **182**; *vociferans,* **182–83**
Tyto alba, 71, 101, 102, **165–66**
Tytonidae, 165–66

Urbanization, 84, 248

Vauquelinia californica, 42, 234. *See also* Rosewood, Arizona
Vekol Wash, 24, 39
Verdin, 49, 69, 73, 84, 86, 100, 108, **203–04**, 206
Vermivora celata, 67, 102, 104, 105, 106, **221–22**; *luciae,* 63, 69, 70, 83, 84, 101, 104, 105, **222–23**; *peregrina,* **221**; *ruficapilla,* **222**; *virginiae,* **222**
Vireo, 83; *bellii,* 63, 69, 70, 84, 101, 104, 105, **219–20**; *gilvus,* **220–21**, 226; *huttoni,* 67, **219**; *philadelphicus,* **220**; *solitarius,* 66, 67, **220**, 226
Vireo, Bell's, 49, 63, 65, 69, 70, 74, 75, 84, 85, 86, 87, 101, 104, 105, 107, **219–20**; Hutton's, 67, **219**; Philadelphia, 51, **220**; Plumbeous, 220; Solitary, 66, 67, **220**; Warbling, **220–21**

Vireonidae, 219–21
Voucher specimens of plants, deposition, 41
Vulture, Black, 87, 88, **130**; Turkey, 72, 83, 100, 106, **127–30**, 135, 194
Vulturidae, 127–30

Warbler, 58; Audubon's (*see* Yellow-rumped Warbler); Black-and-white, 96, **221**; Black-throated Gray, 91, 97, 101, 104, 221, **224–25**, 226; Chestnut-sided, **225**; Grace's, **225**; Hermit, 91, 97, **225**; Lucy's, 49, 63, 65, 69, 70, 74, 75, 83–86, 104, 105, 107, **222–23**; MacGillivray's, 100, 101, **226**; Myrtle, 96, 224; Nashville, **222**; Northern Parula, 164, **223**; Olive, 88; Orange-crowned, 60, 67, 91, 101, 102, 104–06, **221–22**, 226; Prothonotary, **221**; Red-faced, 88; Tennessee, **221**; 226; Townsend's, 91, 97, 101, 102, 104, 105, **225**; Virginia's, **222**; Wilson's, 96, 97, 100, 101, **227**, 236; Yellow, 65, 71, 74, 75, 77, 79, 80, 87, **223**; Yellow-rumped, 59, 60, 63, 64, 71, 91, 101, 102, 104, 105, 217, **224**
Wash, Greene's, 24, 39; Santa Rosa, 39; Vekol, 24, 39 (*see also* River, Santa Cruz)
Wastewater. *See* Sewage effluent; Irrigation, Pima; Pond, Run-off
Watershed deterioration, Gila, causes of, 3, 11, 29
Water table, 2, 3, 16, 26, 33, 34, 39, 78, 83, 196
Waxwing, Cedar, 95, **217**
Webb, George (Pima author), 33
Wheat, 9, 17, 18, 19, 21, 31, 159, 235
Wigeon, American, **158**
Willow, Goodding, 1, 7, 16, 18, 19, 20, 24, 33, 36, 37, 38, 56, 57, 58, 59, 60, 62, 64, 65, 68, 72, 74, 85, 86, 89, 91, 107, 137, 141, 142, 146, 149, 163, 164, 168, 169, 177, 178, 180–81, 183, 188, 191, 195, 202, 219, 220, 221, 223, 224, 225, 227, 230, 236, 240, 251
Wilsonia pusilla, 100, 101, 226, **227**
Wintering status, changes in, 58, 91–93
Wolfberry, 32, 39, 42, 46, 47, 49, 53. *See also Lycium*
Woodcutting, 3, 30, 33, 50, 68, 83
Woodpecker, 169, 219; Acorn, 95, **180**; Gila, 70, 71, 73, 87, 88, 101, 102, 104, 106, 169, **180**; Ladder-backed, 70, 71, 73, 86, 87, 100, **181–82**; Lewis, 95, **180**
Wren, Bewick's 65, 67, 69, 70, 74, **205–06**; Cactus, 73, 100, **206**, 247; Cañon, 72, 73, 106, **207–08**; House, 92, 100, **205**; Long-billed Marsh, 73, 74, 77, 79, 80, 81, 86, 104, 105, **207**; Rock, 72, 106, 207, **208**; Winter, **205**

Xanthium canadense, 47, 54; *strumarium,* 47, 54. *See also* Cocklebur
Xanthocephalus xanthocephalus, 79, 99, 103, 104, 105, **246–47**

Yellowlegs, Greater, **152**
Yellowthroat, Common, 37, 73, 74, 75, 77, 78, 80, 84, 85, 86, 96, 97, 104, 105, 107, 109, **226–27**
Yucca, 42, 242

Zea mays, 47. *See also* Corn; Native agriculture
Zenaida asiatica, 70, 71, 86, 88, 101, 102, 103, 104, 105, 106, **159–60**; *macroura,* 69, 70, 86, 100, 101, 102, 103, **160**
Ziziphus obtusifolia, 33, 35, 36, 37, 47, 50, 203, 209, 211. *See also* Graythorn
Zonotrichia leucophrys, 93, **239–40**; *querula,* **240**
Zoothera citrina 213; *mollissima,* 213; *naevia,* 94, 95, **213**

Index of Piman Words and Expressions

á:ḍo, 162
ákimul aw'awtam, 9
áw'awtórtupwa, 235
áwkakoi, 33, 159

ba:k, 133–34
báifchul, 177
básho sve:k, 212
bet kúishnam, 209
bíchput, 234–35
bit kik, 50

chíkukmal, 182
chívichuich, 150
chuáwgiakam vákoiñ, 126
chúduk ví:sukĭ, 136
chuf chíñgam, 209
chúhugam, 181
chúhupĭ, 160
chukud, 166
chúkut, 166
chúvul ú'uhik, 153

é: 'et váhudam, 165

gáw'uguf, 201
gídval, 195, 197
gí:sop, 203
gúgu, 161

haiñ júlshap, 147
háupal, 132
hávaiñ, 200
háwhi, 160
háwkut, 206
háwshañ, 34
hé'et kávult, 229
hévachut, 195
hévaichut, 195
hé:vel maws, 184
hiá:lak, 156
híkvik, 180

jujupul, 130

kádgam, 163
kákachu, 139
kó:logam, 172
kókoho, 170
kómakĭ ú'hik, 210
kó:makí vakoiñ, 142
kud vik, 209, 210
kúdat, 178
kú:kul, 168, 169
kú:kvul, 168
kúli huch, 209, 211
kwígam, 217

ñúi, 127, 130, 135
ñúput, 173

oam ñúi, 135

s'bában mákam, 235
schuk máw'awkum gí:sop, 216
s'chuk ñúi, 130
schuk ú'uhik, 199
se:p wéhadam, 134
sháwshaiñ, 245, 249
shú:dakĭmat, 151
shu:g, 208
sípok, 229
sísik, 136
sívichich, 150
s'náwkadam u'uhik, 162

támtol, 239
tash cúnam vákoiñ, 146
táwdai, 164
tóa vákoiñ, 142, 143
tósif, 244
tóva, 140

unmú'uhik, 197

váchpik, 149
váchukukĭ, 242
vá:kaf, 133
vákoiñ, 142, 143, 146, 147
vákoiñ ba:k, 134
vápkuch, 78
vápkukĭ, 155
vátop, 78
vávas, 208
vípismal, 174
vísukĭ, 135
vúhigam, 160